From Runway to Orbit

NASA SP 2004-4109

From Runway to Orbit:
Reflections of a NASA Engineer

Kenneth W. Iliff

and

Curtis L. Peebles

The NASA History Series

National Aeronautics and Space Administration
NASA History Office
Washington, D.C. 20546   2004

Library of Congress Cataloging-in-Publication Data

Iliff, Kenneth W.
    From runway to orbit : reflections of a NASA engineer / Kenneth W. Iliff and Curtis L. Peebles.
        p. cm. -- (The NASA history series) (NASA SP ; 2004-4109
    Includes bibliographical references and index.
    1. Iliff, Kenneth W. 2. Aerospace engineers--United States--Biography. 3. Dryden Flight
Research Facility--History. I. Peebles, Curtis. II. Title. III. Series. IV. NASA SP ; 4109.

    TL540.I357A3 2003
    629.4'092--dc22
    [B]

                                                                                    2003059308

For my parents, Warren and Dorothy Iliff,
And
My wife's parents, Dean and Katheryn Shafer

Table of Contents

Acknowledgements viii

Preface x

Foreword xi

Introduction xiii

Apprenticeship of a Young Engineer 1

Birth of the Lifting Body 31

Building the Heavyweight Lifting Bodies 57

Flying the M2-F2 and Other Adventures 79

Flight Research in the 1960s and Early 1970s 105

Origins of the Space Shuttle 133

Getting Ready to Fly 153

The Approach-and-Landing Tests 167

Counting Down to Launch 189

STS-1 201

Analyzing the Data 219

Becoming Operational: STS-3 through STS-5 239

STS-6 to the Loss of Challenger 259

Return to Flight: The Shuttle Program in the 1990s and Beyond 283

Going Nowhere Fast: The NASP 303

Hypersonics in the 1990s 323

Stairway to Heaven 339

Appendices 361

Acronyms 380

References 383

Index 392

About the Authors 400

Acknowledgements

The quest for higher and faster flight started when the Wright brothers first flew 100 years ago. This is my version of going higher and faster for the past forty years, while I worked at NASA Dryden Flight Research Center. During those years at Dryden I have been in the privileged position of analyzing data from over ninety aircraft configurations, including over thirty first flights. I was fortunate to work on, among others, all of the early lifting bodies, the X-15, the XB-70, and the Space Shuttle.

This document began as an oral history of my experiences and still bears definite signs of its origin. This is my personal story, as a working engineer. Because of this it's told from my point of view. When I started working at the NASA Dryden Flight Research Center in 1962, I worked with senior NASA employees who could remember first-hand the excitement of the Wright brothers' first powered flight in 1903. The lifetimes of those senior employees plus the forty years of my career span one hundred years of powered flight, from the Wright Flyer at Kitty Hawk to the Space Shuttle Orbiter and beyond.

This story, personal though it may seem, involves the contributions of thousands of dedicated and talented people. I have only mentioned the few with whom I worked closely and with whom I remembered as I recounted my version of these past forty years.

The loss of the Space Shuttle Columbia and her crew happened when this document was in the final stage of editing. Like everyone else, I had a very emotional response to the loss itself, as well as the subsequent investigation. I originally felt that I should redo the Shuttle-related material to emphasize the dedication of those who worked so hard to make the Shuttle, a very complex vehicle, fly so well for the past 25 years. On reflection, the publishers of this book decided to leave the discussion of the loss of Columbia to later writings.

There are many people to whom I owe much for the help they gave me. The first is my co-author, Curtis Peebles, who did all the hard work of turning the original taped narration into a coherent document. Dill Hunley, retired Dryden Center Historian, suggested that I make the narration and then encouraged me to turn it into this book. Michael Gorn, the present Dryden Center Historian, has been dedicated to getting the account from draft form to publishable document.

Many people assisted in the transformation of the extremely rough draft into a finished book. While I can't name them all I can list the major contributors. The process of publication is very demanding and sometimes tedious, but these people persisted, driven by their professional pride, desire for accuracy, general obsessiveness, and goodwill toward me.

These individuals include Bill Dana (retired Dryden research test pilot), Ed Saltzman and Mary Shafer (both retired Dryden research engineers), Darlene Lister, and Sue Henderson. All of them read the entire manuscript and made innumerable suggestions and corrections that turned the document into an accurate historical account. (Mary Shafer's motives also included the understandable desire not to have her husband, me, look any more foolish than necessary.)

In additions, Tony Landis, Ed Saltzman, and Joy Nordberg spent many hours helping me find the photographs that I used. Mary Shafer also provided a number of Dryden photos from her collection. Once we found the right pictures, Carla Thomas scanned and perfected them, a process necessary for the high quality images she demanded.

Barbara Rogers and Sylvia Dolber ensured that all my references are available to the public. Muriel Khachooni compiled the list of references and checked the accuracy of every citation using the original documents. Denny Gonia contributed to the creation of the appendices.

Dryden DC-8 pilot Mark Pestana painted the beautiful jacket pictures. Justine Mack created all the line drawings, just as she has done for years for my technical publications. Steve Lighthill laid the book out in its beautiful final form.

However, even the most gifted specialists sometimes miss things. All errors and omissions in this book are my responsibility.

Preface

In his remarkable memoir *Runway to Orbit*, Dr. Kenneth W. Iliff — the recently retired Chief Scientist of the NASA Dryden Flight Research Center— tells a highly personal, yet a highly persuasive account of the last forty years of American aeronautical research. His interpretation of events commands respect, because over these years he has played pivotal roles in many of the most important American aeronautics and spaceflight endeavors. Moreover, his narrative covers much of the second half of the first 100 years of flight, a centennial anniversary being celebrated this year.

Dr. Iliff's story is one of immense contributions to the nation's repository of aerospace knowledge. He arrived at the then NASA Flight Research Center in 1962 as a young aeronautical engineer and quickly became involved in two of the seminal projects of modern flight, the X-15 and the lifting bodies. In the process, he pioneered (with Lawrence Taylor) the application of digital computing to the reduction of flight data, arriving at a method known as parameter estimation, now applied the world over. Parameter estimation not only enabled researchers to acquire stability and control derivatives from limited flight data, but in time allowed them to obtain a wide range of aerodynamic effects. Although subsequently involved in dozens of important projects, Dr. Iliff devoted much of his time and energy to hypersonic flight, embodied in the Shuttle orbiter (or as he refers to it, the world's fastest airplane). To him, each Shuttle flight, instrumented to obtain a variety of data, represents a research treasure trove, one that he has mined for years.

This book, then, represents the story of Dr. Ken Iliff's passion for flight, his work, and his long and astoundingly productive career. It can be read with profit not just by scientists and engineers, but equally by policy makers, historians, and journalists wishing to better comprehend advancements in flight during the second half of the twentieth century.

Kevin L. Petersen
Director, NASA Dryden Flight Research Center
October 2003

Foreword

Dr. Kenneth W. Iliff has had a long and distinguished career in aerospace at NASA Dryden Flight Research Center. The retired Dryden Chief Scientist has now written a history of the activities that comprise his life's work. The scope of his activities were extraordinarily broad and he has a fascinating story to tell, partly because he was born at the right time and chose the right career, but also because he possesses a highly original vision of the future, penetrating observational skills, and a storyteller's knack for recounting a good tale.

Ken W. Iliff was born during World War II in rural Iowa. When Sputnik orbited in 1957, Ken was in high school. He was captivated by the space program, and shortly after the launch of Sputnik, began building and launching rockets of his own design.

Ken attended Iowa State University, graduating in 1962 with degrees in mathematics and aerospace engineering. He received offers of employment in aerospace engineering from the government as well as from private industry, but he chose to work for NASA Dryden Flight Research Center at Edwards Air Force Base in California.

NASA Dryden's major program in 1962 was the X-15 research airplane, then in its envelope-expansion phase. Iliff quickly gravitated toward that program, where he analyzed vehicle stability and control. He remained with the X-15 in a number of capacities until it was cancelled in 1968. He also worked with the advanced Mach 6.7 version of the X-15 and the proposed delta-winged design.

Ken's career also included the M2-F1 lifting body, the first of its kind ever to fly with a human aboard it. Ken's work with the lightweight M2-F1 led him to the M2-F2, a lifting body almost identical in shape to the M2-F1, but built of aluminum and stressed for supersonic flight.

The M2-F2 had an instability in its roll axis which caused it to oscillate in bank angle when flown to low angle of attack. In From Runway to Orbit, Ken covers in detail his frustrations with the M2-F2.

His next project involved the XB-70, a large Mach 3 prototype bomber acquired from the U.S. Air Force as a research airplane. The XB-70 shared with the lifting bodies and X-15 a common problem: it was difficult for the research engineers servicing these research airplanes to separate and identify the aerodynamic and other parameters that define their behavior in flight. Ken, challenged by this difficulty, developed a theory that he termed "maximum likelihood estimator" to assess parameter values. A revised version, known as the "modified maximum likelihood estimator," or MMLE, is used by flight-test organizations worldwide to extract estimates of aircraft aerodynamic, structural, and performance parameters.

After a break to attend graduate school, Ken returned to NASA Dryden in 1971 and began to work on the Space Shuttle, still in its formative stage. Dr. Iliff observes that although technology existed to build a modest two-stage-to-orbit spacecraft (the second stage of which could serve as a space-station supply vehicle) Congress mandated requirements that inbibited its pursuit. Ken notes with insight the congressional decision and its result.

Ken worked on the Space Shuttle for 15 years—from the early stage, through the approach-and-landing tests, to orbital flight test and post-Challenger operations. He also contributed to several advanced airplane design studies, including the now-cancelled National Aerospace Plane and several multistage concepts which proposed to use the SR-71 reconnaissance airplane as the first stage.

From Runway to Orbit makes observations about the United States aerospace program that will not be found elsewhere. The chapter about the National Aerospace Plane is in itself worth the price of the book, as is the story of a trip Ken took to the Soviet Union during the height of the Cold War.

This volume will satisfy the most ardent aerospace-history enthusiast, and will inform the thinking of the present generation of engineers, students, and policy analysts alike.

William H. Dana
NASA Dryden Flight Research Center
September 2003

Introduction

The first thing I noticed was an incredibly bright flash at the base of the Saturn V. The flame got brighter and larger as it spread up the side of the thrust trench. Slowly, the 363-foot vehicle started to lift off the launch pad. At first, I could see very little acceleration as Apollo 11 rose up along the gantry. About 12 seconds after launch, I started hearing the sound. It was absolutely overpowering. I've never heard a recording that did it justice. It was as if the air were discontinuous, the sound coming in crackling waves.

My body resonated with the power of the sound. The further the vehicle lifted, the more energy I got from the sound. I was completely awed by what I was observing. As Apollo 11 got higher, there was a small, thin cloud above the launch pad. I could see through it, but it was there, nonetheless. As Apollo 11 passed through that cloud, the engine exhaust left a round hole larger than the vehicle. I could even see a change in the light from the hole in that small cloud. I've never seen a film of the Apollo 11 launch that even approaches showing the extreme contrast between the blue mid-morning sky and that long bright flame. I could see the white vehicle perched on top of an exceedingly long yellow flame. The edges of the flame were so sharp and bright that it looked like an animation. I, like the other people in the VIP area on July 16, 1969, continued to watch this spectacle, which I've used in evaluating all later engineering accomplishments in my career. Also in the VIP area were Vice President Spiro Agnew, high-ranking NASA managers, various entertainers including Johnny Carson, and a few people who, like me, were lucky enough to get passes. I sat in the VIP area because I'd had the good fortune of knowing astronaut Fred Haise when he was a test pilot[1] at the NASA Dryden Flight Research Center.[2]

For the two days prior to the launch, I observed the three Apollo 11 astronauts in simulations. Watched them rehearse the various failures and emergencies. It was obvious that the astronauts were bored with this process, because they were flip in some of their responses, although Neil Armstrong was very quiet, as he always was. A pilot at the Dryden Flight Research Center, he had been selected as an astronaut just as I arrived there. The astronauts' bored response to the emergencies being simulated was something that I had frequently observed in the X-15 simulator, the project on which I had just completed work prior to the Apollo 11 launch.

At Dryden, I'd had the opportunity to work a great deal with Fred Haise in the simulator and in other studies, primarily with the lifting body program [1]. In addition to being a very smart, easygoing, nice guy, Fred was an outstanding engineer, so it was natural for me to work with him. We could converse engineer-to-engineer sorting out the problems that we were trying to resolve. After Fred was selected for the astronaut program, I'd kept in infrequent touch with him. I hadn't expected him to be able to get me a pass to watch the launch because I'd asked late, fairly close to the Apollo 11 launch date. But Fred managed to get VIP passes for a friend, Lowell Greenfield, and me.

This event is pivotal to the story I will tell because for the previous decade, much of the United States' technical resources, like the U.S.S.R.'s, had been devoted to the space program. Most NASA engineers were big supporters of the space program. We felt that going to the Moon was well worth doing, even if it meant that some of the things we thought needed to be done in aeronautics would be delayed somewhat.

After the launch, the Vice President made a speech on future space policy. Most of us hoped he was going to say that now that we'd shown that we could go to the Moon, we were going to take the next step, landing on Mars. Unfortunately, he didn't say that. What he did say was that the United States had many problems and needed to use the resources of the space program here on earth for problems like the Vietnam War. I had always wanted us to go to Mars to explore the possibility of life there and learn what the environment was like, so I was disappointed.

The Road to Space

My interest in space began when I was a child in West Union, Iowa. I remember hearing about the first supersonic flight when I was six years old. It was a big deal but it wasn't announced until a while later. Nevertheless, everybody I knew who had an interest in aviation was quite excited about it, as I recall. By the time I reached high school, I'd read science fiction, by Jules Verne and others, about going into outer space, but the books and stories usually involved voyages to the Moon, a planet, or other stars. I'd also seen some science fiction movies in the mid-1950s. I knew that there were real rockets, but there wasn't much film coverage of them then. Occasionally, you could see film of an actual rocket launch, but that was about it.

The concept of humans actually leaving the Earth never occurred to me, as I remember, although it was accepted widely in science fiction. In later years, I read that Konstantin Tsiolkovsky of Russia was probably the first person to look seriously at these concepts, at the end of the 19th century. Everything changed, of course, in October 1957, when the Soviets orbited Sputnik I. For the first time, it occurred to me that orbiting the Earth had real significance. I had read things that mentioned it, but it hadn't really sunk in, for me or most Americans, until the Soviets did it with Sputnik.

Having been good in science in high school, I decided (in the wake of Sputnik) that I'd like to make some rockets. I did know a little bit about them. Zinc and sulfur rockets were fairly straightforward and I knew the mixtures needed to make them work. With my science instructor's permission, I ordered the mossy zinc and the sulfur I needed for the rockets.

Together with Larry Enderes (a friend since the age of seven) we started our own rocket program. We began with pinched-off copper tubing and launched the rockets from my back yard. They'd go up a few hundred feet and come tumbling down. We very quickly learned how to make rockets that would go up a little ways. We also learned that they got very hot, that if we went over and picked up the rockets when they first came down, we'd burn our hands. So we came up with welding a nut on the end of the thin-wall tubing. We would load the tube up with the zinc and sulfur mixture and put fins on the back of the rocket to stabilize it, but we never launched those rockets in town. Another friend, Jim Grimes, lived in the country. Larry and I found one area on Jim's father's farm that was a little hillier than others, with a gully we could get down in to launch the rockets. We actually had our own little natural bunker. We could duck down behind a little ridge in this gully and not have to worry about explosions.

We learned to launch some fairly big zinc and sulfur rockets. Then we tried other kinds of propellant. We made bigger rockets, with bigger tubes, bigger fins, and more zinc and sulfur. We tried two-stage rockets which we got to work by using a technique that we had stumbled across. We found that if we put gasoline in the zinc and sulfur mixture for the second stage, it wouldn't ignite until all the gasoline had burned off. So we used gasoline as an internal fuse for the second stage. Obviously, it didn't stage optimally, but both stages did fire. We sent up mice and parachuted them back down to earth. Sometimes the parachute failed, which was very bad news for the mouse. When we couldn't find a mouse, we'd send up a cricket or something like that. Larry and I were emulating the space programs as we understood them, trying to see for ourselves how things worked. We were not the only high school kids doing this sort of thing, naturally.

We thought that we'd gotten a rocket up over nine thousand feet once, calculating by triangulation and timing. We could see from the ground that when the parachute came out, the rocket was at its apex. Timing from parachute release until the rocket hit the ground was our most accurate way of determining altitude. Thinking about it years later, I realized our technique wasn't all that accurate. That rocket could have reached only six thousand feet, or gone as high as ten thousand. The Chicago Tribune printed a story about our success, saying that it may have been a record altitude for an amateur rocket.

These rocket flights probably focused my interests toward space. Many other Americans my age with a bent toward physics, chemistry, engineering, or mathematics were also attracted to it because it seemed important to the nation, as well as being a field where one might make a useful scientific contribution.

I briefly thought of starting my career after I received my degree in mathematics from Iowa State University. I was 20 years old, and it seemed like a good time to start making money and quit accumulating college debt. However, because my roommate, John McElrath, seemed to be enjoying aerospace engineering so much, I decided to also finish a second degree in aerospace engineering, which I did the following year.

After graduating from Iowa State University with Bachelors degrees in math and aerospace engineering, I wanted to be involved in aircraft or space activity. My two primary choices were going to work for NASA—at Kennedy Space Center in Florida or Dryden Flight Research Center in California. This was during the Mercury program, shortly after John Glenn had orbited the Earth. At the time I felt that I probably could make a bigger contribution and have a more exciting career if I went to work at Dryden, a smaller place and not as much in the public eye as Kennedy. It also meant working on the very important problem of getting aircraft from the ground into high altitude and high speed regimes.

Other factors influenced my decision. Having spent most of my life in Iowa, California in the early 1960s had that special attraction of being where life was really happening, where all kinds of exciting things were going on. There was an ocean and mountains and California girls, as the song said. As a young bachelor, I felt California to be a good decision.

Dryden and Kennedy offered the lowest pay, with the exception of the the Army Corps of Engineers. My initial salary, I believe, was about $6,300. I had also received offers from Rockwell, Lockheed, and McDonnell in the $8,000 range. But work, not money, was the issue.

I had read and studied about what had gone on at Edwards Air Force Base and Dryden. I was particularly attracted to the X-15 program, which had had a fair amount of publicity in popular publications. I thought it sounded like what I might want to do.

Getting Started: From X-1 to Dyna-Soar

Dryden and the Air Force Flight Test Center (AFFTC) made the first step in going from the runway to orbit in 1947 with the first supersonic flight. The goal in 1947 was to penetrate both the supersonic and the transonic regimes to try to understand them. The X-1 was mated to the B-29 mothership and flown to an altitude where it could be dropped. The X-1 could then either glide back and land or fire its rockets and accelerate, which it ultimately did on October 14, 1947, in the first supersonic flight with Chuck Yeager.

The advantage of the X-1 team's approach was probably obvious to the team members and , as a young engineer, it soon became obvious to me. By using an air-breathing first stage (the B-29) and a rocket-powered second stage (the X-1), they separated the problem into two well-behaved, well-understood regimes. The rocket part was less well understood, but they expanded the flight envelope incrementally, so they understood the subsonic and transonic regimes most of the way to Mach 1.

This approach also relieved the problems of energy management and abort contingencies. By using the well-tested air-breathing propulsion system in the first stage, they could position the vehicle at a point where it could launch, knowing that if the rocket didn't light, the pilot could still land the vehicle on the dry lakebed. If it did light, the pilot could accelerate, and the energy and maneuverability of the vehicle would enable it to glide (once the rocket was out of fuel) to a safe landing on the dry lakebed. Another advantage of air launch is that at any point prior to the drop, the launch can be aborted. Aborting the launch preserved the system and bringing back both the mothership and the X-1 to earth allowed the problem to be sorted out. The two vehicles might have to be demated for some major modifications, or the problem might turn out to be something small, caused by the weather or an instrument failure.

In my opinion, the technique chosen in the 1940s to launch the X-1 is still the optimum way to fly from the runway into orbit and beyond. The D-558-II and X-2 aircraft also used the same approach of going to the launch point, being able to abort at any point prior to that, and then flying with incremental increases in velocity and altitude. Throughout all of this the pilot knew that he could get back to land on the dry lakebed at Edwards.

Energy management was probably the primary problem on Mel Apt's fatal flight in the X-2. He went faster than the flight planners had intended and could see that his landing site was going by too fast, so when he elected to turn the vehicle while it was still above Mach 2, he got into trouble. My opinion, based largely on what Richard Day – a retired NASA engineer and inertial coupling researcher – has said that Apt had gone beyond his normal landing point because he'd gone faster and, therefore, further. As he decelerated, he continued to go further from his landing point.

Apt knew the vehicle was predicted to be unstable above Mach 2 at a significant angle of attack. Pilots in such an airplane tend to be sure they can sort out those things out in terms of stability and control and handling qualities, get the vehicle pointed the other direction, and then proceed with the flight as planned. In Mel Apt's case, the inertial coupling was so large that he was unable to compensate for it with his piloting skills and he lost control.

As the NACA and AFFTC pursued the X-15 program, the idea was to understand the hypersonic region up to Mach 6 and the high-altitude region of rarified air where reaction controls had to be used to control the vehicle (because there was not enough dynamic pressure at the low atmospheric density there to use standard aerodynamic controls).

They used the same approach of going from the runway on the B-52 mothership to some predetermined launch point uprange. If the X-15 pilot had to abort, he might not be able to reach Rogers Dry Lake, but he could land on one of the alternate lakebed sites. Doing that meant energy management became a bigger issue. However, the idea was to select a specific launch point so that if the rocket failed to work – did not start, did not work long enough, or did not get sufficient power – the pilot could always abort to one of the uprange lakebeds along the ground track. If the rocket did work, the vehicle would be near Rogers Dry Lake and, as with the X-1 and D-558-II, the pilot would be able to make the intended dead-stick landing.

Using this approach throughout the X-15 program[3] we had many aborts. After takeoff, we would find something on the "no-go" list. We would abort the flight, return, and analyze the problem on the ground until we were fairly sure that everything was going to work. Then we would again mate the X-15 to the B-52, refuel, and fly back to the launch point. There's always a risk in these things, so we'd try to take care of all foreseeable difficulties while still taking a big enough step forward that we would continue to expand the flight envelope learning even more.

When I arrived at Dryden in 1962, the X-15 program was involved in envelope expansion, using the XLR-99 engine to get the vehicle going as fast and as high as possible. During the next few years, Joe Walker set the altitude record, above 350,000 feet, for a winged vehicle. After that we concentrated more on issues of high-dynamic-pressure heating in the hypersonic (above Mach 5) region.

To get the X-15 to go a little faster, tanks were added, making it a three-stage vehicle—the first stage being the B-52, the second stage being the reusable tanks that would be dropped and parachuted up range, and the third stage being the X-15 itself. We were still using the air launch as the first stage and the rocket, in this case, as both the second and the third stages. Some might prefer to call it two-and-a-half stages instead of three, but I always viewed it as three stages.

Another Road to Space

As the X-15 underwent development during the late 1950s, the future was taking a different shape. Since the late 1920s, it had been assumed that spaceflight would be an extension of atmospheric flight. This assumption was implicit in the so-called Round One, Round Two, and Round Three concept. The X-1, D-558, and X-2 aircraft constituted Round One, designed to reach higher Mach numbers and altitudes within the atmosphere. The X-15 (Round Two) was intended to reach hypersonic speeds and make brief forays to the edge of space. The X-20 Dyna-Soar (Round Three) was planned to test the ability of a winged spacecraft to reenter from space and make a controlled horizontal landing. Beyond the X-20, the Aerospaceplane concept envisioned taking off from a runway, fly into orbit, and then return to Earth. It was conceived to be the ultimate achievement of aviation technology. The X-20 and Aerospaceplane were cancelled in 1963 because of cost and other constraints.

With the launch of Sputnik I by the USSR on October 4, 1957, space became an arena for the Cold War rivalry. While a winged spacecraft was still a distant possibility, a capsule could be developed and launched using existing technology and boosters. As a result, both the U.S. and the USSR selected capsule designs for their early manned spacecraft

A capsule launched by a ballistic missile supplanted the winged spacecraft because the United States was in a race, first of all, to get people into orbit and, second, to go to the Moon and return safely. But this approach did not seem to me to be an impediment to ultimately going from a runway to orbit. It had a different objective, but it shared some of the risks and physics involved in transitioning from an air-breather on the runway to a vehicle that ended up in space, came back, and landed.

The initial capsule launches occurred during my college years, and I followed them closely, finding them exciting and interesting. These were the Vostok launches of Yuri Gagarin and Gherman Titov in the U.S.S.R., followed in the U.S. by the orbital flight of John Glenn after two suborbital flights by Alan Shepard and Gus Grissom.

The Vostok was a zero-lift capsule in the shape of a sphere. One side of the capsule had ablative material on it, and the capsule's center of gravity was positioned so that the Vostok would automatically reenter with the ablative side forward. The Soviets conducted their launches with elliptical orbits so that even if the de-orbit burn failed, in a week or so the orbit would decay, the capsule would re-enter, and the cosmonaut would land by parachute somewhere on Earth. The cosmonaut had enough oxygen and other supplies to survive that long.

In those days, we thought the Soviets lacked concern for the lives of their cosmonauts. Years later, it became clear that the Soviets had the same fears and hopes for their cosmonauts that we had for our astronauts. The Soviets' very simple approach of a spherical capsule with no lift and an elliptical orbit was actually one of the most risk-reducing approaches to putting a cosmonaut into space.

The U.S. picked a similar concept for the Mercury program, a cone-shaped capsule that was parachuted into the ocean for recovery by the Navy. Our capsules also did not generate any lift, so astronauts ended up spending a very short time in reentry but were under very high g's, which they found to be very uncomfortable. Of course, the cosmonauts endured the same discomfort in the spherical Vostok capsule.

In the early 1960s, nobody had flown a lifting entry in a capsule. In contrast, the later Gemini and Apollo capsules had low hypersonic maximum lift-over-drag ratios (L/D max)[4] in the neighborhood of .4 or .5. These were the first capsules to generate any lift, which, by rotating the vehicle, presented the opportunity to change not only where it landed downrange but also to change the lateral (crossrange) displacement during the re-entry, or the "footprint."

After the retros were fired on a Mercury capsule, the footprint was fairly small. For Gemini or Apollo, the capsule generated enough lift to increase crossrange and modulate downrange for a more precise landing. For Gemini and Apollo capsules, the landing was in water, in the Pacific and Atlantic Oceans. When things went as planned, the primary recovery ship would be nearby.

At that time, some of us at Dryden were working on a different concept. Partly it was an outgrowth of the earlier work done with winged vehicles like the X-1 and X-15. It also involved a hybrid concept, a wingless vehicle which could develop lift. The "lifting body" idea had come from research on warhead designs. Cutting a cone in half created a shape which could develop a small amount of lift. With fins and other control surfaces, the object could be steered and maneuvered. All of the lifting-body shapes of the late 1950s and early 1960s had hypersonic L/D max ratios of 1.2 up to 1.5, much greater than the L/D max ratio of the Gemini or Apollo spacecraft. With that much L/D max in hypersonic regions, the footprint was increased so a vehicle would have a crossrange that allowed it to return roughly to its launch point after one orbit.

Some fairly elementary orbital mechanics show that a hypersonic L/D max of 1.2 or greater will give sufficient crossrange to launch from and land at Kennedy. It also improves the downrange picture. By modulating energy with bank reversals,[5] as the Shuttle does, crossrange and downrange can be controlled to land a vehicle where intended.

The idea of an air-launched, two-stage orbital vehicle was more or less in the back of my mind during the 1970s and afterward. I'm not sure whether somebody told me the idea or whether it was my own extrapolation, a unique combination of things that I was working on then. In any event, these concepts used technologies that we really weren't ready to send into space in the early 1960s, when I started work at Dryden, but working on them was a good exercise in terms of getting us focused on the right problem. They led ultimately to the Shuttle[6] and will, I hope, lead to future vehicles as well.

[1] See Appendix A for the various terms for pilots used throughout the text.

[2] We will refer to the Dryden Flight Research Center (Dryden or DFRC) throughout the text, regardless of the name used at that time. See Appendix B for the actual names at the various parts of the chronology.

[3] Three-view drawings of the aircraft that I discuss in detail are given in Appendix C.

[4] The glideslope is roughly proportional to the inverse of the L/D ratio, or D/L. The higher L/D is, the shallower the glideslope is and the larger the downrange and crossrange.

[5] See Appendix C for the terminology used for various Shuttle configurations.

[6] We use the terms "Shuttle," "Orbiter," and "Shuttle Orbiter" interchangeably, except where context makes it clear that we are referring to the Orbiter, external tank (ET), and two solid rocket boosters (SRBs). The term "Shuttle stack" refers to the Orbiter, ET, and SRBs.

I

Apprenticeship of a Young Engineer

As graduation from Iowa State drew near, Don Kordes from NASA Dryden interviewed me and made me an offer. So with my degrees in math and aerospace engineering, I started work at Dryden in 1962. My primary assignment was to support the X-15. [2-7] (Fig. 1) I'm sure my job description said more than that, but that's how it was explained to me. Being fresh out of college I didn't know exactly what this meant. My first supervisor, Gene Matranga, was quite busy at the time, shepherding the maturation of the Lunar Landing Research Vehicle (LLRV) concept. Gene had a lot of experience working on the X-15 and other vehicles, and occasionally made himself available to me.

Fig. 1. A 1962 photo of the X-15 number 2 with the XLR-99 engine burning after launch. (EC88-180-1)

Data Processing

But since Gene was so busy, I mostly talked to my actual supervisor at the time, Ed Holleman. Ed said, "Well, just pitch in where you're needed." That was the sort of mentoring that was done at Dryden then. I found engineers who had some lower level analysis to do that was time-consuming or just not much fun, and they'd let me volunteer to do it for them.

The data systems on the X-15s were very simple by current standards: a film that recorded the movements of the galvanometers as they reflected light from tiny mirrors onto the film.[8] If you looked at the data from one today, it would look like a bunch of random squiggles, lots of them, all over a piece of film three or four inches wide. We read the traces from copies of the film, or we worked from Ozalid copies. They had a film scale that looked like a clear ruler with various scales. For identification most channels had a uniquely-placed gap in each trace. There was a zero point and a scale for each signal. We would read off the scale from the zero point. A person could get pretty good at using the film scale and those people who had been doing it a long time always got the best answer. With some practice we all picked up the knack and would then would perform various calculations, converting the position of that trace on the film to engineering units.

The conversions were done using what today would be called a spreadsheet. Above each column, I would write down each step of the calculation – such as add A to B, or divide A by B – as I worked my way across this sheet. The time of the flight was on the film, so I could read it off for the interval I wanted. Most of what I was doing were time histories, so each trace moved throughout the maneuver that I was reading up. I usually read the traces at intervals of ten data points per second. I think the more skilled people could do a little better, but that was the standard in trying to read traces. As can be imagined, there were a lot of opportunities for me to volunteer to do those calculations. Sometimes it wasn't volunteering, though. People would get behind, trying to get results calculated in time to make the input for the next flight plan or to understand something before the next flight and it would be all hands and the cook helping.

Doing these conversions could become tedious, but as a new engineer, I found the task interesting. If the data needed to be very accurate we used the Frieden calculator, a noisy one-of-a-kind mechanical device that multiplied, divided, subtracted, and added to many decimal places accurately. A few of them, the very special ones for the very privileged people, would take a square root. If you felt like creating a little turmoil, it was fun to go into a quiet office, plug a number into their Frieden, start it up taking the square root, then leave the room. I have no idea how the Frieden took the square root mechanically, but it made a lot of racket for a long time. Everybody in the office knew what you had done, and later they would berate you for creating all that uproar. We didn't have a Frieden in the first office I was in, and there were five engineers in that office when I started.

In any event, I worked my way across the spreadsheet until I had the number both calibrated and corrected for all known effects, putting it in the desired form. I'd do this for each time slice. If each time slice was for ten seconds, there would be a hundred data points. I needed to read off the data and go through all of those calculations until I had the final correct number. As long as the spreadsheet was set up right, it was self-checking. Because it was a time history, a little engineering judgment would show if one of the numbers was too far off from the others.

If it was, another engineer or I would check to see if it had been read wrong on the film scale or if a calculation error had been made.

The primary tool used for these calculations was a slide rule. In a TV show at the time an engineer explained what he did and how a slide rule did additions and subtractions. We only used the slide rule for multiplication and division, though. We usually did additions and subtractions the old-fashioned way, writing the numbers down and adding or subtracting them. Or we used the Frieden. Calculating with a slide rule is a lost art now, but three-place accuracy using a 10-inch slide rule was possible for very complicated calculations if you were careful. Working on experimental data, three-place accuracy was enough in most cases because other sources had errors larger than that.

Doing these calibrations from the onboard recorded film was the primary way of putting the flight results into the form needed by the research analysts. One of the obvious things to do after we'd made these spreadsheets was to plot the data. This was a very tedious job that wore out many of the engineers as well as the people we called math aides. There was a group of four or five math aides, usually women, and often called "computers" – when I started at Dryden in 1962. They read the film, did selected segments (either more accurately than the rest of us or with different parameters), and sometimes did the plots as well. We didn't have automated plotters at the time, so numbers had to be plotted by hand, usually as time histories. Sometimes there were cross plots of various parameter combinations, such as Mach number and angle of attack, useful to an engineer to examine simultaneously.

These tasks were my main work assignments at the time. When the X-15 was flying frequently, they kept me busy. When it was not flying so frequently, I'd more or less get caught up, as would everyone else.

X-15 Handling Qualities Studies

My first assignment at Dryden was to work with Ed Holleman on the X-15's handling qualities – that is, how the pilot and the aircraft interacted and how we could make the vehicle better, easier, safer, and more comfortable for the pilot to fly. That's really what I was doing with stability-and-control analysis on the X-15. Stability, in this case, refers to the tendency of the vehicle to fly more or less in the direction the nose is pointed and to not oscillate, spin, or swap ends. The control part has two aspects. The first is the pilot's interaction directly with the aircraft dynamics. In other words, as the pilot moves the controls, he expects a predictable response from the vehicle. The second aspect is the control system itself, used primarily to augment the handling qualities of the aircraft so that they are predictable to the pilot. The control system also stabilizes the vehicle, a key aircraft requirement as one gets close to the edges of the flight envelope.

In addition to the handling qualities and the stability and control of the X-15, I was also doing some work assisting other engineers in analyzing heating data and studying the boundary layer of the X-15. In fact, we flew the X-15 in large part to see if we could change the heating rates on the aircraft's skin, which explains why I became involved in studying the boundary layer at hypersonic speeds..

I also looked at was proposed modifications to the X-15. These were ways to make it go a little faster, as well as different configurations. The primary one at that time, which was classified, was converting the X-15 into a delta-wing airplane, making it a little heavier, a little longer and able to fly up to Mach 8 (Fig. 2). I worked on various air-breathing propulsion ideas and what those modifications might do to the stability and control of the X-15. An air-breathing engine needs an inlet for the air and adding that would be quite a large change to the exceedingly smooth shape of the X-15.

Fig. 2. One of many proposed follow-on X-15 designs with a delta wing modification. (E-13857)

After collaborating on the X-15 for a short time, I realized that the key enabling technology for higher speeds was an air-breathing propulsion system. To get to those speeds with a rocket meant being limited by the amount of oxidizer that the rocket could carry. An air breathing vehicle doesn't have to carry the oxidizer—it uses the oxygen in the atmosphere. There were many studies at the time about how we might integrate a ramjet engine into the X-15 and how we might use it to accelerate. These proposals were not usually very well defined. They were preliminary paper studies, looking at different aircraft configurations, all starting with the X-15. [9] (Fig. 3).

I also worked on obtaining stability and control information from the X-15 flight data. We looked for areas where we needed more information and then

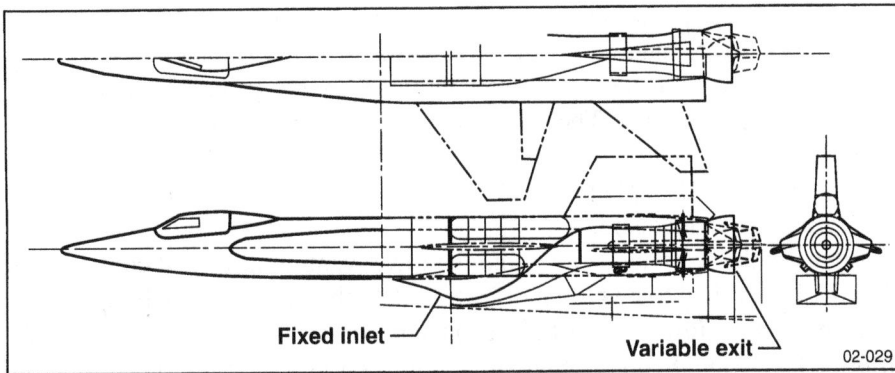

Fixed inlet — **Variable exit** — 02-029

Fig. 3. Marquardt ramjet proposal from 1967 that would have enabled the X-15 to cruise near Mach 5. (Graphic 02-029)

specife the Mach number, angle of attack, altitude, and dynamic pressure that we needed for the maneuver we requested. If our request was approved, the maneuver was put into the flight plan. During this planning process, we worked in the simulator with the pilot and other engineers. The flight planner and test pilot would "fly" the draft flight plan and ensure that enough flight time was available to do the various maneuvers requested by all of the disciplines—heating, structures, loads, aerodynamics, stability and control, handling qualities, propulsion, and performance.

We needed to obtain the data for two reasons. One was to improve our understanding of the X-15 as we continued to expand its envelope. The classic NACA approach to envelope expansion was to fly just a little faster or higher than we had flown previously. We analyzed the data that we obtained at one condition, then asked the pilot to get more data at a slightly faster or higher condition. As we did this, we gained confidence. We always inched close to the edge of the flight envelope with the faster or higher condition. However, if a problem occurred, we could always slow or descend to the previous point where we had good information about how the vehicle flew, how much heat it was exposed to, and what other issues might arise. The other reason was really the primary objective of the research program: to obtain data in a flight regime where nothing had flown before. This included local measurement of heating, loads, boundary layer, and pressures on the vehicle itself, so that we could find out if our understanding of the interaction between an aircraft and the atmosphere was correct and complete.

X-15 Envelope Expansion

In the first year I was at Dryden, we repeated the envelope expansion done earlier on the X-15 because, beginning in 1962, the lower ventral fin was removed for all flights. We were going back through some of the conditions that the airplane had been flown at before, looking for any indication of controllability issues. We used the Cooper-Harper scale to measure the handling qualities [10], in which a rating of 1 is perfect, and a rating of 10 is awful. We'd draw contours of Mach versus angle of attack to show where the ratings were better than $3\frac{1}{2}$, between $3\frac{1}{2}$ and $6\frac{1}{2}$, and worse than $6\frac{1}{2}$. Then we'd define those contours as flight regimes, whether the vehicle was flyable or not. Any rating worse than $6\frac{1}{2}$ means a bad airplane. A rating of $3\frac{1}{2}$ or better means a very good airplane and a rating between $3\frac{1}{2}$ and $6\frac{1}{2}$ is adequate.

Originally the X-15 landed without the lower ventral because it couldn't land with it in place. It had to be blown off the X-15 with pyrotechnic bolts and parachuted down a few seconds before touchdown, right after the X-15 crossed California Highway 58, on the edge of Edwards Air Force Base. The crew would go out and pick it up later. Most of the X-15 flights I worked on had no lower ventral on for the whole flight. We compared the handling qualities without the fin to those with the fin (at the same flight condition) by performing the same stability-and-control maneuvers that were flown before.

We also had "dampers" back then, which now would be known as a control augmentation system (CAS) or a stability augmentation system (SAS). Dampers fed back the velocity rotation rates to the control surfaces to reduce, or damp, the motion. The dampers improved the handling qualities, the apparent stability and control, and the flight dynamics for the pilot, who almost always flew better with the dampers on than with them off. However, for short periods the dampers could be turned off so the pilot could evaluate the X-15 without them. We did stability and control maneuvers and we also asked for pilot ratings. At the postflight debriefing, the pilot's comments helped us interpret the ratings. For example, if the pilot said it was a 4, it might be because it had a little lag in roll or because the pilot couldn't stop a motion on a dime. Pilots at the postflight listed what they did or did not like about the vehicle. Using a bare number in a rating sounds like an oversimplification, but the pilot comments that accompanied it helped us determine what we needed to do to improve the airplane, so it would fly better at that flight condition.

Just to summarize, what we mean by envelope expansion is that we've flown at one point in the flight envelope with a vehicle, and on a succeeding flight we fly at a little higher Mach, a little higher angle of attack, or a little higher or lower altitude than we have before. Engineers abhor extrapolation. We do interpolation and know its pitfalls, but extrapolation can be very dangerous if we go in large steps. Doing flight envelope expansion, we expanded or extrapolated in small steps; every half a Mach number, every two degrees angle of attack or every few thousand feet. We felt we understood that step after we had analyzed the stability-and-control maneuvers, the pilot ratings, and the effects of dampers on and dampers off. On the next flight, we increased the Mach number another half or the angle of attack another two degrees, one at a time. Each step meant that we knew the airplane could fly higher, faster, at more conditions, until we knew everywhere the airplane could and couldn't fly. That's the envelope.

Engineers like to build a little, test a little, and then propose the next test that's not very far from where we've been. That's how we engineers "bootstrap" our way from a place where we've flown to a place a little bit beyond it. Once we've understood all of those points, we feel we've expanded the envelope to its limits. Once we've flown all of the points many times with many different pilots, we start to be confident that we understand the vehicle. Understanding the vehicle means understanding the envelope and its limit. The envelope may not be risk-free but the risks are understood. The placards (restrictions), comments, or flight rules for that region are added to the flight plan.

The original X-15 envelope expansion went just over Mach 6 and up to an altitude of 250,000 feet or so (Fig. 4). The third X-15 had a new flight-control called an adaptive system that did a better job of switching smoothly from the pure reaction-control jets used above 300,000 feet to the aerodynamic controls, used around 100,000 feet. The same concept and technology were used for the Space Shuttle two decades later. The Minneapolis Honeywell adaptive flight control system that had been developed for the Dyna-Soar (which was later cancelled). It flew every flight with this flight control system.

In order to fly faster than Mach 6, it was necessary to modify one of the X-15s. The second X-15 was rebuilt after an engine failure forced pilot Jack McKay to land on Mud Lake in Nevada on November 9, 1962. The landing gear collapsed, flipping the X-15 on its back and damaging the vehicle extensively,

Fig. 4. The number one X-15 just after being launched on June 8, 1959. This is Scott Crossfield's first glide flight flown with the X-15 nose boom, North American Aviation livery, and the XLR-11 engines. (E88-013-15).

seriously injuring McKay. The aircraft, rebuilt as the X-15A-2 (Fig. 5), flew again in the summer of 1964. Jack returned to flight in April, 1961. When the X-15A-2 returned to service, it was 29 inches longer, to accommodate the fuel for a ramjet experiment. It had external drop tanks, one tank full of liquid oxygen, the other liquid ammonia. We considered in-flight separation of tanks from the X-15, dropping them at about Mach 2 or 2.5. They needed to leave the airplane without hitting it or damaging it, because the pilot would continue to fly the airplane to landing, so we didn't want a big change in stability going from the tanks on to the tanks off. When we flew above Mach 6, the X-15A-2 would be covered in an ablative heat shield for protection from higher temperatures. But first, before we could fly fast, we had to expand the envelope and have the pilots shake down the airplane to make sure it was ready to fly beyond where the X-15 had flown.

We reconfirmed the flight envelope in bigger steps than before, as I recall, because the pilots didn't find the X-15A-2 to be that dissimilar from the other X-15s. They probably noticed some mechanical differences, but the stability-and-control derivatives and the flying qualities were not markedly different. The nose slam-down on landing was definitely harder in the X-15A-2. The most important point of the envelope expansion was to verify the flight envelopes with dampers

Fig. 5. December 1965 photo of the X-15A-2 and active pilots (left to right) Joe Engle, Bob Rushworth, Jack McKay, Pete Knight, Milt Thompson, and Bill Dana. (E14169)

on and dampers off to prepare for the further envelope expansion for higher Mach number and for testing the ramjet.

As we continued validating the stability and control of the X-15 with the lower ventral off, we began more testing of the heating on various parts of the skin and of the aerodynamic boundary layer on the vehicle. Near Mach 6 and altitudes around 100,000 feet, the dynamic pressure could exceed two thousand pounds per square foot, the highest for any program that I worked on in my entire career.

A lot of early understanding of hypersonic boundary layers came out of the X-15 program, done mostly by other groups at Dryden, but when you work on something like the X-15, you soon become aware of the other key disciplines, such as structures, materials and loads, and propulsion, in this case, rocket propulsion. Others included aerodynamics; the flow of air around the vehicle; the atmospheric conditions (temperature, pressure, and density); the character of atmospheric turbulence; the controls and avionics of the airplane that the pilot interfaced with; and the artificial stability.

Instrumentation consituted the final discipline that I considered to be distinct. A lot of instrumentation development occurred during the X-15 program. I think instrumentation is usually not given the same attention as some of the other fields get.

The ball nose on the X-15 for example, represents an instrumentation development. Many of the strain gauges and the mounting techniques used to get sensors to work in the exceedingly hot environment with high stresses were properly research areas. Those of us working in research disciplines benefited when those instruments worked properly. The ball nose told us the angle of attack, angle of sideslip, the dynamic pressure, the Mach number, and the altitude. The other sensors told us the state of the boundary layer.

One of the big questions was whether the X-15 boundary layer would be turbulent or laminar. We found out from the flights that it was overwhelmingly turbulent [11, 12, 13]. We also found that the heat generated by the turbulent boundary layer was 30 percent lower than predicted [13, 14, 15]. The first use of PCM telemetry occurred in one of the rebuilt X-15s, a development that alleviated the tedious job of reading film. I think the first PCM system was developed for Dyna-Soar, and the X-15 benefited from it. The first PCM system we used was a CT-77, a system I used on many programs, up through the F-15 Spin Research Vehicle (SRV) program in 1981.

Everyone benefited from the instrumentation experiments, not just the engineers working on them. All engineers who work on airplanes become somewhat multidisciplinary. If we don't, we might propose things that the vehicle can't or shouldn't do. We all learn a little about the other disciplines, to know how they interact, particularly with our primary disciplines.

I met Larry Taylor, a fellow engineer who would become my primary mentor for my first five years, through some of this interdisciplinary work. He was primarily interested in the control system, and I was primarily interested in the aerodynamics and stability and control. We found a lot of common ground, common interest, and ability to communicate. I knew FORTRAN, the computer programming language, and Larry was learning it, so we started working together. We looked at using the computer to do the tedious work we had been doing manually to calculate all the parameters for the vast number of data point so we could examine many different maneuvers on many flights. This would mean we could do hundreds of times the analysis with the digital computer as we could ever do by hand. Once we had the program checked out, just as we checked our hand calculations on our spreadsheets, it would give the right answer every time.

In those days computers only made one kind of mistake, when everything was absolutely outrageous. Any other mistake was an error in the coding, and that was the programmer's fault, not the computer's. I was programming the computers to do the calculations that I wanted. The digital computer I was using was mostly the domain of the programmers. They were in the Data Systems Division (I was in the Research Division). There was always a little conflict involved in getting them to let Larry and me use their IBM 704 computer, because they really owned it. We finally arranged it so that we could use it on a noninterference basis with some oversight (and, I must admit, some prejudice). But we did get to use it enough that we developed programs for both the X-15 and the lifting body data and we got huge benefits from them.

Larry and I also would use the Army's IBM 1620 computer, mostly for code development. Because we had it all to ourselves, being the operators as well as the programmers, we could get a lot done. In those days, there were many steps that you had to go through that made you aware of how to make programs more efficient. Program efficiency was of paramount importance back when computers were slow and had very little memory.

As we did the envelope-expansion flights, we analyzed the maneuvers from the last flight. Perhaps we were trying to increase the Mach number or the altitude, or maybe we wanted a combination of lower altitude and higher Mach number for higher dynamic pressure. We were always working on several different types of flight maneuvers. Once we did the analysis, we plotted the data so we could see the story the airplane had to tell. Then we told that story. Since I was a junior engineer, I told it to the more senior engineers. If it was important, they might suggest I do something a little differently. I would do so and bring it back. Eventually, we would convince the people in charge of the X-15 program that we understood the aircraft characteristics at the current edge of the flight envelope, so we were ready to go a little faster, a little higher, or to a higher angle of attack.

We'd define the new flight conditions and the kinds of data that we needed. For stability and control, it might be half a Mach number higher or flying with the dampers off. For lateral-directional data, we'd want rudder and aileron pulses, and we would want the pilot's comments as well. All of these requests would be added to the other requests from other disciplines and the other engineers within our own discipline. Then, the flight planner would start laying out the next flight on a given airplane to include as many of these proposed maneuvers as possible, figuring out how to integrate them into a smooth flight plan so the pilot didn't waste time going from one flight condition to another. While integrating these requests, the flight planner also needed to consider energy management and flight safety rules.

X-15 Flight-Planning and Tech Brief

The flight planner and some of the senior engineers would put together the flight plan. In the early days, the people I remember being flight planners were Warren Wilson of NASA and Johnny Armstrong of the Air Force Flight Test Center (AFFTC). Johnny Armstrong, who I still see from time to time, was more senior than I. I think he came to AFFTC several years before I got to Dryden. In later years, the flight planner was Jack Kolf. Jack came to Dryden after I did, and he and I were, for a brief time, in the same office. We were both working on the X-15. By then, I was also working on the lifting body program. I remember that Jack always used to give me a rough time about working on something as peculiar as the lifting body when there was something as exciting and important as the X-15 around. After the X-15 program started to wind down, Jack found himself going from flight planner on the X-15 to flight planner on the lifting bodies. It was fun over the years to pick on him for working on lifting bodies just as he had picked on me earlier.

Those of us at Dryden and AFFTC were a fairly integrated team. The engineers at Dryden also had their counterparts at AFFTC. At the disciplinary level we weren't as integrated as we were at the operational level. Once we got to the flight-planning stage, it became very much a joint operation. The flight planner was the connection between those two levels, because he planned not only the maneuvers on the flight, but also the resources available on the scheduled day. The pilots and the flight planner had a lot of help, too. The flight planner didn't operate in a vacuum, but got lots (maybe too much) of input from engineers, the pilots, and the project managers. Once the flight plan was completed, the flight planner had the pilot practice it over and over in the simulator. Sometimes pilots would discover that they couldn't get quite what you wanted, although they always could come close. They would ask, "Is this close enough, or do you want to wait for another flight?" In many cases it was close enough but in some cases it wasn't, or else you preferred to get a really clear look at something on a later flight just to make sure you got the data you wanted.

In a nutshell, this was how the flight plan came together. A request for additional data was justified if it would increase the knowledge within the discipline, if it would examine a local phenomenon, if it would expand the envelope, or if it would increase safety for a specific X-15 problem.

The one reason we always flew the simulator with dampers on and dampers off was to practice backup procedures. We had to be prepare for the possibility that the dampers would not function properly in flight and the pilot would have to fly the X-15 dampers off. Therefore, the pilot had to practice an alternate flight plan in case the dampers kept kicking off or didn't work properly. This gave a safer profile to fly getting the vehicle back. Of course we verified everything we studied in the simulator in flight. We flew with dampers on and dampers off to examine the difference and to give the pilot practice. This also was very useful in defining the X-15 flight envelope, as well as demonstrating the contingencies for various failures.

John Perry, at that time, was the primary implementor and operator of the X-15 simulator, the primary test and training tool. It was used by the engineers to sort some problems out, it was used by the flight planner to lay out the flight plan, and it was used by the pilot to train on both the normal profile and with various failures and contingencies. I think the pilots probably flew 20 hours on the simulator for every mission that they flew (the missions were about 10 minutes long), so they kept the simulator fairly busy in the daytime when all three X-15s were on flight status. Any work that we wanted to do on the X-15 was done off-shift, with others supporting the simulator. I remember Gene Waltman doing some work then as well.

John Perry and I were good friends, so I always enjoyed being there when they were practicing the flights. That wasn't often encouraged, because they tried to maintain a very professional environment without a lot of external chatter. Consequently, although we were familiar with the simulator and knew that it was the primary tool, we didn't usually work directly with the pilots on the simulator unless our presence was required to make sure a maneuver was being performed adequately or to help them understand some phenomenon they were seeing in the simulator.

After all the flight planning was done, we had the tech brief. The research people and the pilots primarily determined the technical objectives, the constraints-that is, the "go/no go" set of parameters-and the value of getting this data. In other words, they prioritized the requirements. The tech brief was basically a peer review as each engineer presented what he had and what he wanted, then answered the question of others.

It was really a very healthy environment, one that I miss at Dryden, because it seems as if proposals are signed off almost automatically now. Those tech briefs not only made you defend your request, but they were learning opportunities because there were discussions of every discipline. They included reactions from different pilots and from different technologists, for the same people weren't always in the room. In a tech brief, you could pick up the overall technical approach and the operation of the airplane itself from the research point of view. As the program matured, there were discussions of local aerodynamic activity occurring somewhere on the X-15, aside from the vehicle aerodynamics, in these tech briefs. I, as a young engineer, found them to be very useful. I don't know, some of the older engineers may have found them overdone, but I enjoyed them and I learned a lot from them.

After everything was resolved in the tech brief, there was a crew brief, usually the day before the flight, that included virtually all the operations personnel, as well as all the research and instrumentation engineers who would be monitoring strip charts during the actual flight. In the crew brief there might be discussion of something that wasn't quite up to snuff, as well as discussions of the limitations and flight rules. We went through the flight plan, which were on the flight cards that the pilot carried in the X-15 during the flight. There weren't very many blank spaces in the flight cards. The pilot basically went through the cards, from performing one maneuver at a given Mach number and altitude to another, until he needed to focus his attention on terminal energy management for landing properly.

X-15 Flights - The View From the Control Room

The following day we would make the X-15 flight itself. Which personnel were in the control room and what they did varied, depending on what maneuvers were being flown, which airplane was flying, who were making the requests, who was traveling or on vacation, and who would be monitoring various strip

Fig. 6. Strip chart recorders that were used to monitor X-15 flights. This photo appears to show data from Scott Crossfield's November 5, 1959 crash on flight 2-3-9, caused by an engine fire. The vehicle's fuselage structure failed due to a hard landing on the Rosamond lakebed. (E-5031)

charts during the flight (Figs. 6 and 7). The monitoring was primarily for flight safety. We kept track of what the airplane was doing. Frequently I looked at roll rate, yaw rate, pitch rate, angle of attack, angle of sideslip, the control positions, and whether the dampers were on or off. (See Appendix D for definitions of these and other parameters.) I wasn't there for every flight, as different people would monitor these strip charts on different flights, based on availability and expertise. The control room environment created for the X-15 was the predecessor of the mission control center used for the later capsule and Shuttle flights.

Most of the engineers assigned to the X-15 program, at one time or another, would monitor strip charts. We had to be in the control room (Fig. 8) prior to B-52 engine start. As I was looking at stability-and-control parameters, I would begin seeing movement in the strip chart traces of some telemetered signals as the B-52 started to taxi. Before we got the PCM system, we had just the film on board and also had fewer things to watch in the control room.

We could hear the chase pilots as well as the B-52 pilots and the X-15 pilot on either the loudspeaker or the headphones in the control room. The controller,

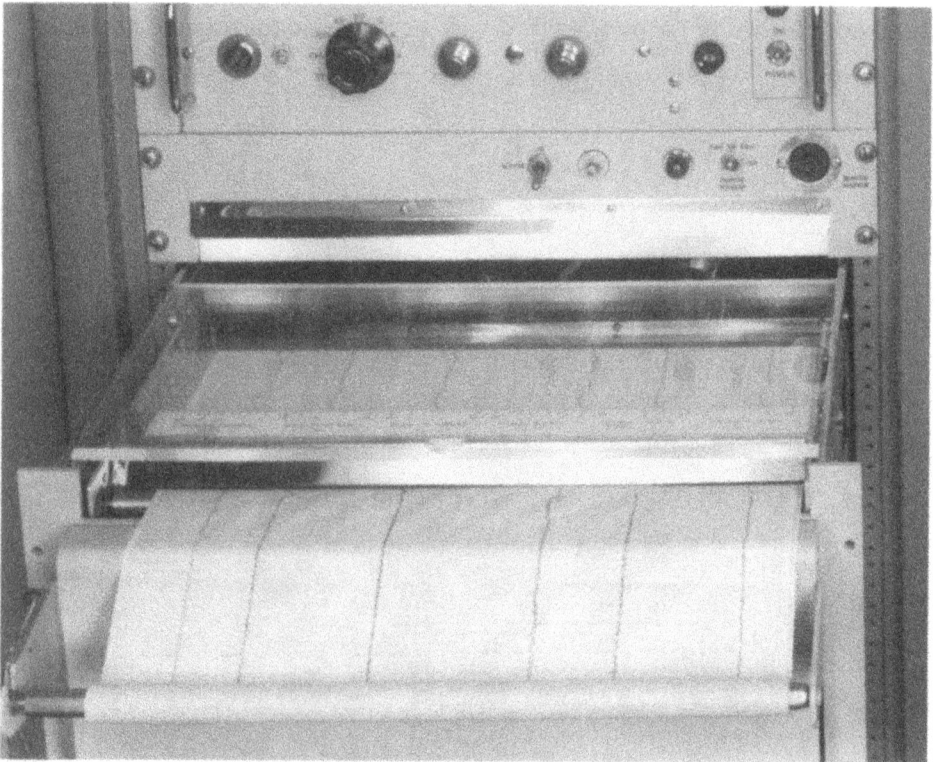

Fig. 7. Close up of the left strip chart recorder in the previous photo. The traces, from left to right, show the malfunctioning dynamic pressure, angle of attack, angle of sideslip, normal acceleration, upper vertical stabilizer position, and right horizontal stabilizer position. The instant of launch from the B-52 can be seen as a dip in the angle of attack and then normal acceleration appearing on the lower portion of the strip chart paper. (E-5031).

14

called NASA-1, was the only person in the control room who talked to the pilot, so if you had any input, you made it to the controller who then decided whether to ignore it as unimportant or communicate it to the X-15 pilot, the B-52 pilots, or the chase pilots. That kept 20 or 30 engineers from making their own comments directly over the radio to various pilots. This kept the radio chatter down, but the engineers on the strip charts always conversed at a low level.

Fig. 8. The control room during X-15 flight 3-55-82. The September 14, 1966 launch was over Delamar lake and Bill Dana reached 254,000 feet and Mach 5.12 on the flight. The personnel at the consoles are clockwise from the left: Don Olson (background left, back to the camera), Ken Iliff, Jim Adkins, Jack Kolf, Pete Knight, and Joe Vensel (foreground, back to the camera). Ralph Mayes is at the plot board behind Knight and Perry Row is at the other plotting board. The B-52 ground track can be seen just in front of Row's face with Delamar lake near the B-52's semicircular turn. Lake Meade is the heavy dark area on the map. (E-15774)

Once the B-52 mothership took off, the X-15 might have to launch early, so we had to be there for the whole flight. We engineers were responsible for monitoring the data available in the control room. If we saw spiking or drifting in any parameters, we might need to tell someone on another strip chart or the flight controller, who in turn might convey it to the pilot, especially if it was something the pilot might not be aware of, such as a symptom of an impending problem.

The maneuvers would be called out by the flight controller as we went through the flight cards. There would be talk between the people on various strip charts as the airplane came south from its uprange launch point, doing various maneuvers and test points. It all went fairly quickly. An X-15 flight, even a long one, lasted only a little over ten minutes. We were very attentive during that time because we were pressing from point to point, putting aside any other distractions.

In its way, an X-15 flight was quite exciting to a young engineer, although I don't recall ever pumping adrenaline during X-15 flights the way I did during lifting-body and early Shuttle flights. I'm sure my respiration and heart rate went up during an X-15 flight, because this was my chosen work. A flight was where the rubber met the road, and it was exciting. I suppose I just don't remember the adrenaline rush I likely had during the X-15 flights. After all, I was only a junior engineer. There were usually 10 or 15 people in the control room who knew all about what to do and how to handle emergencies. I figured they knew what they were doing, because I knew I didn't. In most instances, I would have been useless in an emergency.

Test maneuvers normally ended before the X-15 reached high key, because the pilot had to set up for the landing. High key was around 35,000 or 40,000 feet, and that's what made it so much fun to watch from the ramp behind Dryden. We could usually pick the X-15 out as it came down from the north or the northeast. Sometimes we could see contrails, the X-15 venting fuel, or the chase planes. Lots of people, who either didn't work on the X-15 or didn't have any duties for that flight, would also come outside to watch the X-15 land. If you

Fig. 9. Landing of the number three X-15 on September 28, 1964 (flight 3-35-57) flown by Joe Engle alongside the F-104 chase plane piloted by Pete Knight. (EC88 180-9)

weren't in the control room, there wasn't too much you could do except listen to the radio and go watch the approach and landing. There were good places to view the X-15 if you weren't in the control room at Dryden in the mid-1960s.

As we watched it come over at high key, the X-15 would turn left and head out toward the rocket site and Boron in a fairly large circle. It was descending all the time, of course, because it was unpowered. As it got to low key over the dry lakebed, the X-15 was three, four, or maybe even five miles away. It would then head directly for the runway it was going to land on. The X-15 always landed on lakebed runways because it had skids, not wheels, for the main gear.

The view was better on some flights than on others, depending on the lakebed runway selected (Appendix E). Sometimes we used a more distant runway just because it was the first one to dry out when the lakebed was wet. Most of the time, though, the runway was one that we could see well from the Dryden ramp, so we could watch it come in, float over the ground, and lower its gear as it settled onto the lakebed (Fig. 9). Sometimes it would pick up a little dust before it actually touched down. Then the skids at the back end would touch down and very soon after that the nose would slam down. It was an ungainly landing and a little hard, but I think the pilots didn't complain because it was a very positive way to land. The X-15 would skid along the lakebed runway for a mile or so, depending on how fast the landing had been. If we waited a while we were sometimes rewarded with a flyover of the X-15 on the lakebed by the B-52 and the chase planes as they returned to land (Fig. 10).

Once the X-15 was secured after a flight and returned to the hangar, the flight party would begin. In the early days, we usually gathered at Juanita's in Rosamond. As people got off work, they would stop there for a few drinks. As time passed, more and more people joined in from Dryden, including people who were not working on the X-15 program. We drank, talked, and had a fine old time.

Usually the crews were the first to reach Juanita's. They had to start work very early, so their work day basically ended when they got the X-15 back in the hangar. Next to arrive would be all of the other X-15 pilots, as well as the chase pilots and other Dryden pilots. Finally, the engineers and technicians would show up, along with other people who may or may not have been working on the X-15. I was young when I started working on the X-15, barely of legal drinking age, and I was in a car pool with several other engineers. We would stop at Juanita's, have a few beers, relax, and have a good time. I wasn't a big drinker, and I don't remember ever getting drunk, but having a few beers on an empty stomach tended to be noticeable. For me, it was always beer, often bought for me by someone else.

I'd get to Juanita's by 4:30 p.m. or so. I lived in Rosamond for the first year and a half I was at Dryden, so it wasn't out of my way to stop and celebrate with everybody else. We really didn't seem to need a good reason for gathering at Juanita's. It was simply a part of the culture of Dryden and AFFTC. I think the flight parties were a holdover from the old days when there wasn't much to do

Fig. 10. X-15 shortly after landing. A B-52 and two F-104s are about to do a flyover. This is probably Rushworth's flight 2-22-40 on May 8, 1962. (E-14792)

after work but celebrate flights. Flight parties continued clear through the X-15 program. While there were many flight parties for other programs as well, I remember the X-15 parties as being the largest, the longest, and the drunkest. Drinking was not looked at upon in the 1960s the same way it is now, although I suppose it has always been considered bad for the liver. You could hurt yourself while drinking or be hung over and not get enough sleep, but other than that drinking was accepted by most people back then. Driving after drinking was a bad thing, but many people did it. Nevertheless, the Dryden flight parties weren't meaningless drunken brawls. That isn't at all the impression of them that I had then nor the one that I would want to create now. They were the celebration of an important event.

On the X-15, the crews worked many hours to get the airplane ready for flight. They dedicated years of their lives to this airplane and they were well aware that every connector they put in and every bolt they tightened were vital to the success and safety of the flight. They also kept track of every tool they used to be sure it was put back in their toolboxes, not left in the aircraft. They worked under a lot of stress for two, three, and sometimes four weeks preparing the X-15 for flight. There was a similar level of tension in all the people involved—the operations (ops) engineer and, of course, all the support people. The stress wasn't just from working long hours. Most of it came from knowing that if they screwed up, they would lose an airplane and maybe a pilot, a friend who trusted them, with it. That awareness was shared by everyone working on the airplanes, especially the technicians and mechanics.

The crews had a great deal of pride in their work and competed, informally, to be the quickest and the best. On the day of flight everyone, especially the pilot and crew, pumped a great deal of adrenaline. For the pilot, that was so from suiting up and cockpit entry, to B-52 takeoff and the air-launch point of the X-15, to racing through the flight cards getting all of the data in a ten-minute flight, and to getting energy management under control to make the landing on Rogers Dry Lake. When things didn't go well, of course, there was a lot of extra tension as well, as when the engine didn't light, the SAS didn't engage, or some of the instruments weren't working right.

The research engineers, monitoring safety of flight, probably weren't as stressed as the X-15 crews, but we were stressed as we looked for things such as actuators that weren't working right. When the task was accomplished, where all of the hard work that everybody had put into it was successful, we knew our contributions had been well worth the effort.

A feeling of euphoria is common whenever a significant goal is reached, and the X-15 flight parties were how those of us working on the program released our tension. Everybody gathered together and told stories about other flights, near misses, and funny incidents. The more we drank, the less relevant the stories became and, probably, the less true. Nonetheless, I found the two or three hours that I would spend at them to be more enjoyable than many of the other parties I went to where people bought beer and snacks and invited people over.

I think the flight parties also encouraged a spirit that I miss now. After I'd been at Dryden a couple years, I felt I knew everybody there, at least who they were and I was on speaking terms with almost all of them. I had met them at the flight parties. That's not true anymore, partly because I don't get around as much as I did, but also partly because Dryden seems to have a different spirit today. It's less of a family than it was in the 1960s. Those flight parties gave me the opportunity to meet people I normally didn't work with, including technicians on other programs, skilled machinists, and management. Center Director Paul Bikle frequently would be at the flight parties, having as good a time as everybody else. I don't remember Bikle as being much of a drinker, but he definitely enjoyed the parties, making these little bets that he was always making with everybody, which added a lot to the parties. I think it was clear to everyone attending them that the parties were social occasions where you could let your hair down. They gave everybody an opportunity to speak their mind to others in an atmosphere that was non-threatening.

The X-15 Pilots

In 1962, when I got to Dryden, Neil Armstrong was headed for the astronaut office in Houston and Bob White was headed off to a new assignment in the Air Force (Fig. 11). Although everyone remembers Neil for being the first man to walk on the Moon, in the times that I've been around him, having a conversation with him, he's always wanted to talk about airplanes. I think his heart is really in aviation. The X-15, of course, was the best thing in aviation around at the time. Neil was the first pilot to fly the third X-15 in December 1961. He studied the X-15 and some of its peculiarities on his own. Unlike any other pilot that I associated with in the 1960s, he had used one of Dryden's TR-10 analog computers for some time. I remember stories I heard back then of him examining a control or dynamic problem by programming it himself and taking a look at it. Neil was shy and modest, but there was little question that he was also very capable and knowledgeable. All of the NACA and NASA pilots had college degrees in science or engineering but he stood out. Although he didn't make a big deal out of it, Neil was probably the most innovative engineer among the pilots on the X-15 program. Since Neil left the program about when I arrived at Dryden, I never worked directly with him on the X-15.

All of the new engineers heard about Neil's April 1962, flight to Pasadena in the X-15 and how he barely made it back to the south end of Rogers Dry Lake. From studying that flight later, it was clear to me that he learned something from that flight that we needed to know: that if the pilot doesn't let the airplane settle

Fig. 11. X-15 number one on the left and X-15 number two on the right with three of their early pilots. From left to right are Scott Crossfield, Bob White, and Neil Armstrong. The photo was taken right after X-15 flight 1-21-36 made by White, which was the last to use the XLR-11 engines. (E-6471)

into the denser atmosphere, it will skip back out. That is, if the pilot pulls too much angle of attack too early, the vehicle will not keep coming into the atmosphere and building up dynamic pressure. When we did the flight planning, we assumed the airplane would continue to settle in. Part of this incident was that, I believe, the Minneapolis-Honeywell MH-96 flight-control system was so good it gave the pilot few cues what was going on. Neil felt that he could start his turn, but he actually didn't have enough dynamic pressure to pull the g's and turn as he wanted to. He was commanding it to turn and the X-15 was doing the best that it could, but his flight-path direction wasn't changing appreciably. By the time he had sorted it out and built up enough g's so he could use the lift vector to turn the vehicle, he was over Pasadena. At Mach 3 or 4 a vehicle goes a long way in a few seconds. Neil overshot by 50 miles, going a mile every two seconds. I doubt that anyone was confident during the flight that Neil and the X-15 would make it back to Edwards.

Neil could have landed at El Mirage Dry Lake, off to the east, or at USAF Plant 42 in Palmdale, but he would have had to turn to get to either one of them. He made the proper choice, in my opinion, to fly the X-15 at L/D max, which gave him the absolute maximum range. Even so, he barely made it to the end of the lakebed. Some say he was lucky he made it and he probably was, but whatever else he did probably would have ended up much worse. Even if he had landed short of the lakebed, he would have been on Edwards AFB, avoiding risk to the nearby communities.

One of the stories I heard at the time was that sagebrush was found on the X-15's skids. I don't think that was true. I believe one of the chase pilots said that Neil came in lower than the Joshua trees, but I think that was said just for effect. If it were true, he wouldn't have made it as far onto the lakebed as he did.

Although that experience was scary when it happened, it taught us lessons in flying the return from space, where a pilot has virtually no aerodynamic control, into the denser atmosphere, where a pilot does have aerodynamic control, and how to put those two together into a successful landing. The pilots changed the way they practiced on the simulator after that flight, and the lessons that we learned from that flight of the X-15 were passed along to the Shuttle program, where the reentry was automated and probably a little less likely to skip out of the atmosphere. Nonetheless, the X-15 made all of us aware that aggressive turning required a large lift vector. If the vehicle didn't have a large lift vector, it would roll, but the direction of flight wouldn't change.

The three X-15 pilots that I mostly worked with my first few years at Dryden were Joe Walker, Bob Rushworth, and Jack McKay (Fig. 12). Walker was the head of the pilots' office at the time. He was serious about his job, and I think he made sure the other pilots knew that he was in charge. There was never any question as to his competence. Walker was an excellent research pilot. There were times, though, when you really didn't want to talk to him because he was in a bad mood because of his other responsibilities.

You could stop him in the hall and ask him a question, or go into his office if you had something specific to ask about a technical, flying, or experience issue. It was always clear that lobbying for anything regarding the pilots wasn't something you did with Walker. Some of us had a maneuver or a test point that we would like a particular pilot to perform. We each felt there was one pilot we would prefer because of our previous experience with that pilot or his ability to fly that type of maneuver. However, it was never a subject that I would ever discuss with Walker. He was friendly and treated me well and I certainly never had any complaints about him. I was shocked when I heard of his death a few years later because he wasn't old, as pilots go. He seemed old to me only because I was so young myself. Seeing him in the hall one day and having him gone forever the next was quite an eyeopener for me. It was the first time that had happened to me with a pilot.

Fig. 12. X-15 number 3 in September 1964 with its pilots. (left to right) Jack McKay, Joe Walker, Milt Thompson, Bob Rushworth and Joe Engle. (EC64 421)

The other thing I remember about Walker was that his last X-15 flight was the record altitude flight to 354,200 feet. I don't know what the politics were in those days. How pilots were selected for the X-15 and for the astronaut program in Houston was all a mystery to me and, I think, to many others. However, it was fairly clear that Walker was the chief pilot on the X-15 program.

If you look at the flight logs of the X-15, you'll discover that Walker was the first pilot to fly Mach 3 and live. Mel Apt obviously had been the first, but he died in the attempt. After that, Bob White led the way in terms of Mach number. Joe Walker ended up flying a little slower but a little higher than White. It was my feeling that White had been picked to set the Mach records, and Walker had been picked to set the altitude records. It may not have been true, but it seemed that way to me at the time. Looking at the flight logs it still seems true.

Scott Crossfield, of course, was the first pilot to fly the first two X-15s, in keeping with the contractual agreement between NASA and North American Aviation (NAA). He was also in the third X-15 when the rocket blew up during an engine run. A former Dryden pilot, Crossfield was an NAA pilot when he flew the X-15. I heard criticisms of Crossfield for leaving Dryden and going for the money and power of industry, but I never saw any of that in him. He always seemed to me to be a competent and knowledgeable pilot and a good engineer who liked to talk about engineering details. I didn't work with Crossfield on the X-15 because he was out of the X-15 program long before I got to Dryden. He made considerable contributions to the specifications of the X-15, especially the cockpit and systems. Over the years I have found that he's an engineer at heart. I haven't always agreed with some of his ideas, particularly on the National Aerospace Plane (NASP), but I have never argued with him either.

Bob Rushworth was another pilot who was easy for me to deal with as an engineer. We didn't have much direct contact with each other regarding flying the X-15 as there were some three hundred people working on the X-15, and the people who needed to talk to the X-15 pilots went up through the system. The pilots dealt with program management, the flight planners, and people like that. However, I did have several conversations with Rushworth that I thoroughly enjoyed. He seemed like a really busy guy, very professional and to the point, but occasionally, when we were both waiting for something, I found him easy to engage in conversation about issues that had nothing to do with airplanes. An excellent pilot, Rushworth was, in many ways, one of the most proficient and experienced X-15 pilots.

Another pilot highly experienced and proficient on the X-15 was Jack McKay, the third X-15 pilot I would work with when I arrived at Dryden in 1962. Jack was really special to me because he was always approachable. He was always kind and always very much a gentleman. He was playful when he wanted to be playful. I never saw a really serious side of him, although I know he had one. I've heard about other people who got on the wrong side of Jack and how he let them know very quickly what he thought about that, but I never managed to do that with him.

Jack's twin brother, Jim, also worked at Dryden. Jim McKay was not a research pilot, although he did fly as a private pilot, having been seriously injured in an accident when they were both quite young. They both pursued careers in engineering and Jack became a test pilot and Jim became an engineer. They sure did look a lot alike. They seemed to have quite different personalities, but that could have been because their jobs were quite different. I worked primarily with Jim on the X-15 landing and slide-out distances and landing gear interactions. Jim was working at Dryden long before I got there and worked there for quite a few years after the X-15 program ended.

Jack was almost like an uncle in the way he treated me, and, I suspect, the other junior engineers. If we had something to say, he let us say it. Then, he'd chuckle, pat us on the shoulder, and give his opinion. There was never any confrontation in any of it. My experience, working with him on the X-15, agreed with the reputation Jack had as an outstanding stick-and-rudder pilot. This was true even when things weren't going well, as I recall in a number of instances. Jack had a lot of problems with aircraft systems that shut down. He had one crash and another emergency landing, but he climbed right back in the cockpit afterwards. I noticed in the flight log, for instance, that the accident where he seriously injured his back happened in November 1962, but he was flying again in April 1963. Jack didn't let crushed vertebrae stand in his way. I never noticed any differences in his piloting techniques in the entire time that I dealt with him on the X-15.

Jack was in the X-15 program almost from its beginning, and he was still flying the X-15 after Bill Dana joined the program in late 1965. Jack had also made a lot of flights in the X-1 and the D-558-II, so he was quite an experienced rocket pilot. I was always glad when Jack was going to be the pilot, because I got a little extra chance to talk with him that I wouldn't get with either Joe Walker or Bob Rushworth.

At one point in about 1963, we were missing a piece of data for the X-15 at around Mach 5 and 120,000 feet. I had requested that the data be repeated, as I had a specific reason for wanting to look at it a special way. I think it was sideslip data that we were looking at to understand some of the stability-and-control derivatives. At any rate, it got added to the flight plan.

Having championed this data, I was in the X-15 simulator when they were practicing the flight plan, laying it out. They were making sure this maneuver was going to suit my needs, probably just humoring me, since I seemed so driven by this one point. Later, when Jack was in the simulator, I got to come back in and watch him practice the flight, looking at my data points. I think this was one of the earlier things that I had found on the X-15 that I thought was important, and I was trying to verify it.

When we got to my maneuver at the crew brief, Jack looked at me and said, "Yes, we've really practiced this maneuver." I probably made some comment about how important it was, not that everybody else's weren't equally important, but I was especially fond of mine. Jack said something typical of him. He said, "Don't ya worry, young feller. We'll get yer dater fer ya." He was from the part of Virginia where "fellow" and "data" had r's on the end of them. His comment was friendly, not at all a putdown. It reassured me.

He made the flight the next day. When we got to the point where my maneuver should have been, Jack was half a Mach number too slow and 30,000 feet too low. I knew something was amiss. I didn't know what it was, but there was a lot of talk on the intercom about what was going on. Something had failed on the airplane. They either abandoned the flight plan or went to an alternate one. As a result, nothing even close to what I was looking for appeared on my strip chart. I had gone through all that requesting and politicking to get my maneuver, and I didn't get my data. I was a little disappointed.

However, for flying an airplane with a failure, there wasn't a better pilot than Jack McKay to have in the cockpit. Besides being very experienced and a good stick-and-rudder man, Jack was really strong. There were stories about him bending the stick in the X-15 while trying to demonstrate to somebody that it wasn't working right. I don't know if that story is true, for I've never seen it in writing, but that was the story at the time. Nevertheless, no matter what the failure had been, I suspect he could have flown the X-15 even if it hadn't had a hydraulic system.

At the postflight, Jack went through the various events that happened, mostly a discussion of the failures, and what he had seen and done. Toward the end, Jack turned and winked at me, with a big grin on his face, and said, "Don't ya worry young feller. We'll get yer dater fer ya next time." I've always remembered this experience with Jack because it sums up how neat a guy he was and what a pleasure it was for me, a young engineer, to work with him. He and Milt Thompson were my favorite pilots from the early days. Jack always treated me like a valuable team member, even if it was clear that I wasn't terribly valuable or knowledgeable. I worked a lot with Milt on the M2-F1 and the M2-F2 lifting bodies, and he put up with a lot of things that he might not have liked, just because I was inexperienced.

Milt Thompson and Joe Engle were added as pilots to the X-15 program in late 1963. Joe was quite different from the other pilots that I'd worked with up to that point. Right up through the Shuttle program, he had a reputation for being an outstanding stick-and-rudder man. I don't think he ever let the X-15 buffalo him. I'm not sure that his research maneuvers were any better or worse than anybody else's, but his ability to fly anything was demonstrated with the X-15 as well as the Shuttle.

On his first X-15 flight Joe did a 360-degree roll. He initially denied it, at least publicly. We could see it on the strip charts in the control room and it was clear what he had done. I'm not sure why it's nearly always mentioned tongue in cheek, because it really did happen. Management may not have wanted to publicize a pilot having done that when it was not part of the flight profile. I think Joe got chastised for doing it. He became a very serious and dedicated pilot after that, although he was a very friendly and outgoing guy as well. (I believe that Joe and Larry Taylor had belonged to the same fraternity at the University of Kansas.)

When they joined the X-15 program, Milt Thompson probably replaced Neil Armstrong and Joe Engle likely replaced Bob White. They continued to fly with Jack McKay and Bob Rushworth until Milt suddenly decided to quit flying the

X-15 in 1965 and Engle was selected to become an astronaut. Their positions were filled by Pete Knight and Bill Dana. I knew Bill Dana to be a good guy from other programs. He was a good pilot and he gave us good data and he's still a good guy.

Pete Knight, also an excellent pilot who got good data, ultimately set the speed record of Mach 6.7. He had a couple of very serious problems while flying the X-15. On one flight, he lost all electrical power but managed to somehow get the thing back down on the ground. He also experienced the dummy ramjet burning off on his Mach 6.7 flight.

Mike Adams joined the X-15 program after Bob Rushworth left; unfortunately, he was killed after seven flights. I didn't know him very well. Bill Dana and Pete Knight flew the rest of the flights in the X-15 program.

X-15 Ground Crews

In 1962, when I got to Dryden, there were three X-15s and three ground crews. I think each airplane had a different ops engineer and a different crew chief. The crews were each dedicated to a particular X-15. When the crew of one airplane was getting ready for a flight and was short handed, the other crews pitched in. Nevertheless, each crew was partial to its own airplane. Although the ground crews were always competing to see which could do the best job, have the fewest failures, turn the airplane around the fastest, or diagnose a problem and fix it the fastest, they always kept safety at the top of their minds. All processes and procedures were well documented, and the ground crews adhered to them when they did modifications, repairs, or inspections.

The crews were protective of their airplane, too. You could only tell so much from looking at drawings, and you could get a lot better idea from looking at the actual airplane. Looking at the doors, access panels, rivets and connectors on the vehicle, you could get a clearer idea of what was there. Sometimes you needed to go down and check the airplane out. However, the crews were always watching to make sure you didn't touch anything you shouldn't. In fact, before you approached the airplane, you got permission from the crew chief, or the acting crew chief, before you got very close to the airplane. That was for a good reason. The crew wanted nothing randomly or accidentally switched or shifted on the vehicle. Nor did the crew want any extra equipment along on a flight, such as a pencil that had dropped out of your pocket.

Everybody, not just the people working on the X-15, liked to look at a 50-foot long airplane that went Mach 6 and 70 miles up. The X-15 hangar was a great environment to visit on a break or when escorting visitors. Sometimes you just wanted to see something specific on the airplane or talk to somebody on the crew or to the ops engineer about a procedure for a test or experiment you were thinking about.

The research engineers didn't do anything with the airplane without the crew involved. For such things as limit cycle tests on the control system, led by research engineers, the aircraft had to be powered up. Everybody was in the loop- the person in charge of the test, the technicians, and the ops engineer— and there was a procedure that we followed. When hydraulic systems were on and powered in the hangar, a bright red light was flashing to let you know that this wasn't a good place to be if a hydraulic line should break. When that red light was flash- ing, anyone not directly involved in the test was encouraged to leave.

My recollection is that as soon as the airplane landed from a flight, those who worked on the X-15, primarily the operations people, would do what they needed to with the airplane out on the lakebed, bring it back into the hangar, attend the postflight debriefing, and then go out to the postflight party. The next day, unless something needed to be looked at immediately after the flight, the crew would tear large parts (primarily panels and components) of the airplane down for inspection to make sure pieces weren't ready to break. I won't say they tore the whole airplane apart, but it sure looked like it to a young engineer. They tore the X-15 further apart than I'd seen auto mechanics do with my car when I took it in to a garage for repairs. I could see that the crew members—the ops engineers, the mechanics, and the technicians—were all very dedicated to making their X-15 as good as it could be. They would spend most of two weeks tearing it apart, inspecting it, refurbishing it, fixing the things that needed to be fixed. They would put each component back together and make sure it still worked before putting it back in the X-15. Then, they would do an end-to-end test to make sure the X-15 was ready to fly.

I believe there was a requirement for an engine run between flights. My recollection is that we had engine runs once a week or so. That brought the rocket shop in. The program team assembled out at the test stand, a special area with bunkers and a hold-down area where the X-15 was secured. The crew would bring the X-15 out and go through the engine ground checks. Then they would start the engine and shut it down, then restart it and shut it down again. The crew would run the engine several minutes. You could hear it anywhere in Dryden. Some of us actually liked the noise of a rocket engine, a noise quite unlike that of a turbojet engine. We could go out to the rocket test stand, as long as we didn't get too close, and look at and listen to the engine run. The engine tests were probably bad for our hearing, but they were very enjoyable.

A huge amount of work went into the upkeep of the X-15. With a high- performance production airplane like an F-18 or an F-15, periodic inspections are performed and any squawks are dealt with; however, in general, the assumption is if it flew fine in the last flight, it will probably do okay in the next flight. That wasn't the philosophy on the X-15, partly because it was flying in an environ- ment completely cut off from the rest of the world, from the time the B-52's engines started until the X-15 was brought back into the hangar. The X-15 also used one-of-a-kind systems, in many cases. It was also high technology for those days, when electronics in general were less reliable than they are now. Today, if something breaks, we replace a circuit board and press on. In the 1960s, we still had individual components, soldered in place, and we had to be sure the component was in a proper environment, as the skin would get to 1,200 degrees Fahrenheit or so. The internal part of the airplane, where the heat soaked through was, I think, limited to 200 degrees. It made a difference in where components could be put. That's a harsh environment with high vibration. The slam down on the landing was also rough on components.

X-15 Follow-on Ideas

On the X-15A-2, we planned to use external tanks to increase the vehicle's top speed. We also had plans for a delta-wing X-15, which many of us thought would probably be the next step. We were going to take one of the X-15s and do substantial modification, probably lengthening it and putting a delta wing on, to get up to Mach 8. This would be even more interesting than the Mach 6 that we had already achieved. As the Dyna-Soar had been canceled in December, 1963, the X-15 was our opportunity to examine the much higher heating at Mach 8.

Along with these discussions, we were talking to the people at Marquardt who had successfully built the ramjet for the Bomarc surface-to-air missile program and later, unbeknownst to me, the D-21 drone launched off the M-21 (a modified A-12, the precursor to the SR-71) and the B-52. These were two ramjet vehicles that had successfully flown in the Mach 3+ speed range. Marquardt personnel were big believers in air-breathing supersonic flight, and the ramjet was their propulsion system of choice. There were studies showing you could actually do supersonic combustion. A scramjet would probably work, at least until it ran out of atmospheric oxygen or had incurred so much heat that the vehicle melted.

So there were ideas about air breathing at that point, but only the ramjet had been demonstrated. There were also many small studies going on at Dryden, Langley, and Lewis (now Glenn) Research Center; as well as those made by proponents in the DOD side at Wright-Patterson AFB, the Air Force, and Marquardt. I was mostly working on ramjets and I made some visits to the people at Marquardt. What they could show me and tell me was limited, because projects like the D-21 were Top Secret at the time. Despite this, I had great confidence that this could be applied to the X-15, either with or without the delta wing. I liked the X-15 delta wing because, with the rocket, it was going to be able to go to Mach 8. With a ramjet alone and no rocket, it looked as if we could get to Mach 5 and demonstrate the air-breathing technology. If we went to methane for fuel, we could go over Mach 5.

The feeling that I had at that time was that it had to be easier and cheaper to make a two-stage vehicle with the air breather as the first stage and the rocket as the second stage. This would let the air breather work where air breathers worked and the rocket work where air breathers didn't. My idea was that the first stage would be a yet-to-be-defined air breather, largely based on the research that we would do on a ramjet experiment on the X-15. The second-stage vehicle would be different. At the same time as I began working at Dryden, Dale Reed and a few other engineers were looking at the new concept of a lifting body. Here was an ideal vehicle for the second stage of a runway-to-orbit vehicle.

II

Birth of the Lifting Body

The X-15 program had been going on for several years when I arrived at Dryden, with about half of the Dryden engineers, about 50 or 60 people, working full-time on it. When we weren't as busy as we were at other times, we were encouraged by Dryden management to poke around and find useful things to do related to Dryden's mission. In doing that, I came across the group of people who were starting to firm up the idea of a manned flight program for a lifting body called the M2-F1. This vehicle had no priority or visibility, but it sounded really interesting to a young engineer.

Dale Reed, who was leading the effort in the lifting-body flight program, encouraged me to make a contribution. Being 21 years old and just out of college, I wasn't sure exactly what I could do that more experienced engineers couldn't do. Although I lacked experience, I didn't lack confidence. I looked around and saw several important areas where nobody was yet working. One area involved the wind-tunnel data. We had a lot of it, but it was really for a wide variety of vehicles. Someone needed to figure out which data best represented the M2-F1 that we were in the process of building. So I decided to volunteer for that in my spare time.

Dale took the lead in constructing the vehicle and preparing a flight proposal to find out how viable a lifting-body vehicle might be for reentry. Bertha Ryan, another engineer, was spending a lot of time working on the vehicle. She was an experienced sailplane pilot and had experience in flying, along with Dale, who was also a sailplane enthusiast. Bertha wanted to work with the real data in the traditional fashion of those days, doing calculations and plotting data by hand to understand it, then presenting the information for other people to use.

Another sailplane enthusiast involved with the M2-F1 was Dick Eldredge, who was heading up a good part of the actual structural design of the vehicle and the construction going on at Dryden and at the glider facility. Harriet Smith, a very competent engineer, was also helping out in many areas. She focused on getting a simulator so that we could work with a real flight plan and examine how we might configure the vehicle, how we might get it into the air, and how the pilot might fly it. Other engineers, including Dick Klein and Vic Horton, were working on the construction and design of the vehicle. Many skilled Dryden technicians were working on the construction of the vehicle itself.

Invention of the Lifting Body

The M2-F1 design originated in a study done by Al Eggers and others at

Ames Research Center. This study was done in the late 1950s, before Sputnik I was launched by the Soviets. Other people had helped Eggers, including Clarence Syvertson, for whom I later worked when he was the center director at Ames. They had gone through various designs, trying to decide what would be a good concept for a lifting-entry vehicle.

The Mercury capsule, for instance, had no lift at all. The Gemini and Apollo capsules had a small amount of lift, which made the entry a little bit easier in terms of the g load and gave a little bit more of a footprint for bringing the vehicle down in the ocean for recovery by a U.S. Navy ship. The people at Ames were looking for something that might be able to maneuver enough to land on a runway, rather than coming down on parachutes into the ocean.

They had started with larger cones, gone to half-cones, and then blunted half-cones. I think one of the reasons they decided on a 13-degree half-cone with a blunted nose, the shape on which the M2 was based, was that it would be large enough to be mated to one of the existing rockets. As they got to planning for lower speeds, they decided they needed a boat-tailed afterbody for drag reduction, vertical fins for maneuvering and landing, and horizontal control surfaces for controlling the vehicle in pitch. Little by little they refined the configuration into one they called the Cadillac configuration or the M2-F1 (Appendix C). I think they called it that because it had all those fins on it, Cadillacs of the time having some fairly obvious tail fins. Dale Reed took the half-cone concept and built a small model about 18 inches long or so. Towing it behind one of his model airplanes, he discovered he could make it fly. So he sold the idea of building and flying the M2-F1 to the Ames guys, and they helped sell the program to Paul Bikle, our director.

There were bits and pieces of wind-tunnel data on the different M2 configurations. Looking at the data, I thought that I could augment the data by inferring the difference between the tested configuration and the one we were building. At the time, digital computers were just starting to appear in the workplace. Dryden had a computer that was used only by people in Data Systems back then. They used the computer to put data in a form for being plotted, so that the engineers could study it.

I had written my first computer program in 1960 while still at Iowa State. I was not an expert programmer, but I realized what a tremendous tool it was for getting a lot of work done quickly. Once I had a program checked out, I had very good answers that were less inclined toward error. At the time, there was some controversy whether engineers not in Data Systems were allowed to use the digital computer at Dryden, but with some arm-twisting I was given access to the machine to run on a low-priority basis. I started to write FORTRAN programs to augment our data set. My primary idea was to use some very simple panel methods, methods that I had been taught at Iowa State and actually had used prior to coming to Dryden. These would be considered nonstarters in today's world of computational fluid dynamics, though. I could get an answer within ten to twenty percent of being right by using these methods, particularly since we were going to fly in the incompressible subsonic region.

Initially, we were going to tow the M2-F1 with a custom-built, high-powered Pontiac until we had an idea of how it flew. One of the first things I did was calculate the rotation speed while the vehicle was being pulled by a cable on the nose, the way gliders are towed aloft. I also spent a lot of time calculating the speed at which the vehicle would lift entirely from the lakebed into the air. I also worked with Bertha and Harriet, trying to help their work with the wind-tunnel data that we had. I did any tasks I could to help Harriet get the simulator for the vehicle working. I was a very junior engineer, but I could take the lead on some of the things because there was nobody else working on them. In others areas, I needed more experienced engineers to help direct me.

Back then I thought the engineers I was working with were very experienced and quite old. In looking back, I realize now that Bertha and Dale could not have been much older than 30. Harriet was probably 27 or 28. That now seems quite young to me. At the time, I thought Harriet had a lot of experience and really knew what she was doing. She was one of my early mentors, on the lifting-body and other programs early in my career.

Computer Analysis and Control Problems

The volunteers that Dale had gathered for the M2 didn't include anybody to do the controls. Every aerospace engineer knows a little bit about controls, so it wasn't that they were totally clueless; it was just that nobody was concentrating on controls. I had enjoyed the controls courses I'd taken at Iowa State, and I was also working in the controls area on the X-15, so I decided that I probably should be the controls engineer for the M2. I talked to Dale about it. Dale has always been very accommodating, allowing people to do what interested them, and helping them do it. We both knew that I was inexperienced, but I guess he figured it would be okay and that somebody else would bail us out if I got us in trouble.

At Iowa State, I had learned what, at the time, were the state-of-the art control analysis techniques. These were Bode plots and root locus. I won't go into what those are, but they involved a lot of calculations and had been around for a few years. Because I could program the computer, I could do the calculations quickly. It looked as if I was really working, but I was actually just making the computer work. This initial work with the root locus techniques gave me very strong prejudices against some control strategies and a better feeling about others. In looking at the root locus [16] and how the pilot might interact with the M2, I could see some areas where the vehicle would be very difficult to fly with certain control strategies. Consequently, after I had calculated the rotation and lift-off speed of this lifting body, I started working on stability and control, particularly how we might configure the M2 controls to interact with the pilot's stick and rudder pedals.

About the same time, we were putting together a simulator that we could use to put a pilot in the loop. That was a great help because the root locus technique, although it was very powerful and gave us a lot of insight into the problem, was not a real-time technique. We had to put a pilot in the simulator, flying the vehicle, to visualize how problems might present themselves and how they might be resolved. From the very beginning of my career, the simulator was one of the most important tools that I used. This was true on the X-15 and on the M2-F1, and it continued to be true throughout my 40-year career.

In practicing in the simulator in those days, the engineers and pilots worked closely together, especially on a low-priority program like the M2. When a pilot wasn't available, the engineers did the flying. I did a lot of my own flying in the simulator at that time. Milt Thompson, who was the "chosen one" - the volunteer pilot who was going to try to fly this very strange vehicle - also spent time in the simulator.

The roll control on the M2-F1 was obtained primarily through the movement of the outboard elevons, also called "elephant ears." They looked as if they should be effective for doing exactly what we wanted for the control of the vehicle, but they were actually kind of sluggish in controlling roll. Every test pilot likes to have very quick and predictable roll control, because the vehicle needs to be kept level near the ground. If it isn't, one gear hits the ground before the other or the vehicle goes into some sort of unplanned maneuver down the field, perhaps tumbling or rolling the vehicle. Test pilots don't like that.

Pitch control was done by the split flap at the rear, which was only used symmetrically for pitch, not differentially for roll. Some of the earlier wind-tunnel studies and other analysis had shown that the split flaps would be ineffective as roll controls, so we didn't plan to use them for roll control in the flight program.

As the plywood outer shell delivered by the glider manufacturer was being integrated with the vehicle's internal structure that had been designed and built in the shop at Dryden, I realized we were getting close to flying the M2-F1. This was the first flight program that I'd been involved in from, roughly, the very beginning. Although other people were already working on the program when I started, I still felt I got in on the ground floor. I was quite excited about flying it. I was also excited by the thought of contributions I could make and how I could make the M2-F1 more successful and as safe as possible or reasonable. I don't think anybody would say that the first lifting body was a safe endeavor, though. However, it was all done with informed risk. We all knew what the risks were and the pilot was highly aware of the risks to himself and the vehicle.

The lifting body was lighter initially because it didn't have an ejection seat in it. When we first started towing the M2-F1 with the Pontiac, it was going to go very fast. It occurred to me that these ground tows were going to be dangerous, even before the M2-F1 left the ground. Many parts could fail as the vibration was fairly harsh for a plywood lifting body at those speeds. We were using the dry lakebed runway, not always glassy smooth.

All of these things made me greatly respect Milt's bravery and his confidence that he could fly this thing. I've used Milt as a sounding board on programs he wasn't even involved in throughout my career, mostly because of my M2-F1 experience and because he was the first test pilot that I actually worked closely with on a one-on-one basis. In the X-15 program, although I had worked with the pilots directly, there always had been 10 or 12 other people in the loop and my inexperienced opinion was easily overruled by any of the other more experienced people, especially the pilots.

Bertha Ryan, Harriet Smith and I had worked together as a team over several months and we believed we had accomplished something by building the simulator. We then used it to do the flight-planning for the M2-F1. We had other consultants, both within the project – Dale Reed and Dick Eldredge, in particular – and outside, with people who had more experience in each of the disciplines than we three had.

The flight-planning primarily involved working out how do we get the M2-F1 into the air? Then, once we get it into the air, what do we do? For this program, the first method was towing it with a car to get it into the air. We needed to figure out how we should increase the towing speed until the pilot could first rotate the vehicle, then lift off, and finally keep going to higher altitudes.

Eventually, when the pilot felt comfortable with the vehicle while towed, the M2-F1 would be released from the car towline, allowing it to glide to a landing. After doing this numerous times, we would tow it behind NASA's R4D, the Navy version of the Air Force C-47 or DC-3 Gooney Bird. The R4D could tow it to higher altitudes and release it so Milt could maneuver the vehicle like any traditional aircraft. The three of us each flew the simulator to get our own perspective on whether it was flyable at all and to teach ourselves about energy management so we could help Milt.

In addition to using the database that the three of us had come up with for the simulator, I also had written computer programs to do some independent analysis. This involved such things as using the aerodynamic data and our estimates of the mass characteristics of the vehicle-the weights, the inertias, and the position of the center of gravity-to calculate frequency and damping. If an airplane isn't well-damped, if it diverges in an oscillatory or exponential fashion, it will depart from controlled flight very rapidly, with grave results.

So the first criterion, of course, was to make sure that we had a stable vehicle. I don't recall seeing anything in the simulator that led me to think, with any of the data that we used, that the M2-F1 was not both statically and dynamically stable in both the lateral-directional axes and the longitudinal axis. I also used these analysis programs to calculate the root locus characteristics, which showed that if there were no control inputs by the pilot, the airplane would fly at exactly the open-loop stability characteristics that were demonstrated with these analysis programs, as long as the pilot didn't make any significant inputs.

As the pilot increased his gain and started controlling the vehicle with inputs, the root locus would move toward (close) what are called zeroes (roots of the numerator of the transfer function) of the vehicle. A simplified sketch of this root locus is shown in Fig. 1. The closure on the zeroes means very high gain in the pilot/aircraft combination is possibly unstable. When we talk about stability for aircraft, we're usually talking about what I call open-loop stability, where the pilot isn't making any inputs. For a high-performance fighter, we might also be talking about the closed-loop characteristics, with the control system active. The poles of the root locus show the open-loop behavior and the zeroes show the highest gain closed-loop behavior.

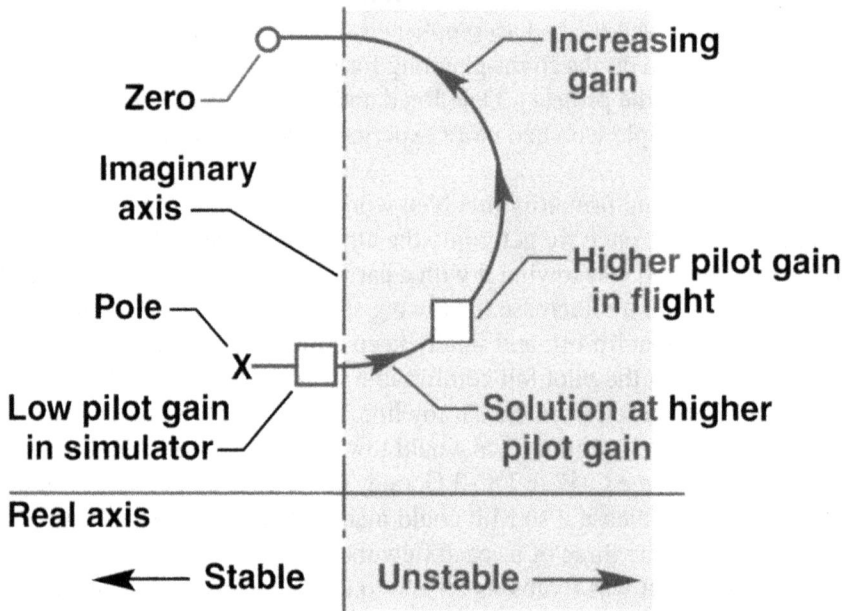

Fig. 1. Sketch of the root locus of using the rudders for roll control on the M2-F1. At low pilot gain the vehicle is stable while at higher gains the vehicle showed an instability (oscillatory divergence). (Graphic 02-027)

The M2-F1 didn't have an artificial stability augmentation system, so the root closure, from the open-loop poles of the Dutch roll mode to the zeroes (which are approached at very high gain), was an approximation to the pilot's input being a pure gain, as the pilot reacted more and more to what the aircraft was doing. Pilots can't really be regarded as a simple gain, but that's the approximation we make in the root locus. The pilot would be moving towards the zeroes, away from the poles. If, at any time, the root locus closure-not nearly as complicated as it may sound here-goes into the right-hand half of the plot (the right-hand half of the real and imaginary plane), the vehicle/pilot combination is unstable (Fig. 1).

What happens then is a pilot-induced oscillation (PIO). If the pilot continues to stay in the loop, to put in inputs and react to what the airplane is doing, the oscillation will grow until, eventually, the airplane crashes. If the pilot recognizes the inputs are part of the problem, he will reduce his gain. The best way to do that is to let go of the stick and put your feet on the floor, not on the rudder pedals. Then the aircraft will immediately revert to the original, stable poles. For the M2-F1, we did this analysis and looked at various control schemes, whether we used the trailing-edge flaps for part of the roll control or whether we used only the outboard elevons.

The two things I looked at initially were the control characteristics and, therefore, an assessment of the pilot-in-the-loop handling qualities using just the outboard elevons or using the outboard elevons and the trailing-edge flaps. I don't remember looking at the trailing-edge flaps alone because the results of earlier wind tunnel tests at Ames predicted that they would provide only low roll rates.

The outboard elevons had quite desirable characteristics as a roll command. When the pilot used the inboard elevons with the outboard elevons, the result was a higher initial roll rate. Because the inboard elevons also tended to put the nose against the roll, the pilot ended up with a quick initial response, followed by a slowed response from the sideslip reducing the roll rate.

We looked at those two characteristics, including Milt in our discussions as the simulation was built. We held meetings spontaneously, whenever somebody had a problem. We also had weekly meetings where, as a group, we quickly discussed what was going on. We didn't have formal meeting rooms then, so we just held our meeting in the largest of the team member's offices or in an office convenient to a large contingent working on the vehicle. There was a lifting-body program office for Dale, Harriet, Bertha, and hardware designer John Orahood and it was usually available for a meeting. We held our spontaneous meetings wherever we were at the time, often in an office but sometimes on the fly, when we met in a hallway or at lunch.

Milt was always available for discussions with any of us. He was quite excited about flying the M2-F1, even though it seemed that only a crazy pilot would do so. Milt was convinced it could be done and that he was the right crazy pilot to do it. I had analyzed two control systems that we had Milt evaluate in the simulator. I was sure he would pick the one with outboard elevons only, not the one with both inboard and outboard elevons. Actually, he didn't like either of them. He said neither of them had a quick enough initial roll response for coun-teracting a gust. I think Milt's real worry, from day one through his last lifting-body flight, was having a side gust hit the vehicle when he was near the ground. Since the lifting body didn't have wings, it didn't have any significant natural roll damping. Instead, it had a low moment of inertia and nice big fins for the gust to hit, producing a roll upset. This is caused by the dihedral effect, which is the amount of rolling moment resulting from a given sideslip (see sketch in Appen-dix D). A side gust causes an instantaneous sideslip, resulting in an instantaneous unwanted rolling moment and, hence, an uncommanded change in roll or bank angle. With that configuration, Milt needed a very quick response to a roll upset, meaning good roll control power. Consequently, he didn't like either control system, even the one that used both the outboard and the inboard elevons, giving a larger initial roll response at the expense of a poorer response later.

Our primary simulation evaluation had the center fin on, which Milt didn't like because it was one more vertical surface for the wind to hit. All our data showed that the vehicle did have a large dihedral effect, because of the vehicle's flat sides and large fins. A gust hitting one side of the vehicle would roll it in the opposite direction. The larger the dihedral effect, the greater the rolling motion from the gust. Milt was concerned that he would not be able to roll the M2-F1 back to "wings level" after a modest side gust. Dihedral effect is something the pilot feels immediately and viscerally, and it's probably the primary parameter a pilot worries about in terms of side gusts upsetting the vehicle and rolling it. I certainly didn't have the experience to disagree with Milt, nor did I have any reason to believe he could be wrong.

Having flown many hours in biplanes, light airplanes, and crop dusters, Milt started flying the lifting body with his feet on the rudder pedals. He discovered that by using the rudders, he could get the roll response he wanted. After he started flying the simulator that way, he asked us to reverse the controls in the simulation, connecting roll control to the pedals and yaw control to the stick, which the simulation engineer did. Later at our request Milt got back in and took a look at the way it flew with the rudders on the stick. He still didn't think the M2-F1 was a particularly good-flying airplane, but he thought the control configuration with the rudders controlled by the stick and the outboard elevons controlled by the rudder pedals was somewhat better than the original configuration.

Because this was worrying us, I started doing some independent analysis on the digital computer, closing the new zeros with the same old poles. The stability of the vehicle (the poles) wouldn't change, but the zeroes changed because we were using the rudders to perform a roll-command task. It was terrible. As I looked at my first results, I thought I'd made a mistake, because at a very low gain the closure would go into the right-hand half of the plane, indicating a pilot-induced oscillation (PIO) caused by small, quick inputs (Fig. 1). I tried to troubleshoot what I'd done, including variations of the stability-and-control derivatives in my analysis program to see what would happen. It looked very dangerous.

If the pilot kept his gain very low, the vehicle continued to fly like the basic airplane. But if the pilot let his gain get high-which meant the pilot had become very reactive to any tiny motion in the vehicle-then the vehicle started performing very differently and, it appeared, very dangerously.

There were two problems with my assessment. First, I had virtually no experience and, second, I was working with data of marginal quality. I didn't have a whole lot of credibility, but I did have the strength of my convictions. I discussed my assessment with Milt, and he listened to me, basically understanding what I was saying. Then, he made the decision that, based on his simulator experience, he still preferred the rudders. My job was to tell him what I thought and it was his job to make the decision about what to do.

A little unsure whether my assessment was wrong or I hadn't done a good job of explaining it to Milt, I went to Larry Taylor. He was an older engineer-all of 28 or 29 years old-but he had a good technical reputation at Dryden. I showed him what I'd been doing on the M2-F1. I don't recall everything that we discussed, but I'm sure that if I had been doing anything wrong, he would have pointed it out. However, he more or less bought the story that I was telling, based on the data that I had.

Informally, Larry started working a bit on the lifting body, mostly through me but sometimes directly with Milt because Larry did have a lot more credibility in controls than I did. After all of the discussions that we had with him, Milt still felt that the way to fly the vehicle was with the rudders on the stick and the outboard elevons on the rudder pedals.

Milt was exactly right. The M2-F1 was a very difficult vehicle to fly. It was unlike any vehicle anybody had ever tried to fly before, and it was Milt's life and body at risk, not anybody else's. Once he made his decision, I did my very best to support it. He was in charge of the vehicle that would be towed behind that souped-up Pontiac the first time, as he should have been.

Unless I changed the stability-and-control derivatives drastically, I got the same tendency for the vehicle to be unstable, at a very low pilot gain, meaning there would be a PIO. If the pilot didn't know how to sort that out-a strong possibility on the first flight of a new airplane-he might not realize that he was part of the problem and would try to keep flying the vehicle until he completely lost control. I was very concerned about this possibility.

As I was fairly green, Milt courteously told me he didn't think my analysis had much to do with the problem, because he had used the piloting technique in the simulator and saw no tendency at all to get into a high-gain PIO. Probably to humor me, as much as anything, Milt agreed to go back to the simulator and try to increase his gain with the rudders controlled by the stick. The elevons were a non-issue because they were the only other surfaces we had and we put them on the rudder pedals. He didn't really use the pedals very much, although he did do some coordination with them.

On several occasions, Milt went back in the simulator to try various techniques to see if he could make this configuration unstable and provoke any of the PIO tendency that I was claiming was there. Occasionally, I think he did see a little of it, but only by doing things that no pilot would ever do, such as making really large control inputs for very small changes. We didn't have a turbulence simulation at that time, but if we had, I believe it would have convinced Milt to rethink his decision. It definitely would have shown him this problem. But we didn't have that capability then; nor did we have any easy way of adding it.

After many weeks of working with Milt in the simulator, we were convinced that we were ready to tow the vehicle. I had misgivings about the control scheme, but I knew the high regard that other engineers and pilots had for Milt, so I assumed my analysis technique had overstated the issue.

Learning to Fly

As we prepared for the first car-tow, we had questions such as how to hook the tow line and make sure it released both from the tow vehicle (in this case the Pontiac) and from the M2-F1. How was the pilot going to communicate with the car driver? What would be the key words and phrases? Would there be hand signals? After these were resolved, we were finally ready to attempt to fly the M2-F1 on March 1, 1963.

We had a tow rope that was about 1,000 feet long, and we were going to tow on Runway 18 (Appendix E). We could get to it easily as it is less than a mile from the Dryden ramp. It's a nice, smooth runway, 22,000 feet (about four miles) long. The crew went out before the tow and inspected the runway, driving up and down it to be sure there were no soft spots. They also checked for debris. Those of us working on the program gathered on the ramp to watch the first flight as we were finally going to see if the vehicle would rotate its nose off the ground.

The crew took the M2-F1 to the runway and hooked it up to the Pontiac. Slowly the Pontiac accelerated with the M2-F1 in tow. The Pontiac reached a planned speed and was held there. Milt then did some handling evaluation of his own to see how the vehicle performed and to plan what to do if the towline broke.

After several low-speed runs up and down the lakebed, they decided to go to the rotation speed, which I had predicted to be about 70 miles per hour. Milt pulled back on the stick, commanding rotation, and the M2-F1 nose rotated at about 70 miles per hour, as predicted. Then Milt went back to the end of the runway and decided to increase the speed to 85 to 90 miles per hour, where I had predicted that liftoff would be. The plan was, I think, to go to about 85 miles per hour, hold it there, and have Milt hold the nose down briefly. Then he would rotate it when he wanted to lift off and he would be up in the air, flying.

When he attempted to do this, Milt discovered that as soon as the wheels left the ground, the vehicle went into various oscillations, bouncing and rocking from wheel to wheel. This didn't give him any confidence at all in going any higher. He was only a few inches off the ground at any time but he was very uncomfortable with the vehicle. He made several attempts to get the vehicle to do what he wanted but was unable to do so. He decided there was something inherent in the M2-F1 that caused all the difficulty. Of course, he was going 85 miles an hour down the runway. That's fast, even on a dry lakebed with no known hazards or obstacles. If anything were to break on the vehicle you're being towed that fast in, it would be very grim for you and the vehicle. He had his gain up fairly high, partly because it wasn't acting the way he expected and partly because he couldn't make it do what he wanted it to.

As I recall, we were doing these tows with the center fin off. There were several reasons for this, some of them having to do with compatibility with a reentry vehicle. In particular, every extra surface had to have a thermal protection system during reentry. Milt didn't want the center fin on because he could see that it would probably increase the dihedral effect and make the vehicle even more prone to rolling in crosswind.

Since Milt couldn't get the M2-F1 to do what he wanted, we decided to put the center fin on. Even though Milt wasn't convinced he wanted to fly it that way, we just wanted to see if he could get it to lift off. So we put the center fin on, and made a few more attempts, but there really didn't seem to be much difference with or without it.

Figuring Out What Went Wrong

We didn't have an instrumentation system in the vehicle, so we didn't have that way of looking at what had actually happened. We did have a photo chase car that had taken 16-mm movies. We looked at the film and could see that the rudders and the outboard elevons were really flopping around. The control gearing and the various pull and push rods were fairly rigid and those surfaces shouldn't have been able to flop around. We felt that Milt probably had, unknowingly, unconsciously, been putting in control inputs.

We had a team meeting and we also looked at the movies. Milt said that his feet had been on the floor and his hand had not been on the stick except to pull and hold the nose up and that he had waited for the M2-F1 to lift off before he put in an input. He felt that the vehicle's flopping around was causing the controls to flop around, as we could see in the movie.

We used a single-frame analyzer on the movie. This analyzer projected a single frame of film so we could make measurements. Frame by frame we measured positions of the rudders and the outboard elevons confirming what we could see, that they moved a lot. Doing the measurements frame by frame allowed us to measure the motions in each frame, yielding a data point. Then the measurements could be plotted and we had a time history of the motion of the rudders and the elevons. Because we could also determine the orientation of the vehicle, we could measure the bank angle and the altitude in the same manner. We assumed the movie was taken at a constant speed, so we could use the time histories of the rudders, elevons, bank angle, and altitude and see what was really going on.

Larry Taylor, who had been more or less leading the effort of analyzing the movie of the car tow, was convinced that there was a significant correlation between the rudder motion and the bank angle we could see on the movie. We presented this to Milt and he said he still didn't believe it, that our approach was too simplistic. He knew what had gone on because he had been in the cockpit and he thought he had it pretty well sorted out. This left us with the idea that Milt was part of the problem, but we didn't have a very good way of convincing him of this.

Pilots, especially highly skilled pilots like Milt, can be very stubborn. They rely entirely on their own instincts and senses so they believe those over almost anything else, particularly what someone else tells them. For this reason, we weren't surprised that Milt didn't feel that our analysis had any bearing on what we were seeing. This was a little discouraging because we didn't really know what to do next.

By then, we had shown other control specialists how unstable the root locus plots were when closing the roll angle and the rudders. We were even more convinced than before that this was a problem because they had more experience with, and greater confidence in the root locus technique. We'd also used my software to analyze some accumulated X-15 data. We had faith in what we had and we thought we understood the problem, but our confidence was shaken because Milt was dismissing our analysis completely.

After the towing attempts on March 1, we had to get the vehicle ready for transport to the Ames 40-foor-by-80 foot wind tunnel, then the largest wind tunnel in the world, on March 4. We were going to run a very complete set of stability-and-control and performance measurements on the M2-F1 in the wind tunnel. The M2-F1 would be mounted 20 feet above the floor of the tunnel, which was fairly high. Then they'd blow air down the tunnel at over 100 miles an hour. The wind in the tunnel is not quite as smooth as atmospheric air, which would make the tests very interesting.

At that time there was still some question whether the M2-F1 was going to be flown with or without the center fin. We already had a lot of small-scale wind-tunnel data with the center fin on. When the M2-F1 was in the wind tunnel at Ames, we did a complete set of tests with the center fin on and with it off. I think we did that to get good comparison with all the smaller-scale data that we already had. As I recall, the intention at that time-mostly because of Milt's aversion to side gusts-was that we were going to fly with the center fin off.

The wind-tunnel testing took about two weeks and we ran at least two shifts. I remember working 16-hour days hand-plotting points. There were no automated plotters in those days and it had to be done by hand. The wind-tunnel personnel wrote down the data. As we got the data, we plotted them to assess the relation-ship of one data point to the others and to find out if there were holes in our data or if more tests were needed.

Milt sat in the vehicle, holding the control surfaces in different positions, while the wind was blowing. From what I heard then and what I've read since, it was a terrible, uncomfortable environment to be in for eight hours. We had priority in the tunnel and the data kept coming. The hand-plotters were Bertha, Harriet, Dale, and me. I remember Dick Eldredge doing at least some of the plots, as well.

This test was conducted by Clarence "Sy" Syvertson, who later became the Director of Ames when Dryden was a facility under Ames. I met him then and developed a great deal of respect for his management and people skills. It was really a pleasure to work with him. I was a young engineer who didn't know too much and he showed me all the nuts and bolts of the tunnel. He also let me sit in any of the operating areas during the tests, so I'd get a good feel for it. I really appreciated that because I've carried my impressions from those wind-tunnel tests for the rest of my career, of the quality and dedication of the wind tunnel people and the care with which they handled the data.

It was an invigorating environment, just sleeping and plotting, as I recall. We got the data we needed and returned to Dryden with the vehicle on a flat-bed truck.

Fixing the M2-F1

Over several more weeks, we prepared to go back out and do more car-tow testing. While Milt was gone briefly to support another airplane project, I argued for a couple of car-tow tests with the aileron on the stick and the rudder on the rudder pedals. I believed that it couldn't fly any worse than it did already and this way, at least, we could get more photographic evidence of what was going on.

I believe we set it up with the center fin on, because that was the way we had flown it last before we went to Ames. I convinced Dale and the other people in charge of the program to try the configuration that had looked much better in my analysis. Dale has always been easy for me to deal with, anyway, because he always listens carefully and tries to reach consensus among all the people working on the project.

Bertha Ryan, in particular, was upset that Dale went along with me. She felt that Milt was the pilot, the only one of us with everything at risk; and we really didn't have a right to change the configuration without his approval. In the past, when I had made this suggestion, Milt had said very quickly that he didn't want to try this configuration. He'd flown the simulator enough, he said. He felt that trying this configuration would not be worth the effort. So he was against the idea, but probably not entirely adamant about it at that point. However, Bertha-and, probably, others-were very adamant that we shouldn't do it.

Being young and not yet knowing any better, I probably advocated that we change the configuration and not tell Milt, which I think is what set Bertha off, quite rightly. Then, if he could fly it, we'd tell him what we had done. As I recall, I advocated that we do it that way because I knew Milt didn't want to try it.

To get Bertha back on the team, Dale decided he had to tell Milt about the changes and let him make the final decision. It would take only a day or so to switch the vehicle back to the other configuration, so we wouldn't lose much time. Dale made the right decision. I had no right to try to trick the pilot into flying something he didn't want to fly.

I had pitched this with Larry's support all the way. Larry had great confidence in his engineering skills, but he also had a reputation for being a little heavy-handed with people who disagreed with him on technical matters. He wasn't known for his tact, so it was I, not Larry, who worked on Dale to get the configuration changed. Larry had, however, probably been more insistent than I that we should try to change the configuration.

About the time that Bertha got upset, I started to have doubts about the M2-F1. On the simulator I'd watched, the steep descents this vehicle would be making. During airborne tows behind the R4D, at an altitude of 10,000 feet, the pilot would have to release the tow line, push the vehicle over, and descend at a 30-degree angle until he got fairly close to the ground, within 300 feet or so.

Then he would have to pull back on the stick to flare. He would need to do the initial pushover to get enough speed so that, as the vehicle rotated through the flare and the speed bled off, he would still have enough energy so the lift would be roughly equal to the weight.

As I watched the simulations, I recalled that the Ames tunnel tests had shown that the L/D max was going to be more nearly three than four. Consequently, the pilot was going to have to come in even more steeply than we'd thought. It occurred to me that the pilot was going to be diving at the ground in a vehicle with no wings, a vehicle that nobody had ever flown before. The pilot was going to have to somehow arrest that sink rate, control the lateral and the directional axes simultaneously, and counteract any side gusts. This was dangerous thing and I started getting cold feet about whether my idea was a good one and whether I had the right to try, with my limited experience, to convince the project to fly the M2-F1 with the new control system.

Larry told me that not only did I have a good idea, but that it was what my job was all about. Engineers did analysis and shared it with other people, some who were working on the project and others who weren't, who would then make decisions. However, he said, if I didn't advocate the idea, there would be no way that it would be evaluated and a program decision made.

I have carried this experience in my head over the years. I've always been very cautious in dealing with pilots who didn't immediately challenge what I was saying, because Milt did challenge me. I knew that Milt would take what I said with a grain of salt and make his own decision. When a pilot simply says "Sure, we can do that," I always worry that maybe he doesn't understand what I'm asking. I like pilots who argue with me, probably because Milt did and he was the first test pilot I worked closely with.

The changes I'd suggested were made to the M2-F1 and the vehicle was configured with the outboard elevons on the stick, the rudders on the rudder pedals and the center fin in place. When Milt came back, Dale told him what we'd done. I expected Milt to be angry, but he really wasn't, as I recall. He did think it was a stupid thing to do, but since we'd already wasted the time making the change, he was willing to spend a day looking at it, so we could get this issue behind us and get back to solving the real problem. I think he was disappointed that we had done something he didn't like, but he knew that we were working hard and maybe we deserved to look at this before we went on to other things.

The second series of tow tests occurred on April 5, 1963. First we towed the M2-F1 at slow speeds to make sure that nothing had come loose in the interim. Then we got, fairly quickly, to the 85-miles-per-hour lift-off speed again. I remember that I thought he got further off the ground on the first try and was actually flying it, but it was still oscillating. He got both wheels off the ground, up in the air a ways, but he had his gain very high with a control system he didn't like, the worry about side gusts, the center fin on, and an airplane that had never been flown before. I think his high gain had him putting in extra inputs. They were more attuned the vehicle, so he did actually get it off the ground.

We all agreed that we should try it without the center fin. Milt did even better. He'd been up in the air once and the offending center fin was now off, so his gain was naturally lower. Because of this successful test the rest of the M2-F1 flights, until its last flight in 1966, were flown with the lateral stick connected to the outboard elevons and the rudder pedals connected to the rudders. Once we got past that hurdle, we went on to towing the M2-F1 to higher and higher altitudes (Fig. 2) during April and May, 1963. We took it up a couple hundred feet each flight. Milt was convinced he could fly it now.

Fig. 2. A photo of the M2-F1 taken immediately after the first release from a car tow, capturing the moment of the very first free flight of a lifting body. Milt Thompson flew the M2-F1 on April 23, 1963. (E 9829)

We had car chases for the tow tests. These vehicles, authorized to be on the lakebed, would drive beside the towed vehicle to help interpret what was going on and to give Milt any information he might need based on what they could see. I had the opportunity to ride in a chase vehicle a couple of times. It involved going 70, 80, or 90 miles an hour in a tightly sprung vehicle on a lakebed that wasn't nearly as smooth as I had thought it was. The ride wasn't really very rough but what impressed me was how much noise and vibration there was, just trying to follow the M2-F1. This helped me appreciate Milt's environment. Despite being in the secure environment of the chase vehicle, I found the environment very dominating and disquieting.

With the additional equipment the vehicle weight had grown to about 1,250 pounds, resulting in a lift-off speed of nearly 95 miles per hour. We were going to press on, getting the instrumentation, landing rockets, and an ejection seat (Fig. 3) installed and then towing the M2-F1 to altitude with the R4D. This R4D was also used to fly people around to aerospace facilities in California. It was very convenient for us. If the plane were going where we needed to be, we'd just get in the aircraft and go.

We got to fly on the R4D with all of the NASA research test pilots -Joe Walker, Jack McKay, Bill Dana, and Milt Thompson-many times, so we spent a lot of time in the air together. Even though it wasn't the same environment as a research flight, with the pilot in a research aircraft and us engineers huddled in the control room, trying to anticipate problems he might run into, it did build a good relationship between the working engineers and the pilots. Dryden was family in those days, because of such shared experiences. Some family members were more equal than others, but we were all treated very well.

We made additional car tows during July and the early part of August 1963, which finished the initial car-tow tests with Milt. Shortly after we completed the successful car tows, in early August, the M2-F1 was ready to begin the air tows behind the Gooney Bird.

Fig. 3. Milt Thompson using the M2-F1 landing rocket on the lakebed on August 6, 1963. This was ten days before the first R4D tow of the M2-F1. (ECN 205)

M2-F1 Air Tows

We were ready to go. We picked a nice day, August 16. We always started at the crack of dawn, the big concern being turbulence. In the desert the calmest atmosphere is almost always at sunrise, and programs for which it matters try to use that opportunity. We keep flying until thermal activity picks up and the air is just too rough. We're always getting up in the dark, driving out to work in the dark, and getting ready for the flight just when the sun comes up.

I knew the flight was going to be a real crowd-pleaser and I really wanted to watch. After launch, the vehicle would appear to be coming straight down. It wasn't; it was coming down at a 30-or 35-degree angle from the horizontal. The descent always looks a lot steeper, nearly straight down, than it really is.

I didn't get to watch the flight, however, because the members of the project team who had been working with the vehicle's performance, stability and control, energy management, and flight planning were needed in the flight control room. The airplane data system downlinked to the antenna at Dryden and the data was put out on strip charts that we were monitoring. To this day we use a very similar-seeming system, though the technology behind the process is much more sophisticated now. Today we can bring up cross-plots or other plots in addition to strip chart-like plots. We can make these changes in real time, too.

In the simulator we watch the variation of the individual state parameters-the roll rate, the angle of attack, the sideslip, and the control positions. Looking at all of them together on the strip chart is an extension of that and a natural, intuitive display. We were prepared to tell the pilot to eject if we saw anything surprising on the strip charts, anything we didn't recognize.

We were in the normal control room environment but our little program had a lot less staffing and rigmarole than the X-15 had for its flights. I didn't monitor all the X-15 flights, just the ones where we were looking at a control, heating, or handling qualities problem, but that was enough time that I got trained. The control room was a familiar environment to all of us.

NASA-1 controller Bill Dana was the one person who was in contact with the tow aircraft pilot, Milt in the M2-F1, and the chase pilots. We had other aircraft that loitered at various altitudes, off to the side, so Milt didn't have to worry about hitting somebody. Milt was very intent on flying when he left the tow, went to the runway, and set up a good landing.

There were contingencies in the flight plan. We had "go" and "no go" parameters; if certain things weren't going the way we had specified, we canceled the flight right then. No discussion, no negotiation, we just cancelled. We also had "go/no go" rules that, to my knowledge, were never violated by Dryden. Other organizations have violated such rules and they've paid the price.

The advantage of "go/no go" rules is that everybody is familiar with them. Anybody can bring up an issue or an objection in an open forum, prior to the flight test, and have it talked through until all those concerned understand the issue completely. Then they make decisions based on the relevant ground rule. For instance, Milt was very concerned about crosswind controllability, as were we all, so crosswinds would have been a "no go" condition. Certain instrumentation not working also would be a "no go." We would just cancel the flight until it could be fixed because safety was primary.

The plan from takeoff to the M2-F1's landing basically always had contingencies for every foreseeable problem. Whether it was an abort and we returned to try again some other time, an inadvertent disengagement of the tow line by the towing aircraft or the M2-F1, or some other problem, Milt had a place he could reach to make a reasonable landing.

The way the vehicle was towed above the R4D included a "dead-man" zone that ranged up to four or five hundred feet above the runway. If either end of the towrope became detached below that point, or if either vehicle were not controllable at that point, the only possible response was to eject right away. This had been rehearsed and Milt had it all well in mind.

The emergency procedure had all been reviewed by the various organizations at Dryden. In some ways, these reviews were less formal than the regular reviews, but they were more meaningful because all the people involved were there and they all spoke up. We talked through every one of the issues in depth. The discussion wasn't formalized with people going through bullets on charts; it was an open discussion by experts.

On the day of flight, we were all at our stations and the weather was deemed okay. It was a good day, as I recall, with no wind. Of course, we could always have a knot of wind in either direction, just from a little unsteady air, but that wasn't enough to be too concerned about. We'd done all of the checks for the control room and for the vehicle prior to every flight. The vehicle checks were the final step; once they were done nothing could be changed on the fly later.

The various vehicles were given the go-ahead and the Gooney Bird rolled down the runway. The M2-F1, as planned, lifted up above the tow plane before the tow plane actually took off. The tow plane didn't have much of a wake when it was on the ground, because it wasn't generating significant lift from the wings. As a result, we didn't get significant turbulence behind the towing airplane until it actually started to lift off the runway.

The M2-F1 would lift up about one hundred feet above the Gooney Bird, on a thousand-foot towline. Milt would then tell the pilot of the tow aircraft to take off. Together they would climb slowly, always within range of an abort landing site for the M2-F1 (Fig. 4).

The airfield was closed-or supposedly closed-for these flights. I recall a nuisance aircraft appearing once where it shouldn't have, which always makes a pilot nervous. The pilot of the first experimental low L/D glide vehicle doesn't have time to be looking for traffic. Usually when we close the airfield for a brief time like that, it works and there's no one in the vicinity.

The M2-F1 and the tow plane flew to the drop point so the M2-F1 could be released to land on the long lakebed runway. I think Milt's plan was to aim to touch down halfway down the runway. If the vehicle had less performance than expected, it could still land short on the runway. Since the M2-F1 had neither speed brakes nor thrust, we were pretty much committed to a limited flight envelope. Milt had to do the tow release because we didn't want him flying around with the towline dangling from the M2-F1. The tow plane pilot could also release the tow line in an emergency, should the M2-F1 have put the Gooney Bird in jeopardy.

Fig. 4. The M2-F1 being towed to altitude behind the R4D Gooney Bird on a flight in early 1964. (E 10960)

Vic Horton was in the old ex-Navy R4D with its little observation dome. Vic could stick his head up into the dome and look back at the M2-F1. He and Milt were in radio contact. He gave Milt a countdown in terms of a minute, a half-minute, etc, until release and then said "Release" and Milt knew when to release.

Milt could see downward a little through the window in the bottom of the nose of the M2-F1. I think the window was there to allow the pilot to watch the tow aircraft. A nose window isn't really much use for judging height on landing and I think most pilots don't use it for that. In fact, when I asked Milt if he would use the window to judge altitude, he said it would be a blur field and that he wouldn't be able to judge altitude at all because it would be going very fast. He said most pilots get their height-above-the-ground information either from landmarks or from looking out the side of the vehicle.

So, that was the plan for approach and landing. Milt could see the runway he was headed for and see if it was within the range that he felt comfortable flying. So he released the tow and did some small inputs just to make sure everything was connected properly and everything moved in the direction that it was supposed to move. I'm sure he'd been doing the same thing on tow as well. We were watching the strip charts to see if we saw anything unexpected.

Milt didn't say a lot during the flight (Fig. 5), as I recall, perhaps because I was so busy looking at the data that I wasn't listening. The flight was only about two-and-a-half minutes long so I didn't have much time to think about anything. We spent the flight either preparing to dive at the ground or preparing to flare. Milt did a practice pushover and flare at a higher altitude. Then when it got down to an altitude that seemed adequate, he pushed the nose over to around 120 to 125 knots and held it there until he got to the point to initiate the flare.

Fig. 5. The M2-F1 in free flight above Edwards AFB after release from the R4D at about 10,000 feet. Milt Thompson can be seen looking through the nose window ahead of the nose gear as he descends on this August 21, 1964 flight. (EC 64 406)

He knew he needed so much speed because of the weight of the vehicle, because of the speed that would bleed off in the flare, and because he would need enough speed to touch down. He knew the rough profile just from being a very experienced pilot and many of the details based on the results in the simulator. Everything I knew about what he was doing was based on what I'd seen in the simulator.

I don't remember seeing anything scary on the strip charts I was watching. I thought Milt was putting in a lot more inputs than I expected and wondered if it was turbulent or if he had a little oscillation that we couldn't see. He didn't comment about anything, so I assumed he was just doing what test pilots do, putting in a little input and seeing what happens, then putting in a little more and seeing what happens.

When Milt got to the flare altitude, which, I think, was called out by one of the chase plane pilots, he initiated the flare and came down right over the ground. We had altitude information because we had tracking radar and onboard data, so we could see about where he was. We could see, from the information we had, that the flight was going well. Milt glided right above the runway for a few seconds, and then set the M2-F1 down.

I'm sure the landing was a big relief for Milt, more than for the rest of us. Pilots go through these things more often than the rest of us but I don't think that makes it any easier. I'm sure that my heart was beating exceedingly fast even though I personally was in no danger. It just seemed as if there was so much uncertainty.

The excitement of finally doing something that we'd been working on for a year and having all go well was exhilarating. Having the intact vehicle with the intact pilot stop on the runway was quite a relief, with great exuberance everywhere. I know everyone watching the flight felt the same way that I did.

That's how we got through the first flight. There were some issues, some little things that Milt wanted to look at, requiring some modification to fix things that had annoyed him during the flight. That's normal after a first flight. As soon as we took care of them, we proceeded through a series of flights.

Bruce Peterson, the second person to fly a lifting body was the co-project pilot. I don't remember when he did his first flight in the M2-F1, but it was probably several months later. We didn't fly very often. Milt had a lot of responsibilities with Dyna-Soar, the X-15, and other programs, so when he, the M2-F1, and its team were available, we flew.

During those flights, Milt did a lot of maneuvers. He did pushover pull-ups and windup turns for the performance measurements to get better flight-derived lift and drag. He did steady sideslips, a traditional flight maneuver for finding out about the control power and sideslip characteristics-in other words, the directional stability and the dreaded dihedral effect. We also did stability-and-control tests, which consisted of a pulse on the aileron followed later by a pulse on the rudder. Milt did as many of these maneuvers as he could in the few thousand feet available before starting the terminal phase of the pushover. Even with such short flight, we got nearly all the stability-and-control and performance maneuvers during the approximately ten flights that Milt did.

I think we stuck with Milt so long because he was an outstanding pilot in every way. He had the confidence to do the flights and to get the data that we wanted. We did have more formal flight-planning sessions to decide what maneuvers we'd fly, how many there would be, in what order Milt would do them, and when he would quit doing them and go into the landing phase of the flight.

After this phase, we brought in two new pilots. Bruce was the next NASA pilot to fly. Then, Chuck Yeager showed a great interest in the program and Paul Bikle, Dryden's director, wanted to know what Yeager thought about the vehicle. Yeager and Bikle had been good friends since the days when Bikle had worked for the Air Force before he became Dryden's director.

I don't think there was anything that had wings and flew that Yeager wasn't interested in flying. He always made it really clear that he wasn't interested in being a monkey in a tin can, like the Mercury astronauts but that didn't apply to real airplanes, even if they didn't have actual wings. I think he always wanted to fly the M2-F1 and he knew he could fly it better than anybody else. Some of that was typical pilot bravado but not all of it. When I worked with Yeager, getting him prepared to fly, I found him very easy to work with (Fig. 6). He knew I was a junior engineer, but he answered every question I asked him and he treated me with the same respect he gave everybody else. He made it really clear that he was the pilot and that he was going to make all the decisions, as he should.

Fig. 6. A photo taken in early November 1963 of the first four pilots to fly a lifting body. The photo was taken during car tow practice getting ready for subsequent R4D tows that all four pilots would participate in. The pilots from left to right are Milt Thompson, Chuck Yeager, Bruce Peterson, and Don Mallick. (E-10627)

On the first of his five flights, I thought he wasn't doing anything in the airplane. The control motions that he put in were so small that I thought he was just going to wait for the flare and land it. Later, looking at the data, I could see in the time histories that he had made a rather nice study of the angle-of-attack envelope by putting in little inputs.

When I talked to others who had worked with Yeager in earlier years, they said that he never did any more with an airplane than he had to. He was one with the airplane if anybody was. When he put in a little input he sensed in every way what it was doing to the airplane. He didn't need a big input to figure the airplane out. He did lots of startling and scary maneuvers once he understood how a vehicle flew, but he learned to fly the vehicle in a most cautious way, getting the most information in the shortest time by proceeding that way.

Around the same time as he was flying the M2-F1, he went up in the NF-104, reached too high an altitude and angle of attack, and had the airplane go into a spin. He got the plane out of the spin with the drag chute. However, he didn't get the angle of attack low enough before he released the drag chute, so the airplane went back into the spin and he had to eject.

Yeager was burned by the ejection seat rocket during the ejection. When he flew the M2, he was still in pain from this, but most pilots have to be in great pain to not fly something that they're really interested in flying. After all, he exceeded Mach 1 in the X-1 with cracked ribs from a horse-riding accident. I don't think we missed any flying opportunities. It was quite spectacular to me that right in the middle of the effort to get our lifting-body data, one of the world's most famous test pilots gave us an evaluation.

We continued the flight program until we had the data we needed to complete the analysis and certify the stability and control performance of the M2-F1. (Fig. 7) We got most of the data within a few months of Milt's first flight. Then we quit flying for a while, as I recall. Later, we started bringing other pilots in as we

Fig. 7. President Lyndon Johnson was one of many important people to visit Dryden during the 1960s. Milt Thompson explains the M2-F1 to President Johnson with center director Paul Bikle just to his left in this June 1964 photo (ECN 281)

began getting ready to fly the next lifting body, the M2-F2.

There was lots of analysis to do once we got the data, I did some analysis with the digital computer, but at that time most of the analysis didn't lend itself very well to the computer techniques that we had available. Harriet Smith's primary responsibility was to extract the stability-and-control information from the flight data using analog matching. Since she was my senior by seven or eight years, she led that effort, and I helped her. I learned a lot in the process.

I had done some analog matching earlier on the X-15 with Roxanah Yancey, one of the original engineers at Dryden. She was the one who worked up the X-1 flight record and certified the first supersonic flight on October 14, 1947. I wasn't required to do any stability and control derivative extraction from the X-15 maneuvers. I was interested in the data, but I was more interested in the process of getting information. We could use the information to update the X-15 simulator to see where wind-tunnel predictions agreed or disagreed with what we were getting in flight. Roxy spent the better part of a day analyzing a single maneuver. Becoming familiar with analog matching interested me for two reasons. First, I thought that analog-matching was exceedingly tedious. The second reason was that analog matching took a long time and required engineering judgment, which varied from analyst to analyst. Analog-matching just wasn't very satisfying because there was no consistent metric for determining how good a match was.

The maneuvers I worked on with Harriet were much simpler than the X-15 maneuvers, even though we had close to a hundred of them. A problem we kept running into was that the flight control system was a mechanical system with cables. When the pilot tried to hold the aileron steady as he did a rudder pulse, the aileron cable stretched and the aileron surface moved around, too. For analog matching, one parameter needed to remain fixed throughout each phase of the maneuver to give good results.

Even as I was analyzing lifting body stability and control maneuvers and X-15 maneuvers with Roxanah [17], I was looking around for ways to automate the process. Larry Taylor and I initially looked at regression techniques and found that they could add some value, but they had other kinds of problems. We also looked at the time vector technique, primarily used by Chester Wolowicz, for getting some types of information.

The analog matching experience I had with Roxanah on the X-15 analysis and the experience I had with Harriet on the M2-F1 in 1963 led me to develop what turned into the parameter identification techniques that I spent much of my career devising, perfecting, and helping others to use over the years.

There were certain places in the X-15 envelope where the vehicle was marginal in stability-and-control. We really didn't like flying without the SAS control system engaged, which meant that if we tried to analyze those maneuvers, all the control surfaces were always moving. Having them moving made it very difficult to get a consistent match with analog matching. We were adjusting potentiometers to do the matching for each stability-and-control derivative. It worked best if the controls were fixed for a while, so we could work on the rate and sideslip derivatives. Then we wanted to have only one control, the aileron, moving, to get the aileron derivatives with the rudder fixed. To get the rudder derivatives, we allowed the rudder to move and fixed the aileron. This was one of the motivations for some of the work that I went on to do. I started in 1963 with those labor-intensive, difficult techniques and, by sometime in late 1965 or early 1966, I had a computerized technique fairly close to the maximum likelihood estimator in use today.

Vic Horton, Dick Eldredge, and Dick Klein were analyzing the M2-F1 performance data [18] and Harriet, with some of my help, was extracting the stability-and-control derivatives [19]. We continued to get data over the next six months or so. The M2-F1 flew fairly regularly with either Milt Thompson or Bruce Peterson doing additional maneuvers for us. By the time we were done, we were satisfied that we had a very good model of how the M2-F1 performed and how its flight and wind-tunnel data compared. Some of the M2-F1 derivative analysis was a little disappointing because of the mechanical control system and the limitations of analog-matching. It was, though, as good as anyone could do and the report that Harriet Smith wrote was the culmination of that activity.

My final recollection about the M2-F1 program is that I was greatly startled by the first time I saw the movie of the M2-F1 in flight and saw the horrendous angle of its descent. It looked as if we were seeing the whole top of the vehicle in the ground-based camera footage as Milt dove toward the dry lakebed. It gave me a cold chill watching that, thinking about how little experience I had at the time. I was amazed that the contribution I made had actually helped the vehicle. It could have ended in disaster if my analysis had been incorrect. My experience with the M2-F1 has given me a great appreciation ever since for the overwhelming importance of the work that an engineer does in supporting a flight project. This work is far more important than the research data we get because it's about the risk to the pilot, to the airplane, and to the program. If Milt had crashed and severely hurt himself on that first flight, it quite likely would have been the end of all lifting-body programs ever. If it had been caused by anything that I had worked on during that project, I probably would have never forgiven myself and would have chosen some other kind of work. So it was important to me that I had worked with a very skilled group of people on the M2-F1 and had done a good job.

On the M2-F1 program, I was treated as an equal from the beginning. I tried to make a real contribution and afterwards I felt that I did. Everybody listened to me when I had something to say. This helped build my confidence in myself and in the others I worked with on that program.

III

Building the Heavyweight Lifting Bodies

In the late summer or early fall of 1963, we proposed more maneuvers to further analyze and thrash out the M2-F1 flight data. At that point, I was still supporting the X-15 program, working primarily on the handling qualities and the stability and control. There were some interesting stability-and-control problems on the X-15 at some higher altitudes and higher Mach numbers. I developed some FORTRAN programs to help myself and others look at these problems. Larry Taylor was still primarily involved with the X-15 and I used him mostly on an emergency basis for the lifting bodies, whenever I thought we needed somebody with more expertise and experience than I had.

Future Plans

Originally, we had three lifting-body shapes under consideration for research flights: the Ames M2, a half-cone/large-angle shape from Ames called the M1-L, and the Langley lenticular shape. All three shapes were designed to use the same internal steel tube framework in their construction. Dale Reed had argued the M2-F1 shape was the most reasonable, so that was the first shape we built. I assumed that after we studied the flight of the M2-F1, we would probably go to the lenticular and M1-L configurations to see how they flew. That's what Dale had always talked about and I assumed our little skunk works group would build more of these and get some ideas about where we might go.

At the same time, Ames Research Center had developed a strong interest in the M2-F1 and was looking at how to make it a more realistic vehicle for returning from space. The Ames people had been testing models in the wind tunnel, discussing what they would need to do. I first became aware of what was going on at Ames in late summer or early fall of 1963. I found out that the configuration that the Ames people were interested in was something they were calling the M2-F2 (Fig. 1), which had three major differences from the M2-F1. The canopy was moved forward from where it had been on the M2-F1. They had significantly increased the boat-tail properties to reduce the base drag and increase the lift over drag,. They had also removed the outboard ailerons, which they thought were totally unsuitable for reentry (See Appendix C). I couldn't disagree with them, because the outboard ailerons certainly would have burned off with the technologies available in 1963. In fact, I suspect that even today we'd have a very difficult time protecting those kinds of surfaces from inner gap heating on reentry.

Ames had proposed that we go back to the split trailing edge inboard ailerons,

Fig. 1. Dale Reed (right) and Ken Iliff, using a slide rule, pictured with the Langley lenticular lifting body and the M2-F2 models on the table. (E 15468)

using them differentially for roll control. With the body flap and split ailerons, they would be able to do the pitch and roll control. They also modified the rudders slightly to use them for directional control of the vehicle. I could see where the configuration had come from, but I could also see that the ailerons were not going to be very effective for roll. We knew from the M2-F1 that the rudders were exceedingly dangerous to use for roll, which meant they wouldn't be a very good option for the M2-F2 either.

We got the stability-and-control derivatives and the performance numbers from the Ames wind-tunnel studies. I'm sure Bertha Ryan and Harriet were also looking at the data. I had established my own interest in the lifting bodies as being mostly in the areas of stability and control and handling qualities. I took the analysis programs that I had developed for the M2-F1 and the X-15 and used them to start looking at the proposed new configurations from Ames. My recollection is that although Ames looked at a number of configurations, they had two favorites. One was the vehicle with inboard ailerons, modified canopy, and increased boat-tailing. The other one was roughly the same, but it had a center fin as well. The M2-F1 had been tested in the wind tunnel both with the center fin on and off. We ultimately ended up flying most of the M2-F1 program with the center fin off, we had many indications that suggested that as we went to higher Mach number, the center fin would be good for directional stability. This was especially true transonically, down to about eight-tenths Mach number, the vehicle having a tendency to be directionally unstable at the higher Mach numbers.

Ames had a couple of objections to the center fin. One was that it was another surface to get hot in the reentry, another to have to protect. It also increased the drag of the vehicle. Additionally, many people, including Milt, disliked the center fin because it looked as if it would increase the dihedral effect, which was, I believe, the real culprit in Milt's reactions to these vehicles. He thought the dihedral effect was already much too large and a fin would make it larger. As a result, he didn't want to fly with a center fin if he could avoid it.

We knew from the M2-F1 tests that we had done in the Ames 40-by-80-foot tunnel that the inboard ailerons had totally unacceptable adverse yaw. If the pilot put the stick to the right, he initially got a very quick roll to the right, followed by a rapid roll to the left. The interesting thing is which way the nose goes when an aileron deflection is made. If it goes with the roll, it's what engineers call "proverse yaw." If it goes against the roll, it's called "adverse yaw." A little adverse yaw is usually not a bad thing. A lot of adverse yaw is a very bad thing, because you end up with a roll reversal.

We had noticed in the full-scale wind-tunnel tests on the M2-F1 that, although the center fin was ahead of the split elevons, the adverse yaw from the differential elevon deflection was substantially reduced [19]. We suspected that the inboard side of the outboard fin was experiencing a high pressure area, creating a yawing moment, when the differential elevons were deflected upward. This moment caused the nose to swing away from the direction of the roll. Therefore, the center fin experienced a similar high pressure area which counteracted the one on the outboard fin, reducing the overall adverse yaw. It was a small effect, as I recall, maybe a 25 percent reduction, but it was in the right direction. My recollection, looking at the initial Ames data on the M2-F2 with the center fin on, was that it was clear that there was virtually no adverse yaw.

I also remember that we saw the initial M2-F2 wind-tunnel data at least by late 1963. The only information I can find now is a reference to the tunnel tests in an M2-F2 internal Dryden document and an Ames wind-tunnel report by Earl Keener published after the end of the flight program showing center fin data on the same model they had been using for the rest of the M2-F2 data [20]. These imply that the center fin data was run at the same time. The Langley HL-10 and Air Force SV-5 lifting bodies, circa 1963, both had center fins and they were probably put there to at least partially reduce the adverse yaw. The report is circumstantial evidence, at best, for showing when we actually knew about the reduction in adverse yaw with the center fin. But my recollection is we knew that very early. I can remember having discussions and drawings on the blackboard that high pressure areas on the outboard fins without the center fin would cause the adverse yaw in the high pressure area.

My recollection is that we looked at both of those sets of data-with the center fin and without it-and discovered that the M2-F2 had low roll power. We couldn't generate high roll rates with those inboard ailerons, as they weren't out on the wing tips, and we didn't get much rolling moment from them. But they did tend to roll the vehicle in the direction the pilot wanted with little, if any, adverse yaw. However, we ended up with an airplane with a center fin that had the desired roll control effectiveness, although it could never generate very high rates.

The rationale for removing the center fin was that we could fix the adverse yaw problem by incorporating an interconnect that would use rudder to reduce the adverse yaw. In other words, the rudder would move in the same direction as the roll, so that the nose would not swing to the left when the vehicle rolled to the right. This was proposed by quite a few people and it was largely favored by the people in charge of the program at that time, because it did eliminate the need for the center fin. Early on we had a bull session about these things and this didn't seem like a terribly serious issue. We each picked from various options and configurations what we wanted to proceed with next.

Early on, I remember, Milt felt, in both formal and informal discussions, that the M2-F1 was already a handful in terms of dihedral effect. As a result, if he got hit by a large gust from the side, close to the ground, he would not be able to compensate for it, even with the outboard ailerons. Milt was very opposed to the idea of adding another vertical surface that looked as if it would increase the dihedral effect, even though it would also increase the directional stability.

Milt was a good engineer. He'd flown in a lot of vehicles, and he knew what scared him and what didn't. There was very little that actually scared him, but in several conversations during this time and later as we proceeded with the M2-F2, he continued to say that the thing that scared him most about flying the M2-F1, and eventually the M2-F2, was being hit by cross winds or side gusts. He felt that this was a good way to lose the vehicle.

I don't remember where the other engineers on the program stood on the issue of center fin versus no center fin. I think Dale was leaning toward the center fin, partly because it looked as if it would be less complicated to implement without that rudder interconnect.

My biggest concern about the adverse yaw was the almost blind-faith belief that we could couple the rudder with the inboard ailerons and end up with nice roll control simply by having that interconnect. Using the data from every wind-tunnel test I had seen for analysis with various control analysis techniques, primarily the root locus, showed that using the rudder this way was dangerous. The discussion of the M2-F1 rudder characteristics with the simplified root locus discussion of Fig. 1, chapter 2 still applies. It was going to be difficult to finesse the rudder into the control system and not have it always be a looming problem. As I recall, even in the M2-F1, Milt and the other pilots commented that the rudders were tricky and that you needed to keep your gain down when you were using them, because they were obviously a big hammer compared to controlling roll with the outboard elevons through the stick.

Building the M2-F2 Simulator

To aid in thrashing out questions about the M2-F2 control system, we used a simulation. As I said earlier, building a simulation was not a minor task. It took many, many months to get one implemented and checked out. It also took a full-time simulation engineer to maintain it and check it out every day to make sure that it was working right, because it was on an analog computer system.

Lowell Greenfield was a simulation engineer. He'd gone to Iowa State at the same time I did, ending up with a degree in mathematics. As he was graduating in 1962, he wasn't sure where he wanted to go, but he liked the idea of California. He'd heard that I was going to be working at Dryden and asked if I thought he might also get a job there. Since we got along so well, it seemed like a good idea to both of us. We inquired at Dryden and John Smith reviewed his application. Smith, head of the simulation group, was also from Iowa State. Dryden made Lowell an offer, and he and I started at Dryden at roughly the same time.

Iowa State was quite advanced using both digital and analog computers for engineering work. Analog computers had appeared in the Aerospace Engineering Department at Iowa State in about 1961, along with a new faculty member, P.J. Herman, who had worked with the GEDA analog computers made by Goodyear. He had also worked on Goodyear blimp simulations. Quite an expert on analog simulations, he got Goodyear to donate analog computers to Iowa State.

As a result, there was a period of about three or four years when Iowa State was turning out experienced analog computer users. We programmed the computers and used them to analyze the data. In addition, we had the Cyclone digital computer which was, for a few months, the fastest digital computer in the world. I had learned to program on it in machine language with zeros and ones on punch tape. Machine code wasn't easy to write, unlike FORTRAN, and it was very sensitive to little mistakes.

The program I remember best was one where I thought that I had invented a perpetual motion machine. I had built a lookup table in the program for computing drag on a glide vehicle, but when the vehicle got down to 36,000 feet, it started accelerating. While troubleshooting it, I discovered that my lookup table had the wrong sign for drag below 36,000 feet, so rather than slowing the vehicle down, the "drag" sped it up. My career as a designer of perpetual motion machines ended at that point.

As a result of departmental foresight, all aerospace engineers at Iowa State from about 1960 or 1961 until at least 1965 were very advanced in programming and using digital computers. A lot of the early work in computational fluid dynamics came from the use of the digital computers at Iowa State. I learned the panel methods that I used on the M2-F1 and delta-wing X-15 in my junior and senior years at Iowa State. I used those methods to fill in the missing data for various aircraft configurations. As these methods advanced, they became the fully computational fluid dynamics techniques now called CFD. Ames Research Center, a major center for CFD development, had a large number of postdoctoral researchers throughout the 1960s and 1970s who had learned their computational fluid dynamics at Iowa State. The education in both digital and analog computers was a big advantage for Iowa State students joining the workforce.

When Lowell Greenfield received his offer from Dryden in 1962, he had not ever programmed a digital computer or an analog computer. He knew that I had experience on both at Iowa State. Once we found out that he was going to be at Dryden, I took Lowell to the aero lab and showed him how to write a couple of simple programs on the digital computer, like a harmonic oscillator. Lowell picked that up fairly quickly and thought it was neat stuff. At Dryden he was going to work on the analog computer, so I showed that to him. I taught him the strategies and how to use the output as the input to bootstrap his way to a solution. That intrigued him. He designed his own analog computer program that was supposed to produce a simple undamped sine wave. However, it turned out to be a highly divergent sine wave, so Lowell's first analog computer program was not a roaring success.

Lowell became very proficient in programming analog computers at Dryden. He could do everything assigned to him and still have spare time. I said to him, "You know, it would be neat if we could look at some of these advanced lifting-body configurations on an analog computer. What do you think?" He agreed, got some numbers from me, then went off and programmed it. He set it up with a simple cockpit, with a dowel suspended in a coffee can by four springs that moved the potentiometers so that we actually had lateral and longitudinal stick control.

Dick Musick, a simulation technician at that time, Lowell, and I had a good time learning to fly the M2-F2, even if we didn't fly it very well. We didn't have very good instruments, but we had strip charts and Dick eventually came up with some dial faces. I don't know when Lowell actually did the first unofficial simulation of the M2-F2, but I would guess it was in late 1963. Later, it became an official simulation, going through the more traditional developments that all simulator programs at Dryden went through at that time.

Looking at Advanced Lifting Bodies

After looking at the M2-F2 and not really knowing if building it was a possibility, we started talking to Langley Research Center. Just as Ames had dropped the M1-L shape in favor of the M2-F2, Langley had abandoned the lenticular lifting body (Fig. 1) for a series of concepts that they called "HL," for horizontal landing. By late 1963, they were up to the HL-10 shape (Appendix C). We got the HL-10 wind-tunnel data from them, although it may not have been the final set of data. I still have my calculations of HL-10 frequency and damping, as well as pilot ratings from Milt from May 1963. I think Langley was still working out the sizes of the center fin and outboard fins on the HL-10, but it was a fairly complete data set otherwise. In fact, they had more supersonic data than the M2-F2 did, as I recall. One problem with the HL-10 was that it was classified Confidential, which meant we had to keep all data and analysis in the safe, which made it a little more difficult to work on.

The stability and control of the HL-10 was unbelievable. We put the HL-10 on the little simulation that Lowell had made and it flew like an F-104. It had high roll rates and it was very responsive in pitch, making it a good vehicle from many points of view. As an early wind-tunnel data set, it stuck in the back of our minds. It was a lifting body that actually flew like a high-performance airplane. It also had a purported L/D max of 3.5 or 4, considered quite good. However, most of our work was being done with the M2-F2, partly because of its similarity with the M2-F1. I know that Dale, Harriet, Bertha, and I were familiar with the M2-F2 data that we were getting from Ames and the HL-10 data that we were getting from Gene Love's group at Langley. In fact, I remember a group from Langley coming out a couple of times discussing the configuration with us.

The Langley group was always willing to listen to the people at Dryden, because of our M2-F1 experience. In our critique of their vehicle, asking them if they could do more of one thing and less of another, we found them very responsive. I remember working with Jack Paulsen and Bob Rainey in particular. They had taken a different approach to defining lifting bodies. The Ames approach with the M2 shape had been to solve the heating and the hypersonic problem and build add-ons later. Coming into the program a little bit later, the Langley people did a more global design, trying to find a shape that would be good everywhere. Part of the shape would be good for hypersonic flight, part of the shape would be good for supersonic flight, and part of it would be good for subsonic flight and landing. They chose a negative-camber airfoil shape for the body, and gave it a leading-edge sweep, like a delta wing, except it was a lot fatter. They designed the fins and the control surfaces to provide trim where it was needed. They were very concerned about trimming the vehicle hypersonically so the body flap didn't get too hot. A fairly mature configuration, the HL-10 didn't have a lot of the warts that the M2-F2 had.

The only thing that I remember from those early data sets is that the HL-10 was directionally unstable around Mach 2.3. That was going to be a problem because in those days we only had analog systems and we did not intentionally fly an aircraft that was statically unstable. Another problem was the data seemed to be so good that there had to have been an error.

Toward mid-1964, probably, we had some discussions with the Martin Company and Air Force personnel about a configuration of theirs that was very controllable, had a high L/D max, and was better in every way than what we had been examining. They were obviously quite enthusiastic about this configuration, called the SV-5.

Since the SV-5 was classified as Secret, all documents were controlled and all plots had to be stamped immediately and kept in a security container. Its classification meant that there was really no possibility that we could do much in the way of running it on an unclassified computer. Work done on something classified as Confidential could be disguised to a large extent if it wasn't identified and you were careful what you did with the data. But with something classified as Secret, you really did need a classified computer and a classified work environment, so we could work on it only in the offices. My impression of the vehicle was that it didn't look as if it would work. I've heard the SV-5 described as a potato with fins, and I believe that was my initial reaction to it. It was just too ugly to fly. However, the data indicated that it was predicted to have a really high L/D max subsonically and a good one hypersonically. It was stable everywhere and had a higher volumetric efficiency than any of the M2 shapes or the HL-10. A measurement of the internal volume compared to the vehicle's external area, the volumetric efficiency is defined as six times the square root of pi, times the vehicle's volume over the vehicle's surface area to the three halves power. This is a very important parameter for what we were trying to do, because we were looking for something better than a capsule for returning from orbit.

The supposed volumetric efficiency of the SV-5, its unbelievably high subsonic L/D max, about 4, and its stability throughout the speed range made it a very attractive candidate. We didn't do much with it then because of the classification, though the information was being shared with us mostly because we were working with lifting bodies. I don't know if we ever got all of the data, or all of the really good data, but it was clear that the SV-5 shape was in the running. After Martin lost the contract to Boeing for building the Dyna-Soar, it received Air Force money for studying lifting bodies. I think the group working on the SV-5 shape was a spinoff of that relatively old and quite mature effort. As flight tests would show later, the M2 was the big hammer approach for getting something that would work at all, the HL-10 was a global look at how the pieces went together, and the SV-5 was a refined analysis design with a really high volumetric efficiency and a high L/D max.

I was told a few years later that the SV-5 had had more wind-tunnel test time than any other configuration at that time. I believe that this record was broken by the Shuttle, which may be the only vehicle to do so. The SV-5 had thousands and thousands of hours of wind-tunnel time in the early to mid-1960s. The X-23A vehicle, shaped like the SV-5, was launched on an Atlas rocket and went to near-orbital speeds before successfully reentering. Another derivative of the SV-5 became the X-24A (Appendix C).

About then, in addition to what Lowell had done on the M2-F2, I programmed some of the problems that I thought the M2-F2 had into a TR-20. We had TR-10, TR-20, and TR-48 analog computers [21]. The TR-10s didn't have a removable patch panel, so once you patched it up, it was yours until somebody else took it away from you. This was the same type of TR-10 analog computer that Neil Armstrong had used to study X-15 characteristics. Because the TR-20s had removable patch panels, they could pass from user to user. You would clamp on your patch panel and it would simulate your experiment in an analog fashion. Rather than having a full flight envelope simulation on these small portable analog computers, I programmed the landing tasks of the M2-F2 with and without the center fin to get a better idea of what it looked like. You couldn't program a lot of versatility on a TR-20. It remained very focused on one specific problem. I was looking at how we might interconnect the rudders to come up with something that worked with the center fin off.

Building the M2-F2

In early 1964, NASA issued a request for proposal for two heavyweight lifting bodies. The evaluation of the proposals was finished by mid-1964 and the contract was awarded to Northrop. I wasn't much involved in the proposal specifications. I think Dale and the more senior people did that work. I gave them my ideas because by then the M2-F2 was specified as having the center fin off, a configuration I was unhappy with. However, if that was what we were going to build and fly, I would do my best to make it work.

I also wasn't involved in evaluating the proposals, but it seemed quite early on that Northrop was going to get the contract. After the contract was awarded to them, we started meeting with the Northrop people. From my point of view it was a very nice association. I hadn't really worked earlier on configuration design or on manufacturing of airplanes. I'd seen some of it going on and I'd seen some of the modifications that Dryden had made, but other than the M2-F1, I had never seen an airplane built. The internal structure, the skin, and the hydraulic systems were pieces I had seen in F-104s and in the X-15, so the heavyweight lifting bodies were basically airplanes that were being drawn up, designed, and prepared for manufacturing.

NASA Headquarters selected the M2-F2 and the HL-10 shapes for building, rather than a pair of M2-F2's. I think the HL-10 design was still being finalized, but it was roughly what we had looked at initially. The M2-F2 was defined in terms of mold lines, configuration, and control surfaces. We could go down to Northrop in Hawthorne, California, in the R4D, with our pilots, frequently Milt Thompson or Bruce Peterson (the early project pilots), flying it. Later on, Fred Haise was added to the program and he also went with us back and forth to Hawthorne. If the R4D wasn't available, we could fly in the Northrop airplane based at Edwards, an Italian airplane called a Piaggio, a nice flying airplane. The flight from Edwards to Northrop in Hawthorne was fairly short, but sometimes it was a rough flight because neither the R4D nor the Piaggio flew very high and we had to go over a mountain range in all kinds of weather.

As Northrop started building the vehicles, John McTigue took over the project management of the lifting-body program. I hadn't worked with John very much, but I was impressed with his ability to find out what the problem was and get somebody who knew what they were doing to solve it. As Northrop ran into problems, or John ran into problems with what Northrop was doing, the word would spread. Sometimes I'd get called to create a data table or a plot that was required to further the process, but it was really design, not straight stability-and-control research. It was also a study in manufacturing techniques, putting the pieces together making them work. It was very interesting for me, because in looking back at it, it was the only other truly skunk works project that I worked on. The M2-F1 certainly had the goals and objectives that the Lockheed Skunk Works had developed over the years, as did the M2-F2 as well. The project was also interesting because it was a collaboration between a NASA center and a major airplane manufacturer.

We had a working relationship that split fairly easily, at least from my point of view. Northrop knew how to build stuff, and we supposedly knew how to fly it. We knew what it was we wanted, so if something wasn't specified quite right, Northrop and Dryden quickly resolved it and got on with the project. It was a nice working environment. You could watch the various fixtures as they were built at Northrop, how they got everything to fit together, and how they tested it and put it together. It was a good experience for me as a young engineer, seeing how the rest of the aircraft world really worked. I felt that I knew all about how Dryden worked; after all, I'd been there two years by then. The people I thought were experienced engineers, of course, were in their late twenties and early thirties, which shows what people can do at a young age if they're empowered to do it by management. Paul Bikle, then our center's director, trusted us to get it right, challenging us frequently, sometimes a little unfairly, and it resulted in a safe product. It was cheap and it was fun. Bikle was probably not unique then, but he would certainly be unique now.

I'm sure other engineers at Dryden had a different feeling about manufacturing the vehicle than I did. I was busy trying to come up with a control system that would make the vehicle fly without the center fin. I spent many hours in the simulator with first Milt and then Fred Haise, trying to come up with a set of feedbacks and interconnects that worked with our database. There were variations in the database because we didn't know how good it was. It was hard to make a very good airplane out of it, so the days that I did get away to go down to Northrop were always very refreshing to me because I could see that there was a real airplane that was going to be launched from the B-52 at 45,000 feet. It also reminded us that we had better get our work done, because we needed to make it the best possible flying airplane we could to get Milt from the B-52 to the ground.

One of my favorite recollections from that time was the rollout of the M2-F2 (Fig. 2). Invited to attend, I went to Hawthorne on the Piaggio, which was an Italian-built transport. A couple of engineers and I met at the Piaggio and got on. Then Chuck Yeager got on and asked the copilot if he minded if Yeager sat in the front seat. The pilot said, "Sure, go ahead." I knew of Chuck Yeager's fame for the first supersonic flight, and having worked with him for several months on the M2-F1, I had gotten a very healthy respect for his piloting abilities.

As we flew over the mountains to Hawthorne on the Piaggio, I was sitting in the front row. I could see the pilot and the co-pilot. Chuck Yeager asked if he could fly the plane. I think Chuck liked to log flight time in all the airplanes he'd flown in, and he maybe hadn't flown a Piaggio before. He flew the airplane from Edwards to Hawthorne. Chuck spent the whole time talking to the pilot, telling war stories and using his hands to show the attitudes of airplanes and his reactions. He didn't seem to be paying very close attention to what was going on with the Piaggio. I remember thinking at the time, here's probably the best pilot I've ever worked with in terms of a stick-and-rudder man, who's doing a terrible job of flying a business aviation airplane. It was fairly rough going over the mountains that day and he wasn't paying much attention. I'd feel the airplane roll clear off to one side and Chuck would grab the yoke and straighten it up, then it would roll off

Fig. 2. The M2-F2 rollout at the Northrop facility at Hawthorne, California in June 1965. (ECN 738)

to the other side. I guessed he was a good stick-and-rudder man when he has an airplane that's hard to fly, because this one probably wasn't.

It's always stuck in my memory that the only flight I ever took with Chuck Yeager was one of the worst flights I've ever been on in terms of keeping the wings level and the pilot not really seeming to pay attention. Chuck was paying attention, knowing him. Probably he could see there weren't any other airplanes around and he wasn't going to hit a mountain. The rest of it was just a matter of getting from Edwards to Hawthorne and telling his war stories.

Preparing to Fly the M2-F2

I don't remember all of the advocates for no-center-fin on the M2-F2, but I do remember one, and that was Milt. To me he was the only one that mattered, because I had a great deal of respect for him as a pilot and as an engineer. I thought I could convince him not to fly with an interconnect without the center fin. I used all my bag of tricks and showed him various studies. Comparing the M2-F2 and the M2-F1 studies, I told him why the M2-F2 characteristics could result in bad things. He patiently listened to me, smiled at me, and probably joked around the way he always did. When I was done, he replied, saying something like he was sure that's all very true, but I haven't solved his one problem, and that's what he was going to do about the side gusts near the ground. By putting that fin on, I was going to exacerbate the problem.

In those days, like the other working engineers, I flew the simulators. We were trying to understand problems, trying to scope out a flight plan, trying to understand-maybe by limiting the degrees of freedom on the simulator-certain aspects of the flight envelopes that we were planning to fly with the M2-F2.

Milt was a very willing participant in the simulator and very actively involved in the flight plan. He had a pilot's perspective of always having an off-ramp for every problem. He wanted to study problems ahead of time, eliminate those that he could, and learn how to live with those that he couldn't. We spent many hours in the simulator trying to refine the control system. He agreed it was a handful, but said it wasn't something that he didn't feel he could handle, because he knew what to watch out for.

One thing that surprised me in watching Milt fly the simulator was that every once in a while it would get into a residual low-frequency lateral-directional oscillation. I'd never seen that in the lateral-directional axis in an airplane before. Little by little I began to think there might be something wrong with the implementation of the simulator. I couldn't track anything down, though, so I went back to my analysis techniques and saw that the lateral-directional characteristic equation didn't factor into two real roots and one set of complex roots, as expected. Instead, it factored into two sets of complex pairs, which meant that the second pair of complex roots was the cause of the oscillation, an order of magnitude slower than the Dutch roll, that I was seeing on the simulator. There were ways that you could get it going on the simulator and it wasn't very controllable.

Milt didn't like this, but he didn't think it was a big problem because it didn't do much. However, at lower angles of attack, regardless of the interconnect gain, the analysis definitely showed that this long period lateral oscillation was undesirable. In some research that I did, I called it the wallowing mode, because that's how it looked on the simulator. As I looked through the literature, I discovered that it was known in other airplanes [22], although it wasn't common. It was called the lateral phugoid, because it was a long-period mode in the lateral axis, as the phugoid mode is in the longitudinal axis. It made designing a control system a lot more difficult.

I was the primary person defining the control system. I still had the help of my mentor Larry Taylor, but I was doing most of the dirty work. I was a young engineer and needed the experience and I was certainly getting it. Looking at the root locus, I could see that the poles of this second oscillatory mode were quite close to the unstable half of the imaginary plane. If the zeros from the control system with the interconnected ailerons were from poorly predicted characteristics of the rudder and aileron derivatives, the root locus would tend to try to close in the right-hand half of the plane, which would be unstable (Fig. 1, chapter 2), and we would get a pilot-induced oscillation from it.

With the analysis of the lateral phugoid and the uncertainty of the predicted zeros resulting in the undesirable location of the zeroes of the interconnected aileron, we spent a lot of time looking for a set of gains to reduce the tendency of the vehicle to become too sensitive in that region. You could lower the interconnect gain for the low angles of attack and make it a little better, but it was still really a problem if you could get the pilot's gain up, which you couldn't do with Milt.

However, he could try to make his own gain go up to look at things. For most pilots, it's hard to get their gain up in a simulator because neither their life nor the vehicle is at risk. When you get them in a real airplane, that's when their gain goes up. So you try to anticipate the higher pilot gain that you know you're going to see in flight. One of the techniques that I spent a lot of time on was called "electronic compensation." I was going to put electric components in the aircraft to have the phugoid close on some zeroes that I would artificially place near them. Then I was going to move the poles up on the left-hand side of where the zeroes were on the interconnected rudder.

Some of those worked well, but I didn't have a lot of experience with them. I knew what they were and I knew how to design and define them, though. As they're quite sensitive to the uncertainties in the aerodynamic data, there was always a chance you were going to make a bad situation worse if you got extra compensation in there. Initially Milt and I spent a lot of time looking at various kinds of pole zero compensation and at different feedback gains for roll rate into the aileron, roll rate into the rudder, yaw rate into the aileron, yaw rate into the rudder-anything that we could come up with-but we never could make it very satisfactory. I was changing those control gains, and I was also changing some of the primary stability and control derivatives a little bit-10 percent or 20 percent-and then having the system evaluated.

I'd do this in limited degrees of freedom so that we weren't flying the whole flight envelope. The simulator was fixed in velocity and altitude and we were looking just at sideslip, roll rate, and yaw rate-those three degrees of freedom. We'd spend a lot of hours looking. I'd adjust the potentiometers which would either change the gain or a derivative and then he'd evaluate it. The way we did evaluations in those days, as we still do, is to use the Cooper-Harper pilot rating scale with ratings from 1 to 10. A 10 is really bad-unacceptable under any condition-and a 1 is as close to perfect as you're going to see. For an experimental aircraft like this, a pilot rating of 3 or 4 wasn't bad, especially if it was not for the primary task. We'd get some 6's, especially at low angle of attack, because of the lateral phugoid and the poor characteristics of the interconnected aileron with the rudder.

Milt was a fairly heavy smoker at the time. He would sit in the simulator, smoking away, telling stories, mostly in a good mood. I'd keep him in the simulator for a couple hours at a time, but then he'd have had enough and leave. Milt was in the simulator to tell me whether configuration A was better than configuration B. He would give me a pilot rating. If he gave one of the configurations a 3.5, then for the next one he'd say, "No, that was a little worse," and he'd call that a 3.8. Later he'd say, "That one's a little bit better. That's a 3.75." We would keep doing this until he was giving me pilot ratings that were meaningless, because of unrealistic precision, in terms of the scale. However, it was a way for me later, after we were out of the simulator, to track what kinds of things he liked better and what kinds of things he didn't like as well. He would also give me four-or-five-place pilot ratings.

All of the rest of the years Milt and I worked together, right up until a day or two before he died in August 1993, if he and I didn't have anything else to say when we met in the hallway, he'd ask, "Anybody give you six-place pilot ratings lately?" We'd both laugh.

For all our joking, Milt knew that I was interested in making a good airplane. While he was probably humoring me to a certain extent, he was a good engineer and always wanted to understand everything about the airplanes- aerodynamics, mechanical parts, failure modes, and backups. Milt wanted to know all of that before he flew any airplane, including the X-15. This meant that using the simulator was partly about training him, because there was no other way to train him, and partly about educating me to see if we could come up with a better system.

We eventually came up with a system that was compensated and we had it reviewed by various controls experts at Dryden and elsewhere. It made the airplane at least one full pilot rating better everywhere. It didn't make it perfect. It didn't fix it. With higher-order electronic compensation, we probably could have made it better. However, since we'd never done any electronic compensation before, we were uncomfortable with that approach. We decided to fly the best of the mechanical systems that we'd come up with, which was an interconnect gain of about 0.4 or so. Sometimes we found 0.5 was better, and sometimes 0.4 was better. In the following discussion the interconnect gain is defined as how much negative rudder deflection would be commanded for a given aileron deflection, so for a 0.4 interconnect gain, a one-degree aileron deflection (by Dryden convention) would result in a -0.4 degree rudder deflection.

Somewhere in there was the best gain for all angles of attack. Probably the best thing that we could have done at that point for that airplane was to have a gain scheduler, where the interconnect gain lowered as we got to lower angles of attack. The M2-F2 would still have been, in my opinion, a dangerous airplane, but it would have been less likely to have the problems that showed up later.

Milt was busy during this period. He was doing X-15 flights. He was still flying the M2-F1 early in that development. He flew chase on the X-15s, and he was actively involved in many other things. I was glad to get some of his time to do control system evaluations. Later, as Milt got busier or felt he wasn't going to learn any more working with me and my simulations, I got Fred Haise, who was, to my way of thinking, a new pilot then. I think Fred came in 1963. He was the most eager pilot I had ever been associated with for helping you out. He'd fly the simulator, and he'd help with anything that he could. He was also a good experienced test pilot. He treated me as even more of an equal than Milt did. Fred really surprised me when he let me know that he was willing to work on engineering studies after four o'clock (quitting time) or on weekends. We worked some Saturdays, something I had never heard of at Dryden before. The engineers and the technicians worked off-shift, but for the most part the pilots didn't. I won't say none of the other pilots participated in engineering studies in the off hours, because they may have, but it was certainly pushing your luck to ask one to do so.

This is not to say that the pilots only worked the standard eight-hour shift. They worked a lot of hours, had other duties that overrode participating in engineering studies. The pilots flew early mornings, late evenings, and weekends. Additionally, they served as project pilots, chase pilots, and mission controllers for all of Dryden's research flight programs. For the X-15 flights, they practiced in the simulator prior to the mission and sometimes they flew up the range to check the weather and the status of the lakebeds.

Fred spent a lot of time in the simulator and we became good friends. I had a lot of respect for his engineering as well as his piloting ability. He was another very analytical guy. He understood my root locus arguments, my control concerns, and my Bode plots. It was nice having a guy like that around.

As evaluating the various control schemes drew to a close, I managed to modify an existing Air Force contract with Cornell Aeronautical Labs, in Buffalo, New York, which later became CalSpan and is now Veridian. Cornell had a variable-stability T-33 which I had them program with some of my more interesting lateral-directional and longitudinal sets of M2-F2 data so the pilots could fly it in the actual environment. That would drive their gain higher. As a result, I thought, Milt would see the conditions I was concerned about, decide they were dangerous, and we could go back to looking at the center fin. I accused Milt of being stubborn, but I was obviously stubborn myself.

Larry Taylor and I also were doing criteria studies of general handling qualities at the time. [23] We added a dozen or so sets of derivatives for the M2-F2 as configured with various interconnect gains, feedbacks, and flight conditions onto the list for the T-33. The pilots flew those first in the fixed-base simulator at Cornell. When we were there in Buffalo in February 1965, it snowed 30 inches in three or four days, but we went to work every day, somehow getting there. I spent many hours there working with Bob Harper of Cornell (the Harper of the Cooper-Harper pilot rating system) and Fred Haise (Fig. 3), the project pilot for this set of experiments. The plan was that Fred would get it verified and ready to go, then we'd bring the T-33 to Dryden and have Bob Harper, Fred Haise, Milt Thompson, and Bruce Peterson fly it. Each of them would do evaluations and also fly it in some windy conditions. [24, 25]

Milt evaluated the M2-F2 with the T-33 simulation and did not like it in turbulence, but he felt he could probably handle it the way it was simulated. As a result we flew the M2-F2 with no center fin.

By the end of the T-33 program, the control design for the M2-F2 was fairly well determined. [26] We were going to fly it with fixed gain feedbacks, the roll rate into the aileron, the yaw rate into the rudder, and the aileron and rudder interconnected with a lead-lag or a washout on the aileron so we still would have the ability to generate some roll rate in a turn. It was not going to be a very good flying airplane, but Milt felt that he would be able to handle any eventualities. His main off-ramp, I believe, was to increase the angle of attack if he discovered he was getting into anything like a pilot-induced oscillation.

It was during this period that I got to know Fred Haise very well. He had been chosen to be a lifting-body project pilot. Milt was going to be the lead pilot. I think by then that Jerry Gentry and Bruce Peterson had been picked as the other two M2-F2 pilots and the same group of pilots was also planned to fly the HL-10. I was

Fig. 3. The Cornell T-33 variable stability aircraft with the drag petals extended. Fred Haise piloted the T-33 to test the M2-F2's stability and control, lift and drag during this April 1965 flight. (E 13020)

looking forward to working with Fred and Milt on the M2-F2 and HL-10 flight tests.

Some time after the T-33 flights, however, Fred was selected to become an astronaut, as was Joe Engle. I also knew Joe fairly well, having worked with him on the X-15 program. I knew Fred would be an outstanding astronaut and I thought his chances of getting to go to the Moon were fairly good, although the long list of names ahead of his didn't make it look very promising. I knew that, with his abilities, Fred would rise to the top. He was in the backup crews for Apollo 8 and Apollo 11. He's the one that closed the door to the capsule-the last person to see each of those two crews. Fred did well in the astronaut program, and I owe him a debt that I can't repay for arranging for me to watch the Apollo 11 launch and to watch Neil Armstrong, Buzz Aldrin, and Mike Collins practice in the simulator the day before they left for the Moon.

Another thing Fred did for me at Dryden was to demonstrate an unstable phugoid. Most engineers don't get to fly in unstable aircraft because we're supposed to fix them. The AeroCommander aircraft, in the power approach configuration, had an unstable phugoid in the longitudinal mode. Fred and I went up in the AeroCommander and he let me fly it so I could feel in the wheel what was going on. It was quite deceiving, because the airplane would lose ten or fifteen knots in the phugoid, increase to twenty knots, and then stall out on the next cycle. We did this several times on that flight. Fred also flew me by the top of Mount Whitney and down below sea level at Death Valley on that same flight. I still have the recollection from that experience of how compelling an instability is when you're in the airplane and your life is at risk.

It may seem as if I was the only person working on controls for the M2-F2, which of course isn't true. The primary investigator for the longitudinal axis on the M2-F2 was Wen Painter. [26] He did the early work on defining the longitudinal control system, as did Larry Taylor. Wen also got into the more practical part after the M2-F2 arrived at Dryden from Hawthorne

We had to do the systems testing, a lot of which hadn't been done in Hawthorne because we would have to repeat it all at Dryden anyway. Since we had such a good team relationship with Northrop, I think we took the vehicle before all the checks had been done, although we continued to work as a team. When any deficiencies were found, Northrop was right there with us, helping to fix the problems. It was an excellent relationship and a good way to do things. I don't think I've seen anything done that well since then, in terms of a small team doing a good job on a short schedule. We spent another year testing components and getting the systems really wrung out so that we could do our first flight.

Some of that involved captive-flight tests, where problems are always found. That's another advantage of an air-launched vehicle-being able to check the vehicle out in a harsh environment in terms of dynamic pressure, heating from the electronics, vibration, turbulence, cold soaking at high altitude, and aerodynamic load. All of these things get tested as the vehicle flies on the mothership, a good environment for the final shakedown of an airplane. In a traditional airplane, high-speed taxi tests come first, then a short flight, usually with the gear down, to find the problems that always slip through that cannot be simulated in testing on the ground.

Wen Painter was heavily involved in the testing of the systems, much more so than I was. He gained a lot of experience that I think served him well in the HL-10 later in doing the closed loop test and the structural feedback test. I remember watching them in the hangar, not sure why they were doing those tests. Larry Taylor explained it to me, because he had done a lot of the same work on the X-15. This is a part of qualifying an aircraft for flight that doesn't seem to get written about very much. [27] We do individual tests of the components, then we test several components, and eventually we do end-to-end tests. In the case of structural feedback, we try to find out if any of the vibration is getting fed back through to the control system, which can actually destroy the airplane. Wen did a lot of that testing in the hangar. I believe Berwin Koch joined the program about that time.

The HL-10 Lifting Body

We had started getting information about the HL-10 from Langley back in early 1963. We had some preliminary wind-tunnel sets describing what was then the current HL-10, which was in most respects what we flew with some modification, mostly enlargement of both the outboard fins and the center fin. Because the data on the HL-10 back then was classified as Confidential, we had to be careful with the analysis results. If somebody came to visit, we had to cover what we were doing. If we left the room, we had to lock the analysis results in the safe.

Despite the inconvenience, our working environment was mostly the lifting-body group, so we could work a little more freely and discuss the data and how we were interpreting it. At that point, the HL-10 was an easier vehicle to cope with than was the SV-5. I remember that Bertha Ryan, Harriet Smith, and I came up with our own code so we didn't have to tell everybody what we were talking about. "Vehicle K" was the HL-10, because I liked it so much, and I was impressed with it for not having the deficiencies of the M2-F1 or the M2-F2. "Vehicle B" was the M2-F2, taking Bertha's initial because she had a long association with the Ames people and the M2-F2. In our code, we used Dale's initial for the SV-5, "Vehicle D" for whatever form the SV-5 was in when we got the data.

We did the interpretation of the wind-tunnel data for the HL-10 shape that we had by early 1963, looked at it with our various analysis programs, and predicted it would be a good flying vehicle. It didn't have problems with controlling the roll axis and it had a center fin. There was no need to have interconnects. You could develop highly predictable roll rates with it and it was fairly well behaved in pitch. There were some difficulties above Mach 2, but those were being worked on in the tunnel at that time.

I told Lowell Greenfield where we were on the HL-10, hoping that he might be able to help us build a simulator for the vehicle. There was no approved simulation for the HL-10 at that time, because we had not known earlier that one of the flight vehicles built at Northrop would be an HL-10. We had always assumed Northrop would build a pair of M2-F2s and when that program was over that we'd move on to the next project. So in early 1964, we weren't anticipating getting the HL-10 ready to fly.

The HL-10 simulation was our first opportunity to look at how other lifting bodies might compare to the M2-F2. Lowell was resourceful, playing around with a TR-48 computer until he said he had a simplified simulator he thought we could set in some of the conditions for the HL-10 to take a look at them. Once again we used the coffee can with the dowel and springs as the control stick. We were able to get five degrees of freedom on the simulator-the three degrees of lateral-directional and two of longitudinal. The Mach number variation changed the coefficients, which we didn't have the capability to change, so there was little advantage of doing a six degree of freedom simulation. I don't think we had enough amplifiers to consider having a full six-degree-of-freedom simulator. Harriet, Bertha and I came up with some linearized sets of stability and control derivatives for the HL-10 and we used them in my analysis programs.

We could put them into the simplified simulation that Lowell had come up with and we could also put in a simplified control system by assuming the control surfaces were a first order lag and coming up with feedbacks that we could use from the pitch, roll, and yaw rates. It was a linearized control system, so we would look at just the first-order lag to give us a better feel for the vehicle. The results were a little bit surprising. The HL-10 appeared to be not simply a good flying airplane, but a piece of cake to fly. We had a hard time getting into trouble with it, even looking over quite a wide angle-of-attack range and Mach number range. All of the Mach numbers that we looked at, as I recall, were subsonic or transonic, nine-tenths and below.

We had a couple of the pilots take a look at it. They thought the simulation was good too, but also unrealistic, I'm sure. We didn't feel that we were looking at all of the problems at that point either. It was simply encouraging to see a vehicle, the first time you put it on the simulator, fly as well as an F-104. The HL-10 rolled well and it had good pitch response. We didn't look carefully at the drag at that time on the simulator because we were not doing the longitudinal acceleration equation of motion.

We had a good feeling that the HL-10 was quite a different animal than the M2-F2. It flew well, had a predicted L/D max somewhat larger than the M2-F2, and it didn't have any of those dangerous-looking control characteristics that we were starting to study in depth on the M2-F2 simulation. Actually, the M2-F2 was what we were working on and our HL-10 work was mostly a look at another configuration. We probably spent a total of one or two months on the HL-10 simulation. Its ease of flying and lack of problems was why I chose the HL-10 as my favorite configuration.

About this time, people at NASA Headquarters and management at Dryden and Ames were getting serious about the heavyweight lifting-body program. The vehicle would have a propulsion system to take it up in the Mach 2 region for exploring the transonic and low supersonic characteristics as well as the subsonic. That effort was getting fairly well focused and we were writing the request for proposal in the spring of 1964. It was about the time that people at headquarters and Langley suggested that rather than building two identical M2-F2s, we build an M2-F2 as the first vehicle and an HL-10 as the second vehicle. This decision was reached before the request for proposal went out. As a result, we were quite a ways behind on the HL-10. We had put a good deal of work into the M2-F2, a lot of that, of course, because of the concern over interconnecting the aileron and the rudder and the lateral stability issues.

Although we knew we had both of those vehicles on the way, we had only an official simulation for the M2-F2, which was quite a bear to use because it was all analog and the M2-F2 was not a high-priority program like the X-15. We had taken the pieces that we could get and put together a decent M2-F2 simulation, but it took one or two sim engineers to support that effort. At the same time, in addition to the X-15, we were also doing the Lunar Landing Research Vehicle (LLRV) and some other vehicles as well. As a result, the simulation people were overloaded, doing all of this work. Nevertheless, by the end of 1964, Lowell and I were doing simulations of both the M2-F2 and the HL-10. They were simplified, compared to what we would have as the final configuration was defined.

This effort with Northrop was quite different than most others. It was a moving target at both ends. The people at Northrop were finding problems in design and manufacturing and we were reacting by changing our specifications. Then, as we got more wind-tunnel data or evaluations on the simulator or energy management, we were changing what it was we were asking them to do. This is quite unusual, because in most environments when that is done, the price skyrockets, because with every change, everything else has to be changed. However, Dryden and Northrop had a joint skunk works approach that was really more like a single group with a single voice. Consequently, as the changes occurred, both Dryden and Northrop would accommodate them and try to keep our organizations more or less in lockstep.

By early 1965, we had quite improved HL-10 and M2-F2 simulations. The M2-F2, I think, still had priority and took the bulk of our time because we knew it was going to be the first one that we flew. The M2-F2 was similar in many ways to the M2-F1, with the major difference being the controllability and the lateral-directional stability needing to be sorted out on the M2-F2. There was a strong bias against switching back and forth between the M2-F2 and HL-10 simulations, because there was a huge amount of work involved in going from one to the other. I think it took a full crew of technicians and sim engineers several days working two shifts to make that change each time. We didn't like to lose that much time on the simulation. As a result, we did the HL-10 simulations for a couple of weeks, did the M2-F2 for a month or two. We switched back as we had more data and something that we wanted to look at on the HL-10.

Both were all-analog simulations, using the Electronic Associates Incorporated main analog computers that Dryden had at the time. [21] We also had auxiliary computers, TR-48s in particular, that were connected together. So it was a real ball of snakes for the sim engineers and the technicians to get switched from one simulation to the other. I'm certain that by the spring of 1965 we had an HL-10 simulation that represented some of the more important nonlinearities. We had things called "pot padders," which were a type of servo multiplier, and function generators that could get a more complete set of a six-degree-of-freedom HL-10 simulation.

Little by little the simulator matured, so that by the spring and summer of 1965, we were looking at flight-planning for the HL-10, although that wasn't our first concern. We could see it was not going to be that big a departure from what we planned with the M2-F2. The full HL-10 simulation had some of the nonlinearities and a control system mechanized more or less the way we would want to fly it. I think there were still some refinements Langley was doing on the outboard and the center fins.

The HL-10's nonlinear characteristics were difficult to fully reproduce with our early simulator and using pot padders was not a very reasonable way for looking at all of these things. We needed a more powerful computer. I think that John Smith, who was the head of simulation at that time, made a plea in mid-1965 for the lifting body program to purchase a Scientific Data Systems SDS 930 digital computer. We already had one for the X-15, as I recall, and we may have used it a little bit as well, coming up to speed. However, we did get the SDS 930 for the program fairly soon after the request was made.

By late 1965, Lowell Greenfield and Don Bacon were jointly doing the M2-F2 and the HL-10 simulations. Don Bacon was the primary HL-10 simulation engineer from early 1964 on through the program until about mid-1966. He and Lowell were both under a great deal of strain trying to support these two simulations, using the same simulator. They had a horrendous number of changes and checks that had to occur each time in switching between the M2 and the HL-10 sims.

In addition, analog simulators continually needed to be checked out. Unlike digital computers, they had drift in the amplifiers. The potentiometers also sometimes got moved unintentionally, so we were continually running what were called "static checks." Before we could start using the M2-F2 simulation, the engineers had to run through a complete set of checks, then troubleshoot any differences. They looked at the values of each of the amplifiers, and they also could look to see if the test dynamic maneuver that they made that day looked like the same dynamic maneuver made the day before. All in all, there was a lot of effort involved in just bringing up an analog simulation.

The SDS 930 computer was a godsend, because we could move a lot of the nonlinearities-the data lookup-onto the 930 computer, which I think is the first thing that we did so that we could get a more realistic envelope simulation for the HL-10. Because the frequencies are fairly low in the longitudinal acceleration equation, and therefore in altitude and velocity, we could use this relatively slow digital computer to also do that degree of freedom for us. It could look up the atmospheric conditions in terms of density, speed of sound, and pressure, and solve the longitudinal acceleration equation. The other five degrees of freedom were still done on the analog computer.

A great number of things were simplified because the function generation for these nonlinearities and for the atmospheric model was in the digital computer. Unless you made a mistake, it remained constant, so there was a lot less checking to do. Of course, even with the digital computer taking care of those parts and the analog computer taking care of solving the five-degree-of-freedom equations of motion and the control system, the system still had to be statically checked between runs. This was quite an effort. It was my first experience on a hybrid computer system, which meant an analog computer connected to a digital computer. The analog part was the more difficult of the two to check so it's where we spent most of our time troubleshooting.

There were also some other issues on a hybrid computer, because the continuous analog signal had to be converted into a sampled digital format to be used as an input for the digital computer. This task was done with an analog-to-digital converter. Then, the digital computer would come out with a digital number. A digital-to-analog converter was needed later to put it back over onto the analog computer. Of course, this process was done several times a second, so there was a big learning curve on how to use the two converters that interfaced the analog and digital components of the hybrid computer. But it was well worthwhile, because the further we got into the program the more things stabilized. We got a good idea as to where failures were, what they looked like, and how to start fixing them.

IV

Flying the M2-F2 and Other Adventures

In the late fall of 1965, my primary lifting-body assignment was being the stability-and-control engineer for the M2-F2. Because of this, I no longer had as much time to spend on the HL-10. We were undergoing a major expansion in staff. We were checking out the M2-F2 in the hangar and the HL-10 was being built at Northrop. We needed people to check out the M2-F2, prepare the simulators for both vehicles, and do the flight planning. In late 1965, Paul Bikle and the local Air Force commander decided to proceed with the lifting bodies as a joint effort between the Air Force and NASA, much the same way the X-15 was.

There were several reasons favoring the joint effort. One was that the collaboration had worked so well on the X-15 that there was optimism at Dryden that the Air Force could add value to the heavyweight lifting-body programs. Another reason was that the Air Force had a strong interest in the lifting bodies. Chuck Yeager's involvement in the M2-F1 in late 1963 was not merely him poking around to see what was going on. A very active Air Force lifting-body program was already underway, as I realized in 1964 when we started getting SV-5 data. The Air Force also had plans for using the SV-5 in several roles. The SV-5J was to be a jet-powered training vehicle that could take off from the ground for use by students in the Air Force Test Pilot School. The X-24A rocket-powered vehicle would be launched from the B-52 and would accelerate to Mach 2 speeds. The near-orbital launches of the related X-23A would be made using an Atlas booster. Using a lifting-body reentry vehicle to return exposed film from photo reconnaissance satellites was also proposed. At the time the Air Force saw lifting bodies as having significant military applications.

With their interest in lifting bodies and our prior good experience working with them on the X-15, the Air Force started to take part in the M2-F2 and the HL-10 programs. I don't remember the Air Force having much to do with anything I was working on until probably very late 1965 or very early 1966. Air Force Captain John Durrett, a flight test engineer at Edwards, started phasing into the program in the summer or fall of 1965. We had a close working relationship where he started familiarizing himself by using our simulation. I think he was in charge of modifying the Air Force's X-15 simulation to support the flight planning for the M2-F2, while we used our simulation for engineering evaluations with the pilot or an engineer in the loop. The Air Force also signed up to do the energy management flight-planning for the M2-F2. We had some good ideas on these tasks before the Air Force took them over, but since we were still doing X-15 simulations, trying to do them for two lifting bodies, and getting actively involved in envelope expansion on the M2-F2, we were relieved to have the Air Force help. We now had another partner like Northrop with whom we'd worked closely and successfully over the years.

I was doing most of the design and analysis of the M2-F2 lateral-directions control system, including the closed loop and pilot interaction. I worked with the pilot in the simulation, trying to come up with acceptable gains. Larry Taylor was more or less in charge of the overall design of the control system for the M2-F2. Ultimately, he was responsible for the work that both Wen Painter and I were doing on the M2-F2. In describing my work, I'm probably ascribing more freedom to myself in what I was doing than I really had, although I don't remember much interference or objection by Larry to anything that Wen and I did. Bob Kempel returned to Dryden from the Air Force Flight Test Center to take over my project duties on the HL-10 in the spring of 1966, just before the first M2-F2 flight.

Fig. 1. February 1966 photo of the first two lifting bodies to be flown, the M2-F1 and M2-F2. (ECN 1106)

Getting Ready to Fly

By late 1965, the M2-F2 had been through all of the component tests and as much of the closed loop testing as we could do in the hangar. We were looking for limit cycles caused by the interaction between the structure and the controls. Some of the systems were put in environmental testing situations where we could control the temperature and altitude (that is, the pressure environment) experienced by the components, so we could make sure they would work at the lower density, pressure, and temperature that the M2-F2 would experience during the B-52 captive portion of the program.

The pilots we were working with for the first M2-F2 flight were Milt Thompson as the project pilot and Bruce Peterson as the co-project pilot. At some earlier point, Fred Haise was going to be involved in the M2-F2 as a pilot. Fred had done the Cornell studies in early 1965 and had flown the T-33 that late spring and early summer. He had done some car-tow tests on the M2-F1 as well. Then, he and Joe Engle were selected for astronaut training. Joe Engle had been an X-15 pilot and showed himself-in all the work I did with him at any rate—to be a Chuck Yeager type stick-and-rudder man. There wasn't much he couldn't coax the airplane into doing. He was less interested in the research part than was Fred, but was an excellent pilot. I was sure they were both going to be outstanding astronauts. (Fig. 1)

On April 22, 1966, we did car-tows of the M2-F1 for both Joe Engle and Fred Haise. There is no record of the number of flights made, but they probably each had five to ten car-tows. In later car-tows, they would get up to 15 or 20 feet and then release, finish the flare, and land. So they both actually flew the M2-F1 lifting body. Then, of course, in 1977 they were both pilots on the Shuttle approach-and-landing test (ALT) flights. Fred flew the first, third, and fifth flights and Joe flew the second and fourth ones. Joe, of course, went on to fly STS-2 and STS-20. So Joe has flown the X-15, the M2-F1, the Space Shuttle ALT flights, and the full orbital mission from Mach 28 on down, giving us many stability and control maneuvers.

As we approached the flight program, we were bringing two other pilots up to speed to be project lifting-body pilots. One was Bill Dana, from Dryden, and the other was Jerry Gentry, from the Air Force. We also had Air Force Colonel Don Sorlie who had flown the M2-F1 before Jerry Gentry became the Air Force project pilot. Bill did a number of car-tow flights in early 1965. Gentry then made a series of car tow flights in June and July of 1965, and he did fine. He then made his first R4D tow flight on July 16, 1965. As he lifted off the lakebed, he had difficulty controlling the M2-F1 and ended up oscillating wildly, getting inverted, and cutting loose about 500 feet above the lakebed. He rolled out of it beautifully and made a pretty good landing. It was a real crowd-pleaser.

Paul Bikle was concerned about Gentry having this difficulty. Bikle was a sailplane pilot of great renown, holding the world altitude record for solo sailplanes. He believed that Gentry needed more time on tow with sailplanes. Milt, Bikle, and Gentry went up to Tehachapi to fly sailplanes on tow. I think Gentry got quite comfortable doing that.

It was nearly a year before Gentry flew the M2-F1 again. He made several good car-tow flights on July 17 and August 10, 1966, which was actually around the same time as our first two M2-F2 glide flights. On August 16, after our third M2-F2 glide flight, Gentry made his second M2-F1 air-tow, only to end up doing much the same thing he had done the first time. Gentry rolled the vehicle upside down at about 500 feet and released the tow. This time he used the landing rocket, then did an absolutely beautiful landing. There was no question about whether he could fly the M2-F1, but there was some concern about him flying it on tow. Afterwards, Bikle cancelled any further flying of the M2-F1. At that point, we were going to be flying only the M2-F2. I think that the M2-F1 program was shut down because it was not generating that much useful information and was obviously not the very best way to be training the new pilots.

While the pilots were being checked out, the final combined M2-F2 systems tests were being completed on the ground. All the loops were closed, and the telemetry was checked. Then the vehicle was mated to the B-52 and we did more combined systems tests with the M2-F2 attached to the B-52 pylon. Once these were complete, we flew the unmanned captive test flights that put the M2-F2 as close to the flight condition and environment it was going to fly in. It was mounted under the B-52 wing, mated with the adapter, and then flown to altitude to cold-soak it as the systems were checked and monitored in the control room. Each of us was watching a strip chart recorder in the control room, checking various parameters. As a stability-and-control engineer, I was primarily looking for signals from the rate gyros on the M2-F2 when the B-52 maneuvered. In addition, when the systems were on, I was looking for anything in the control deflections that looked suspicious. We were all in the control room, with each disciplinary group monitoring its own strip charts and communicating with the flight controller, who was in contact with the B-52 pilot. On the unmanned captive flights, everybody was looking at what they would be seeing on the first manned flight. In its way, it was a rehearsal for the M2-F2, but because it was so similar to what we did on the X-15, it seemed like an obvious and necessary step to me.

As with all initial captive flights, we found lots of problems that we hadn't found earlier. This isn't a big deal, but it took a while to list the problems and plan to fix each of them. Some involved a lot more work than did others. Right after the captive tests, we also did the taxi tests using the M2-F2 landing rocket so we could actually see Milt driving it around on the runway.

Then we got serious about getting ready to fly. We had in mind to do what we did on the X-15. The system was brought up completely and the pilot was put in the vehicle to make a captive flight. The thing that occurred to me at that point was that we had a vehicle that we'd done our darnedest to make flightworthy, but we really didn't know whether our efforts would be adequate. We were taking it out to go through a practice launch with a pilot in the vehicle. If the adapter failed or some kind of a catastrophic failure began, the pilot's only choice would be to eject.

Ejection over the B-52 wing looked to me like one of the most dangerous things that a pilot could do (Fig. 2). The pilot probably wouldn't impact any part of the B-52. But if I were going to eject, I would rather eject from the M2-F2 in free flight than in the mated setup of the captive flight. The M2-F2 was in an uncertain environment, balanced under the wing, in the B-52 flow field. I've never heard pilots talk very much about how that feels, but I know they're well

Fig. 2. The M2-F2 mated to the B-52 mothership on climb to altitude on Milt Thompson's last flight on September 2, 1966. (ECN 1437)

aware of the risk involved. To me, it seemed more exciting and dangerous than most of the things we'd done in the M2-F1. That was partly because we had spent many hours in the simulator studying the B-52 flow field with the M2-F2 descending through it. This data was obtained from wind tunnels where they use a model B-52 wing and adapter and move a model M2-F2 to various points in the flow field to get the forces and moments on the M2-F2. There were two reasons for doing that. The first was that we wanted to understand the gyrations the M2-F2 might go through. The second was we didn't want the released M2-F2 to be forced into the B-52 and recontact its wing or tail. That happens in missile tests sometimes, so it's not a zero probability event. That would, of course, damage the B-52, possibly affecting its ability to return and land. It also would be very bad for the M2-F2 as the smaller vehicle would sustain a lot more damage.

On the X-15, the launch dynamics had been done using various techniques so we'd already spent a lot of time and gotten a lot of ideas about the initial conditions for the control system to reduce the chance as possible that there would be recontact on launch. The other thing we did was to have the pilot hold the controls in a fixed position for three seconds. This would give three seconds of free fall before the pilot made an input, meaning that the vehicle wouldn't be thrashing around on launch.

During the manned captive flights it was really impressive to be in the control room watching what the pilot was going through and doing the various things we were there to do. We wiggled the B-52 around to check the gyros and the structural integrity of the mated vehicle and the other components. The manned captive flight is one more way that we can get information before we actually go to flight. During it, the vehicle sees the full dynamic pressure envelope and gets cold-soaked at altitude. We do everything but launch, so we go through a lot of the risky and uncertain part for any additional, unforeseen failures. Once the vehicle is launched, though, there's still a lot of uncertainty in how it's going to fly. By then, however, we have spent a lot of time practicing with the pilot in the simulator and the pilot has his own vision of what to do under all flight eventualities.

I think all engineers who work on airplanes identify with the pilot. No engineer wants to injure or kill the pilot by not doing an adequate job of analysis or studying a failure analysis. I started getting the same feeling watching the M2-F2 captive flight that I'd had watching the steep angle at which Milt dove towards the ground in the plywood M2-F1. This surprised me. Because we always want the pilot to be safe, we assume everything will go okay, but we watched our strip charts, looking for an unknown motion, a strain gauge indication of a structural loading problem, or a sign of a systems problem with some electronic part.

After those tests are done, you prepare to launch the vehicle. The B-52 was a non-issue to most of us in the control room, because we knew it was a good-flying airplane with competent pilots and a good crew chief on board. So while the controllers kept track of the B-52, those of us in the control room were focused on the experimental aircraft.

Just as we were preparing for our first glide flight, Joe Walker, the chief test pilot at Dryden at the time, had a tragic fatal accident during a flight of the XB-70 (Fig. 3). Flying in an F-104, Walker got into the vortex flow or touched the wing tip of the XB-70, which immediately threw him out of control, rolling him up over the top of the XB-70 and damaging it so badly the bomber crashed. Walker was killed. Al White, the chief pilot on the XB-70, managed to eject in his escape capsule, despite severe injuries. His co-pilot, Major Carl Cross, was killed when the XB-70 hit the ground. The incident demonstrated the huge amount of energy in the vortex system of a 500,000-pound airplane.

Joe Walker, who I'd worked with closely on many X-15 flights, was suddenly dead because of a tiny problem that turned into a catastrophe. That was my first experience with the death of a pilot I worked with.

The only other airplane crash I'd been involved in before this was in late 1962 when Milt Thompson had a failure on his F-104. He couldn't keep the vehicle from rolling when he slowed down to landing speed so he finally put the airplane over the bombing range and ejected. We didn't have any contact with him, so it was frightening. I hadn't known Milt very long yet, but I was already fond of him and, for all we knew at the time, he had died in the crash. An hour or

Fig. 3. August 10, 1967 flight of the XB-70A number one during a test to intentionally demonstrate the complex vortex flows generated by the vehicle as it experienced natural condensation in the humid atmosphere. (E-17462)

two later, he came wandering in to Dryden, having hitchhiked back. Were we glad to see him.

Flying the M2-F2

We had our first launch of the M2-F2 on July 12, 1966. We had stability and control issues with that first flight as well as a performance issue, about what the actual flight L/D max would be. If the L/D max turned out to be lower than predicted in the Ames full-scale wind tunnel, Milt was going to have to push over even more steeply than planned to pick up enough speed to do the flare, ending up parallel to the ground, and safely land.

There were many issues that I'd been through, practicing with the pilot and working on individual conditions and combinations. I thought Milt was as well prepared for that flight as he could possibly be. Not only was he an outstanding pilot, he was also an outstanding engineer. There was very little on that airplane that he did not understand as well as anybody on the program in terms of the systems, how they were hooked together, how the cockpit was set up, and what the failure modes of the stick and rudder might be. From his prior experience as a pilot, Milt knew how to sense a lot of problems of standard component failures. All of that knowledge was all in his head, along with the flight plan, and the various off-ramps that Milt would instinctively use should any part of the flight plan not go as predicted.

Our first flight plan wasn't very ambitious. It was, essentially, to get from 45,000 feet at seven-tenths Mach number down to a safe horizontal landing on the runway. We'd also have some general sorting out of the airplane's characteristics to see if they were as predicted and, if not, how the plane would fly and how the pilot flew it. Milt had a few minutes to look at that. If all the training we'd given him was incorrect, he would have to retrain himself in just a few minutes so he could land the vehicle.

Our biggest concern on the first flight of the M2-F2 was the tendency for a pilot-induced oscillation at low angle of attack. A low angle of attack—required to get the velocity the pilot needed to do the preflare maneuver—traded potential for kinetic energy so the pilot was at a much higher speed than he needed to land. That way, after the pilot did the flare rotation and the speed started to drop rapidly, enough velocity remained so lift equaled the weight of the vehicle while the pilot tried to find the ground. The pilot needs to be parallel to the ground and still have a little bit of time left for letting it down gently for a nice landing. Since that's the critical maneuver, right after launch, the pilot does a pullup to make sure the vehicle responds more or less the way needed in pitch, so that when he does this rotation on flare, the vehicle will round out and not cause the pilot any problems in stability and control.

First, the pilot wiggles the stick in both axes and checks the rudders out, making sure they feel about the way he thought they would. If they act differently, he modifies the way he plans on flying this vehicle for the next few minutes. Then, he does a pushover to get the speed up. At that point, the pilot ends up at the negative angle of attack where we found there was a tendency for a pilot-induced oscillation (PIO), the most exciting as well as the most dangerous area.

Milt chose to do the maneuver at high altitude very soon after launching so that he could do a practice flare. If he couldn't handle it, he had enough altitude left to eject from the aircraft and parachute safely. We knew that as the M2-F2 got to lower angles of attack, it would fly a little bit better if the pilot lowered the interconnect gain. However, the vehicle still had a real tendency to PIO. It was a little nicer-flying airplane at low angles of attack at the lower interconnect gain and the pilot could fly it a little more aggressively then. On a first flight, even for an experienced test pilot like Milt, the pilot's gain is going to be fairly high. The pilot is going to react to anything he sees or feels. That's what makes for a high gain. It's not something that comes from a pilot feeling something and thinking about it awhile. Instead, it comes from the pilot's instantaneous reaction to counteract something with the controls.

Even though we could vary the interconnect gain, Milt had decided that he really didn't want to try to lower that gain as he did his preflare maneuver, diving at 300 knots in a 30-to 35-degree flight-path angle. This was because if he lowered the interconnect gain for the high-speed low-angle-of-attack portion, he would need to increase that interconnect again to have a vehicle that, during the landing maneuver, didn't have a roll reversal at the higher angles of attack while rounding out in the flare. Milt elected, even though the vehicle was a little sluggish at the high angles of attack and had a little bit of a tendency to PIO at the low angles of attack, to leave it at one gain setting throughout. He thought he could manage between the two, go with the fixed gain, and not have one more thing that needed to be changed.

We had talked about putting in a gain scheduler that would change the value for him, but Milt and some of the rest of us didn't want to do that because we weren't confident that the gain prediction was right. With an experienced test pilot like Milt, who had spent a year and a half practicing for this flight in the simulator with various control systems and wind-tunnel data sets, we felt that if he found that the interconnect gain was too low at the higher angles of attack, he would know to increase it. If he found the PIO tendency was too great at the low angles of attack, he would lower the interconnect gain. There was a wheel at his left side for changing the interconnect gain setting (Fig. 4). With all the adrenaline a pilot is using at that time he can do many things in a very short period of time if he's got a plan and a good test pilot will do mostly good things during that time.

Consequently, we were flying with a fixed interconnect gain around 0.4. Milt did his practice dive maneuver at altitude and decided the vehicle had a little

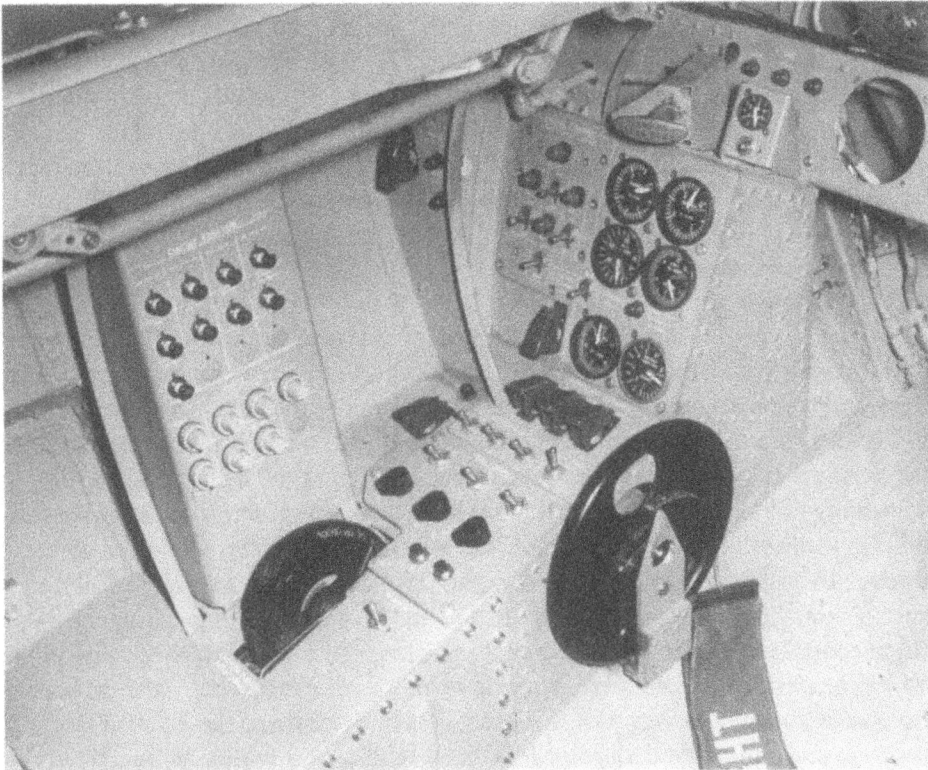

Fig. 4. Left portion of the M2-F2 cockpit showing the aileron to rudder interconnect gain wheel (left) which is defined in percent of maximum gain. (E 13409)

more tendency to PIO than he was used to seeing in the simulator, perhaps due to misprediction of the data or because his gain was higher than it had been in the simulator. Milt was diving so steeply in the vehicle that if he put in any stick control, it had a tendency to oscillate and move more than he wanted. He decided to lower the interconnect gain, according to the plan. He left the gain there as he flew the rest of the maneuvers to set up for the final maneuver prior to approach. As he set up for that maneuver, he was ready to push back over into that low angle-of-attack range. He decided the vehicle was still too sensitive, so he tried to lower the gain some more and the problem continued to get worse until he was in a violent PIO of plus and minus 90 degrees in bank angle.

The response to that would be to increase angle of attack, but he was too close to the ground to do so. The other choice that everybody in a PIO is taught is to let go of the stick. Milt had been in a divergent lateral-directional oscillation in the X-15 in January 1965, which he had initiated by applying the controls too aggressively. He had gotten out of the situation by letting go and stopping his input. In the situation in the M2-F2, Milt quit putting in inputs, but it was apparent from looking at the traces that he still had his hand on the stick. The vehicle quit oscillating. Then he looked down at the interconnect and discovered that he had turned the selector in the wrong direction and, in fact, he had dialed in the full interconnect value, which would make any vehicle PIO. So he put it back to 0.4, went to an altitude of three or four thousand feet, and got ready to initiate the flare.

Milt caused the PIO by trying to counteract the roll. He was actually out of phase just enough that he was adding to the roll. In many cases a pilot can get rid of that tendency by letting go of the stick and the vehicle will revert to its natural poles (Fig. 1 in chapter 2), known to be stable even at the negative angles of attack or the pilot could increase angle of attack. In this case, Milt decided to let go of the stick and see what happened. It was spectacular, considering that he did that at probably over 300 knots at a negative 30-to 35-degree flight-path angle, in a new vehicle he had only a couple of minutes to figure out how to fly. Being a good test pilot, Milt did figure it out, and he did a very nice approach, a very nice flare, and a very smooth touchdown. What to the untrained eye may have looked like a fairly big maneuver in the middle of the flight was a very frightening thing for a minute or two. We couldn't do anything. We can't tell a pilot how to sort something out when we aren't sure what had gone wrong. The only thing we can do is give the pilot information that could be useful in the next minute or so through the flight controller. The only other information that the pilot can get in such a short flight so close to the ground would be a call-out either from the flight controller or from one of the chase airplanes that he should eject. The pilot will also consider his own feeling about where he is with respect to the ground, because it all happens very fast. The pilot is coming down to the lakebed fast and needs to make the right decisions. Milt knew that he was willing to punch out. He'd done it in the F-104 a few years earlier, and it wouldn't take him longer than a blink of an eye to decide he needed to do it.

Instead, he got through a spectacular flight, did a nice landing, and gave us some data that we could then compare with the prediction in our simulation to find out if we needed to change the control system or its gains before we flew again. The most serious discrepancy that we found was Milt turning the interconnect gain selector the wrong way. That was what had caused the PIO problem and Milt was confident that it was not going to happen again, at least not to him. He was ready to fly again, do a little envelope expansion, start to get real numbers on lift and drag, and perform some stability-and-control maneuvers.

And that was what happened on the first flight of the M2-F2 after a couple of years of wind-tunnel work, planning, manufacturing, testing in the hangar, and "what-ifing" in the simulator. A flight of only a few minutes revealing how good a job you did earlier.

Milt's experience of letting go of the stick to stop the PIO was a well-known technique but it had taken a lot of bravery for him to actually let go of the stick. Nevertheless, it was a good way to get out of the loop, let the airplane's natural dynamics take over, and then try to re-engage it again with smaller inputs.

Most of what I was doing at Dryden didn't have much relevance to anything that I would do on my own time, but I remembered Milt's way of getting out of the PIO one day a few months after the first flight of the M2-F2 as I was driving out on Avenue J heading toward the base. All of a sudden, a dog darted in front of me. To avoid hitting the dog, I yanked the wheel one direction. I did miss the dog, but I made a pretty good input, and I found my car heading clear over into the other lane. So I yanked hard as I saw I was headed for the ditch on the other side of the road and my car quickly darted clear across the road. After I did this two or three times, I realized that my Corvette and I were in a PIO. I remember thinking to myself at the time, "Just let go of the wheel." As I came back across the center line one more time, I just let go of the wheel. Lo and behold, the car straightened right out. That was a good application of a lesson that I had learned from working with pilots flying airplanes with PIO tendencies.

Although Milt and I had a serious work relationship, we joked around with each other a lot as well. I enjoyed being with Milt, who was almost always in a jovial mood. Milt was a prankster and tried to set people up. It was done in fun, not in a mean way. He and I had a habit of verbally attacking each other in the hallway or in a meeting as part of our interaction. I think most of it grew from the many hours that we spent in the simulator while trying to refine the interconnect gain and the feedback gains on the M2-F2.

Milt had a Jaguar that he liked to drive fast. I liked to drive my Corvette fast too, but I never did anything while driving it that I considered particularly dangerous. I did see how fast it would go on the lakebed, up to a speedometer reading of 136 miles per hour before the valves started to float. For the most part I was a fairly safe driver. I did speed but I looked well ahead of where I was and I didn't speed if there was significant traffic.

Milt and I occasionally would find ourselves on the road together without much other traffic, especially if we left work at 5:00 or 5:30 p.m., when most of the NASA and Air Force traffic was gone. At one point Milt caught me, going across Rosamond Dry Lake. I sped up. He sped up. I sped up again and he sped up again. Finally I remembered that I had a test pilot with a big ego behind me. I didn't know whose car would go faster, but I knew he was probably willing to drive his car faster than I was mine. So I quit accelerating. We were on the two-lane road across the lakebed at the time (before it became a four lane road). There were no cars coming, so Milt pulled over into the left-hand lane. As he went by me, he turned, gave me a typical Milt grin and flashed me a one-finger salute. I could see him chuckling as he sped past me and pulled over into the lane ahead of me. I slowed down, and I'm sure he probably did, too. I know I'm not the only person that Milt liked to show who could drive the fastest, but he kept showing me throughout the rest of my career at NASA, whenever we were on the road together. I didn't always have a Corvette and he didn't always have a Jaguar, but there was an unwritten agreement between the two of us he would always pass me, flashing the same grin and a one-finger salute. It was just part of Milt's personality. He loved flying, just as he loved the other things that he did. Milt was a fun-loving guy with a lot of talent and the ability to solve some really difficult engineering problems.

As we got ready for the M2-F2's second flight, there were probably some minor tweaks that needed to be done to the vehicle, but for the most part our work was simply to plan the next flight and decide what data was the most important. Milt knew that the incident with the PIO on the first flight was due to him turning the interconnect wheel the wrong way. The simulator had the rotation of the gain changer opposite to what was on the vehicle. It was always a bad idea to train pilots in the simulator with the controls or indicators opposite to what they were on the flight vehicle. "Train like you fly" is a good idea. I don't know if this was fixed before the second flight, but it was fixed soon.

Our flight plan was to do stability and control maneuvers, performance maneuvers, and a further evaluation of the interconnect gains. The second flight would further flesh out the characteristics of the vehicle and validate them. On subsequent flights, we would fly a little different flight profile to expand the envelope. The M2-F2 didn't really have a very big envelope prior to the installation of the XLR-11 rocket engine, since it was all subsonic. Even though part of it was flying in compressible flow, the angle of attack went up to about 15 degrees and down to about minus five degrees in the only region of the flight envelope we would be examining. Since the region was subsonic, we didn't have significant Mach number effects, and the effects of dynamic pressure and viscosity weren't large enough to assess.

Consequently, most of the tests that we were planning were just longitudinal stability-and-control maneuvers, consisting of setting up at a constant angle of attack, pulling back the stick for about half a second, pushing it forward and holding it for about half a second, and then returning to the original state. We wanted to analyze the residual oscillation during the longitudinal maneuver. The change in angle of attack during the maneuver would be about plus or minus three degrees from the trim angle of attack as we started the maneuver. With that envelope, we had five or six angles of attack to investigate. We always liked having multiple, repeated maneuvers so we could get an idea of the scatter or quality of the maneuver we were examining.

The same was true in the lateral-directional axes, although this had another interesting aspect, the interconnect being something that we might like to change. In the lateral-directional axes, we also did stability-and-control pulses. These were a half-second of a given deflection in each direction initiated for the aileron with the stick, then a similar maneuver done with the rudder pedals—once again, at given angle-of-attack increments just as the longitudinal maneuvers had been done. Once we set up for an angle of attack, we did a longitudinal pulse or two, and then a lateral-directional pulse or two at the same flight condition, getting maybe two or three of those per flight condition. That was the plan prior to the first flight, implemented as we went through it.

Sometimes the aileron pulses were done close together with the rudder pulses for the lateral-directional maneuvers and other times they were done separately. That gave us the opportunity to look at aileron alone, rudder alone, or some combination of them, because when the pilot did the aileron pulse, of course, the rudder moved more or less the same as the aileron did because of the interconnect. The only way we could tell the aileron effectiveness from the rudder effectiveness was to do an independent rudder pulse, closely followed by the aileron pulse.

The other thing that we added after the first flight was an investigation of the interconnect. We had the pilot set up a flight condition, start out with the 0.5 or so interconnect gain that we would use for the rest of the program, do an aileron input evaluation, look at the response, go up to 0.8 interconnect gain, and do the same thing, and then drop down to 0.2 or 0.3 interconnect gain and do the same thing. At the lower angles of attack, even with the lower interconnect gain of at least 0.2 or so, the pilot would get roll reversal. The pilot was getting roll reversal at the low as well as the high angle of attack with a low interconnect gain. At the high interconnect gains, there was a tendency for the vehicle to be more responsive at the lower angles of attack and too responsive at the high angle of attack. In the next few flights, we looked at some of those at different conditions to see if we had the interconnect gain right at 0.4 or 0.5. That's fairly much where we wanted Milt and the other pilots to fly.

The second flight was uneventful, at least for the maneuvers that I just described. Milt flew the next four flights and they went well (Fig. 5). As I recall, the M2-F2 was always high on energy, but that was everybody's plan. I think we landed all of the flights on runway 18, about three-fourths of a mile from Dryden, close enough that people could go out on the roof or ramp and watch the flight (Appendix E). We could see the B-52 at launch. People with good eyes or binoculars could follow the lifting body all the way around the pattern to the final dive, always the most spectacular part of the flight.

It came as quite a surprise to me when Milt announced near the first of

Fig. 5. Photo of the M2-F2 followed by the F-104 chase plane taken just before Milt Thompson's final touchdown on September 2, 1966. (EC66 1442)

September in 1966 that he was going to quit flying, including the M2-F2 and the X-15 (which he hadn't flown in the last year). I had no hint of this decision. I thought Milt was in the prime of his career. He was in his late thirties or early forties, really young to me as far as test pilots go. Some pilots at Dryden have been that age or older when they arrived, so Milt seemed really young to be quitting flying. To have him suddenly quit doing what I thought he was probably the best in the world was quite a shock. I probably kidded him about it, but I never got a serious answer from him about why he quit. I know in his later books he said that he didn't see anything exciting lying ahead and that he was interested in getting into engineering and management efforts, but that doesn't sound like Milt to me. He loved flying, and I don't think that a lack of exciting things to fly looked likely. I was looking forward to a future of more lifting bodies, scramjets, ramjets, delta-wing X-15's, and whatever the next thing was that NASA would fly.

I thought Milt's decision came during a very exciting time for flight and it really surprised me. It was also likely the first thing that started to nibble away at my own confidence in the job that I was doing. I considered the M2-F2 to be a fairly dangerous airplane, not that I had much experience with airplanes then. Having watched Milt have a serious problem on the first flight, a problem we had feared since the days when we were trying to car-tow the M2-F1, I was not too confident that the other pilots would do as well as Milt had. Even though Milt was an excellent pilot, he'd had some problems with the M2-F1 and the M2-F2.

Before that, I had developed a good relationship with Fred Haise. I could see Fred was Milt's equal as an engineer and also a very good test pilot, although I'd never worked with him in anything as difficult as the first flights of the M2-F1 or the M2-F2. However, in everything that I had worked on with Fred, he had been superb. After he was chosen to become an astronaut in April 1966, we had to start flying with other pilots. There was a little orientation for each one, which was kind of interesting. Each one handled it a little differently. Milt certainly didn't disappear at that time. He was very active in the flight-planning and the review process.

As most people do after they've done something difficult, Milt became more conservative. I think most of us assume the next person, new to a project, doesn't have quite the experience and insight into a problem that we have. So Milt was a little tentative on investigating the low-angle-of-attack flight dynamics with Bruce Peterson, Don Sorlie, and Jerry Gentry. I think we did those prior to the first flight, but I think Milt had been a little frightened in that situation and knew that he had come close to losing the airplane. I think he just didn't want to put another pilot in the position of risking the loss of the airplane.

We had Bruce Peterson and Don Sorlie fly a few flights. On Sorlie's third flight, he investigated control at around minus three or four degrees angle of attack. In the process, he got into a PIO. He recognized it and, as Milt earlier had done, reduced his inputs and increased the angle of attack a little and recovered from it. Looking at that strip chart with those oscillations, I was reminded of what had happened with Milt. So we had a second pilot who had done a PIO. Although I didn't know Sorlie very well, he seemed very professional and competent. There were no fingers pointed at him for having the PIO, but there was further discussion about how to avoid it on future flights. Next, Jerry Gentry was ready to fly. I knew Gentry had come into the program with people wondering whether he was going to have PIO problems. He was considered to be a high-gain pilot and had managed to invert the M2-F1 twice on air tow already, but obviously he was also a very skillful pilot because he had managed to land it both times. The first time he damaged the landing gear and the second time he had landed without a scratch on the vehicle, so he was a quick learner. Gentry flew four M2-F2 flights before the vehicle was grounded for the installation of the XLR-11 engines for the powered flights.

The plan had, from the outset, called for rocket engines in the vehicle to get it up near the Mach 2 region, because all of the lifting-body shapes that we'd studied had significant changes in their characteristics in the transonic region. As the lifting-body programs were advocated and approved, the intent was to fly the vehicles up to Mach 2. Trying to go faster than that would require a bigger airplane and a lot more engine, and it wasn't clear what we would learn, other than going faster and higher. Consequently, the Mach 2 limit made sense, and that was how the program was sold. We got vehicles manufactured at one or two million dollars apiece, which was a good price even in those days.

As the M2-F2 was to be grounded for the engine to be installed. I left the lifting-body programs for a while. I was still working a little on the HL-10 but I really wasn't part of the team anymore. I knew what they were doing and I knew a lot about their vehicle, having spent a couple of years working on it prior to leaving the primary stability-and-control role I'd had in mid-1966.

The Maximum Likelihood Estimator and Other Projects

Meanwhile, Larry Taylor and I had been trying to extract the aerodynamic stability-and-control derivatives from flight data for comparison with wind-tunnel data and for updating the simulators, resulting in greater consistency between aircraft and simulator. That's very important to do as we expand the flight envelope. It was being done primarily with analog matching but also with time vector, but those techniques gave us, in my opinion, a lot of scatter. [28] They were also very tedious, laborious techniques and it took a long time after a flight to analyze the stability-and-control maneuvers to extract the stability-and-control derivatives so the simulator could be updated.

I'd seen similar problems on the X-15. There was a part of the X-15 profile that Larry Taylor had termed the "hungry eye." This was a little region where the airplane was quite difficult to fly. We'd been analyzing X-15 data, as well as that from the M2-F1, and by that time I had much better M2-F2 data to look at as well. I was quite intrigued by this. Some time early in 1965, Larry and I started looking seriously at what's called a maximum likelihood estimator. For the first time, we felt that we had a technique that had good potential for making the extraction of the stability-and-control derivatives from flight data a much cleaner, easier, and more consistent process. We'd patched together a lot of FORTRAN code, thrown in some X-15 data, and gotten varied results. We were learning how to implement the theory as well as how to hone the art of the extraction. We'd done that on the X-15 and could compare it to numbers that Roxanah Yancey was getting from her analog-matching, as well as comparing it to wind-tunnel predictions on the X-15 [17].

I'd also been doing that with maneuvers that Milt had flown early in the M2-F2 program. Flights two through five had some outstanding stability-and-control pulses. I started looking at those and found they had different problems than those of the X-15, so we learned more of the art. The technique still remained the same, from a mathematical point of view, but there was obviously some art in terms of quality of the data, how various signals were weighted, how start and stop times were picked, and how to help the pilot give us better maneuvers that were more appropriate for this computer-automated technique of derivative extraction.

I was very interested in the problem and I could see that the M2-F2 was going to be down for a while. Although the HL-10 was going to start flying almost immediately, I wasn't directly involved in the project any longer. I intended to support the HL-10 project the same way I had before, in terms of analyzing the stability-and-control characteristics of the vehicle, but I also started looking at the M2-F2 and X-15 maneuvers, trying to use this maximum likelihood estimator to get stability-and-control derivatives. With people such as Wen Painter, Bob Kempel and Berwin Kock involved in the project, I thought this task was a better use of my time.

Still believing the airplane to be dangerous, I asked to be removed from my project duties on the M2-F2. That request was initially refused, and I went to Larry. His advice wasn't always something a young engineer should follow, but I really wanted his opinion on this matter. He said that I should document my concerns regarding the safety of flying the M2-F2, that I thought the vehicle was dangerous and that it was an accident waiting to happen. So I wrote a memo saying that. It was about a page and a half long. I gave it to my supervisor, who asked me, "Are you sure you want to do this?" I said I was. I'd thought about it a long time. I told my story and it got passed on up the chain. I got the memo back. They told me they'd prefer not to have the memo because it would imply that there was something that needed to be fixed on the vehicle and they would have to formally deal with that if they had the memo. Larry advised me to address the memo to files instead of to the Director of Research, which I did, and put it in my files.

Although I think my supervisors probably disapproved and were disappointed, they did take me off the M2-F2 project support. They put me in a support role on the lifting bodies where I would, to the best of my abilities, help with the extraction of the stability-and-control derivatives from the maneuvers.

At the same time, it was becoming clear that the XB-70 (Fig. 6) was not going to be a production bomber, but there was a lot of interest in terms of the propulsion system and the aerodynamics. The vehicle had an unusual configuration, which I remember was described by one reporter as looking like a banana coming out of an orange crate. That wasn't too far off. It was the most unusual-looking fast airplane I've ever seen. It was a lot of fun to hear it fly as it was a very loud airplane. I had been under the flight path of the vehicle at the end of the

Fig. 6. XB-70A number one with the landing chutes deployed. (ED97 44244-3)

runway when it took off from Palmdale on its first flight a few years earlier. It was only a hundred feet or so directly above me.

I was always attracted to it because of its unusual shape and its deep, loud engines. Dryden had two lifting-body programs and three X-15s that were still flying at the time. The XB-70 was coming and there were other smaller programs at well. As a result, there was always a shortage of people. I still had my primary assignment on the X-15, supporting its various activities. After leaving project support on the M2-F2, I was added to the XB-70 project team to look at its stability and control. I thought it was a good opportunity. It was classified at the time, so we couldn't talk about how fast it went. Everybody knew it was a Mach 3 airplane, though. Of course, the X-15 that I'd been working on had gone Mach 6, but the XB-70 was an air-breathing airplane that took off from the ground and had a delta-wing planform. To my way of thinking, that was the way of the future. We had serious plans for adding a delta wing to the X-15, so getting some experience with a delta-wing airplane from the stability-and-control point of view would be good. This would add to the experience I already had on the X-15 at hypersonic speeds and the M2-F1, the M2-F2, the HL-10, and the SV-5 for lifting bodies. The XB-70 (Fig. 7) seemed to me to be a very obvious component to add to that collection of contributors to how we were one day going to be flying from the runway at Edwards into orbit.

Fig. 7. XB-70A number one with the wingtips deflected on its August 16, 1968 flight. (ECN 2125)

For that reason I left the M2-F2 program. I was never unhappy with anybody or anything that happened in the program and I had always been treated exceedingly well. At about the same time, Dale Reed left the lifting-body team as project manager, taking a lesser role. Garry Layton took over in his place. So it was just an obvious time to try something different. I had over four years of experience with lifting bodies and X-15s. I was now going to broaden that experience.

I followed the HL-10 closely because it was an old friend. It was my favorite of the lifting bodies because the simulator version had worked so well right out of the box that it was hard to believe after all the struggles with the M2-F1 and the M2-F2. It was rolled out in January 1966 and moved to Dryden.

My recollection is that when Milt, Fred Haise, and others flew the HL-10 simulation, they felt it flew like an F-104 and they thought that if we could build a lifting body like that, that's what we ought to do. But I don't think they really believed it was that good, because of the lack of maturity of the HL-10 compared to the M2-F2 databases and experience bases. In general, the HL-10 simulation showed that it was an end-to-end design by Langley and that it was a good flying airplane. It had good performance, so the energy management was good. It had good longitudinal and lateral-directional characteristics. It had good response to the roll control on the stick and the yaw control on the rudder pedals.

But it was not to be quite that way. It's well documented what happened with the HL-10. [29] Bruce Peterson was the project pilot on it as well. It flew just prior to Christmas 1966, right after I left the lifting-body project. The flight, to an observer who was now on the outside of the program looking in, did look good, and I knew the vehicle came back intact. I was not involved in the key discussions or the closed room things, but little by little I started to hear rumors about structural feedback. Wen Painter was looking into the problem, was well on the way to fixing it, and was quite qualified to do it, in my opinion, because I had seen him do the same kind of testing with the M2-F2.

Bob Kempel was looking into some of the lack of response to the controls during the higher angles of attack. He couldn't convince himself these were consistent in any way with the wind-tunnel data. He got into a discussion with others at Dryden and Langley. Langley personnel said that they had seen some indication of flow separation on the HL-10's fins during some of the wind-tunnel testing that they had done earlier. Probably some of the nonlinearities that we had been chasing when we were working on the simulator in 1964 and 1965 were the result of this flow separation.

Because of the flow separation at higher angles of attack, the vehicle didn't respond as Bruce Peterson had intended, so he chose to land the vehicle at a lower angle of attack, which meant a higher landing speed. Bruce touched down at about 280 knots indicated, which works out to a ground speed of over 310 miles per hour (Fig. 8). That was the fastest successful landing I am aware of for any flight of any aircraft anywhere. It may be the world's record for touchdown speed.

Fig. 8. Bruce Peterson piloting the HL-10 on its first flight, just prior to its 280 knot landing on December 22, 1966. (ECN 1597)

The HL-10 went back into a wind-tunnel test phase as Langley worked on redesigning the fins. They came up with several proposals that were worked by the various aerodynamicists at Dryden and elsewhere. The shape of the fin was changed. Probably to somebody who doesn't work on airplanes it doesn't look like a substantial change in shape, but aerodynamically it was quite substantial. The airplane was down for many months to have the fins modified and when it came back to flight status again there was never any indication of that flow separation reappearing.

Travels in the U.S.S.R.

About now, a friend of mine—Lee Lytton—who worked at Dryden, had been to Egypt a year earlier, in 1965. He had enjoyed his travel and he started telling me how much fun it was. I'd only been to Canada and Mexico so I thought it sounded interesting. He talked me into a trip he'd been putting together. We would go to the Soviet Union, Rome, and the associated islands for a vacation. Since I'd never been overseas and Lee was already a seasoned traveler, I agreed to go. The Soviet Union and the United States were not good friends at the time, so Lee and I were concerned that we might have some trouble with getting NASA's approval for the trip.

We asked for approval fairly early on, probably early 1966, and we received permission to visit the Soviet Union. In a required briefing we were told what to talk about, what not to talk about, and what to do if somebody captured us and tried to squeeze important information out of us. We would also have to give a briefing when we returned.

We mailed our checks off to Intourist, the official Soviet government travel agency. Intourist was the only way you could travel in the Soviet Union. You had to prepay everything and it was nonrefundable. They limited the amount of cash you could take into the Soviet Union, which seemed strange, but they also limited the amount that you could leave with. You couldn't leave with any Soviet currency.

A couple months later, we got a letter from NASA Headquarters that said U.S. relations with the Soviet Union were deteriorating and NASA would like to withdraw its approval of our trip. This was unexpected, so I didn't know quite what to do. I was still paying off college loans and I didn't have a lot of extra money. Because the Intourist payment was non-refundable, I would lose the money. I don't even remember now what the amount was, but it was a lot of money at the time. So I requested that NASA reimburse me, since I had received NASA approval prior to making payment. I thought that if they were going to withdraw permission for the trip they would reimburse me the money I would lose. The response that I got was basically that NASA didn't have any way of doing that. NASA Headquarters also said that NASA wasn't denying us permission to go, just advising us not to go. That didn't sound like a "no" to either Lee or me, so we went ahead with our trip.

Part of the problem was that tensions with the Soviet Union were increasing because of U.S. involvement in Vietnam. I had no real opinions, pro or con, on that involvement. I was an engineer interested in flying exciting aircraft and getting into space. I figured anything my government did at that time was probably correct. I didn't have a political reason for disagreeing with what the U.S. was doing internationally. At that point, I thought what we were doing in Vietnam was the right thing.

Lee and I went on our trip in late July and early August of 1966, after the first two M2-F2 glide flights. We thoroughly enjoyed the trip and when we got to the Soviet Union we, of course, had to stay in the hotels they had. The visa was good only from the day we arrived until the day we left. All meals were bought with vouchers. You didn't pay cash for meals, hotel rooms, transportation, or anything else. Instead, everything was done with the prepaid vouchers. Although some of the meals were very poor and very late, we always got a meal with the voucher.

The first place we visited was Leningrad. We were taken from the airport to our hotel by train. It was a magnificent steam-driven locomotive, bigger than any I had ever seen before. Lee and I were fascinated with it. The only other passengers were mostly Russians. There were very, very few tourists. The other tourists traveling on the train with us at that time were Jewish, there to visit relatives.

There were about fifteen or twenty of us.

We got to ride on this train with the beautiful locomotive and see all the sights. In Moscow, we got to see a mockup of Luna 9, the unmanned spacecraft that made the first soft landing on the Moon. I hadn't seen photos of it before. I'd been warned prior to leaving the U.S. to be very careful when using my camera. I was also repeatedly warned by Soviet officials that I wasn't to take pictures of anything related to industry or transportation. However, the mockup of the Luna 9 was in the public Exhibition of Scientific Achievement, so I could take pictures of it. I got some poor pictures of some of the early Soviet satellites that I hadn't seen photos of before.

At the Moscow airport, I watched a Tu-22 Blinder jet bomber doing touch-and-go landings on the runway. I had a camera with me, but I didn't take a single picture of that Tu-22. I knew that if I got caught with a picture of the airplane, I might never return to the United States. Just before I had arrived in Leningrad, the Soviets sentenced some Western tourists to twenty years in prison for stealing an ornamental banister bear, so I knew they were serious about their rules. I really wanted to take a picture of that airplane because I'd read about it. However, in everything that I'd read, there had been no pictures of the Tu-22. Here I was, watching one do touch-and-go's in Moscow. I didn't snap a picture of it, but I sure was tempted.

We got on through the rest of our trip, ending in Kiev, Ukraine. That part of the trip went fairly well, too. It was interesting talking to the Soviets. The sense I got was that just about everybody outside of Moscow hated everybody in Moscow because they got all the perks, supplies, and luxuries. This was also very apparent to us, as we saw a few cars in Moscow and even fewer outside of it. In any other Soviet city, almost everybody went everywhere by bus, and all we saw were buses, trucks, and just very few cars. That impressed me. I'd been living in California, so seeing a place with so few cars was quite a change. My German turned out to be more useful than my English most places in the USSR, as only the Intourist guides appeared to speak English. Everybody else either didn't speak very much English or else they wouldn't speak to me at all.

The last stop was Kiev, and when it came time to leave, everyone else got on the airplane but us. The guy who had brought us to the airport tried to explain to Lee and me in his broken English that the airplane was full and it was leaving without us. He said to me that the officials at the airport were going to take our passports so that we could get new visas, because we couldn't legally be in the country beyond that day because our visas would expire. They would also start working through their bureaucracy to get us on another flight out of the country, which I thought would probably be later that day. We were taken back to the hotel and we surrendered our passports. After that, nobody seemed to know anything. All of the people we talked to said that they were unaware of our status, didn't know what was being done and that we should be patient, and eventually something would happen. That was what was worrying me the most.

One of the Intourist people came to visit us at the hotel. Although we were allowed to leave the hotel, neither Lee nor I had done so because we thought that if we were in the hotel, we could somehow make contact with someone if the Soviets were actually up to something. We·suspected that they had found out about some of the things that we'd worked on; we'd been warned by the FBI and others in briefings before we left the country to be alert to that sort of thing. So we were very suspicious that what we were going through was part of a big plan to drag secrets out of us.

We stayed close to the hotel. I remember telling the Intourist person when he came to visit us that I felt very uncomfortable because, without a visa or passport, they could arrest me and put me in prison for being in the country illegally. He said, "That would not be a big problem. Because even though they would put you in jail, it wouldn't be your fault, so it would be okay." That didn't seem like a very satisfactory explanation to me.

We were at the hotel waiting for at least three days, which seemed like a long time. I'd given all but one my books (in English) to a Ukrainian student I had met in Kiev the day before we were originally scheduled to leave, so I had only one book left. There was only one radio station and that was mostly propaganda in Russian. Sometimes there was a TV broadcast, but only to show a ballet or speeches. There wasn't really anything for us to do for the three days we waited. I've always been good at thinking, so I got a lot of that done during that long wait.

Finally a guy in a fancy suit showed up at the hotel. He had our passports, with valid visas, and he was going to take us to the airport. They'd made special arrangements to get us out of the country. To this day, I still don't know why suddenly we started being treated more like VIPs than like spies. At any rate, we didn't object. He took us to the airport and put us on a VIP transport. It had huge seats and was very plush, better than anything but the flat sleepers in first class on airplanes today. The seats in our Soviet VIP airplane were very comfortable and reclined a long way. There was a large distance between seats and between them and the aisle. The airplane had been configured to carry VIPs to important meetings.

Lee and I were put on the airplane with some other people who all looked like officials of the Soviet Union. They were well dressed and carried themselves the way that Soviet bureaucrats did. Then, we went on a very roundabout trip. Lee and I weren't sure where we were going. We went from Kiev to Vienna with several stops in between. At each stop, we picked up people and dropped off people. Of course, a little paranoia had set in by then. We didn't know if we were being taken somewhere to be charged with some serious crimes or if we really were leaving the USSR. I still remember getting off the airplane in Vienna, which was only barely friendly territory at the time, and thinking that I wouldn't be going back to the Soviet Union very soon and that the next time NASA officials warned me that they didn't think my going someplace was a good idea I might listen to them.

Over the years I've thought about this, and I now don't think that I was ever dealing with anything other than incompetent bureaucracy. There was no room on the airplane and they had no way of accommodating us other than by going back to square one and starting over, only then discovering that the only way they could get us to Vienna was to put us on this VIP airplane going a roundabout route.

At any rate, that was my first taste of international travel. I became quite addicted to it on the trip, outside the USSR, and decided to travel while I was young because so many of the tourists that I saw were older and they weren't having any fun. Some only barely knew where they were, their stomachs were bothering them, and so on. So I became a traveler, joining the Travelers Century Club after I visited 104 countries on their approved list.

I'm telling this story of my trip to the USSR because in April 1967, Vladimir Komarov died on the Soyuz 1 spaceflight. He'd had control difficulties throughout the mission. The vehicle was not behaving well. He managed to fire the retrorockets and reenter, but the parachute didn't properly deploy and became tangled. Komarov was killed on impact. We didn't know a lot of these details at the time, just that the Soviets had brought a new vehicle back from space and killed the cosmonaut. They couldn't keep it a secret, of course. They'd made it a big deal that he was in space, so they couldn't have him just disappear. They announced the beginning of a flight once the vehicle was in orbit, anything good about the flight after launch, then the end of the mission. But this time, the Soviets also had to say that, unfortunately, the cosmonaut had died.

I remember being in the hallway at Dryden shortly after that announcement, and having Bruce Peterson and several other people stop me. They were joking that they thought maybe I had given the Soviets M2-F2 information when I was in the USSR and, when the Soviets had used it, they had roughly the same luck that we had with it. That was a running joke around Dryden for a while. Unfortunately, a few weeks later, Bruce Peterson crashed in the M2-F2.

V

Flight Research in the 1960s and Early 1970s

I was working to support the M2-F2 and the HL-10 after November 1966, but I wasn't involved actively in the meetings or flight-planning. My opinion was asked sometimes and I sometimes offered it without being asked, too. I was also analyzing the flight data. The M2-F2 glide flights had been completed, the airplane was being fitted with an XLR-11 engine, and it was to about start flying the rocket-powered phase. I was quite interested in the program as that approached.

I was always looking for tough maneuvers to analyze to see if I couldn't hone the art, as well as the theory, of the maximum likelihood estimator. Larry Taylor and I published a paper about it, which took a while because we were both so busy with all the programs going on then and because the approval process for publishing the data and analysis in an unclassified forum was quite lengthy. I was still working on the XB-70 and the X-15 as well as the maximum likelihood estimator techniques. The paper was published internationally first in September 1968, at San Remo, Italy. The paper did not include X-15 data, because we could not obtain approval to present it in time for the deadline [30]. We finally obtained approval for including the analysis of the X-15, the XB-70, and the HL-10, presenting this version three months later [31]. I was very proud that all these airplanes on which I'd worked were in the first unclassified papers that Larry and I published on the parameter identification techniques.

Jerry Gentry flew the first glide flight of the M2-F2 with the rocket engine installed. It went fairly well, since we didn't have any pilot-induced oscillation (PIO) problems. It was a kind of vanilla flight, but Jerry didn't like the way the aircraft handled with the engine installed. The elevator trim setting and the angle of attack were different than they had been before the engine was installed because the weight and mass properties had changed. He hadn't complained much about the M2-F2 that I recall, prior to that, other than about what all of the other M2-F2 pilots complained about — that it was really sensitive in roll at low angles of attack.

Bruce Peterson made the second glide flight with the engine on May 10, 1967. He got the M2-F2 into a fairly severe PIO on the dive maneuver to the flare. He recovered, oriented himself, and arrested the sink rate but, by then, he was just above the lakebed. The landing gear was coming down and had not yet locked as the vehicle touched down. The M2-F2 rolled across the lakebed several times. Bruce was injured, although not nearly as badly as we thought he was. He lost the use of one eye because he got a staph infection in the hospital. He'd lost his eyelid in the accident and the doctors had reconstructed an eyelid so he wouldn't be blind in that eye. After the staphylococcus infection, Bruce always wore an eye patch. I've always enjoyed working with Bruce. He's a very up kind of a guy, very outgoing. I still see him at Society of Experimental Test Pilot meetings and we chat about the good old days.

The crash ended the M2-F2 program. More importantly, the whole lifting-body program seemed to be on the verge of collapse. The M2-F2 had been badly damaged in the crash and the initial examination suggested that it couldn't be rebuilt. During its first glide flight, the HL-10 had unexpectedly proven itself to be nearly uncontrollable at high angles of attack. There were many critics of the lifting bodies and they seemed to be right.

Toward Mach 8: the X-15A-2 and the Scramjet Program

Although I was involved with the lifting bodies, much of my work was still on the X-15. Midway through 1964 we started flying the X-15 number two, redesignated the X-15A-2, which had its fuselage extended 29 inches and was to fly with two external tanks. The analysis that preceded those first flights started sometime in mid-1963, with looking at wind-tunnel predictions for the added external tanks. How would the effect of the external tanks affect the X-15? Would lengthening the airplane have an appreciable effect?

When the tanks were mated with the X-15, they represented 37 percent of the combined cross-sectional area at the position of the maximum cross-sectional value. The tanks were predicted to increase the supersonic drag by 41 percent at

Fig. 1. X-15A-2 being flown on May 8, 1967 by Pete Knight with a dummy ramjet fitted with a 20-degree nosecone. (EC67-1731)

zero lift, as was subsequently verified in flight. Ultimately a ramjet was to be put on the back end. We always called it a ramjet then, but it was really a scramjet. I think that wasn't really defined at the time, but we were looking at the forces and moments caused by putting a structure back where the lower ventral fin had been. (Fig. 1) The ramjet would be less streamlined and more blocky than the ventral fin, so it would potentially have an aerodynamic effect.

We'd done the analysis on the airplane and we flew it. On each of the first few X-15A-2 flights, something on it would pop open. The nose gear came out on one flight and, as I recall, a landing skid deployed on another flight. It made me a real believer in aerodynamic heating and loads. The gear extension was caused by our adding six inches to the nose gear strut to accommodate the landing dynamics of the longer and heavier vehicle. The thermal expansion of the new, longer strut caused the normal release mechanisms to open prematurely, which hadn't happened when we flew at the same flight conditions before. We also had plans to put an ablative heat shield on the vehicle to take care of the higher heating rates and associated heat load. We'd been to a little over Mach 6 by then on a couple of flights and were concerned about heating at high Mach. The vehicle had originally been designed for the heat loads of Mach 6. We were planning on going to Mach 8 with the X-15A-2 to get additional hypersonic data, using the reusable external tanks (Fig. 2) to study the accompanying experimental effects of greater heating at increased Mach number.

Fig. 2. X-15A-2 in August 1967 prior to the first flight, equipped with the full ablative coating, eyelid, 20-degree dummy ramjet, and external tanks. (E 17339)

The primary justification for increasing the Mach number capability was to test the ramjet, which was really a scramjet. It would have supersonic combustion because we were planning on flying it up to Mach 8. This was a Langley-driven experiment, one in which I was exceedingly interested, because it seemed to me that the hydrocarbon-fueled turbojet was probably limited to Mach 3 or so. If we wanted to go faster, we had to go to the next level of air-breathing engines, the ramjet and the supersonic combustion ramjet-the scramjet.

Most of the early flights were flown with the tanks off meaning, of course, that they had peak Mach numbers below 6—that is, within the Mach range for which we already had X-15 experience. As a result, the changes we saw were nearly imperceptible. Nevertheless, we went through all the checks to make sure that there weren't any surprises. We also flew several flights with the lower ventral fin back on. Later, for the first flight with the tanks on, the tanks would be empty and the flight would be at an even lower speed. Also, we had to jettison the external tanks before we could land the X-15A-2. This requirement gave us the opportunity to evaluate the vehicle stability and control with the tanks on and to look at jettisoning the tanks. For the people interested primarily in the tanks, it was their opportunity to assess recovering the tanks for reuse, just as the Shuttle would later do with the solid rocket boosters.

After the two tanks were installed and we saw how big they were and how much they weighed, we started worrying about aerodynamic interference and the effect on stability and control, both longitudinal and lateral-directional. We did have some wind-tunnel predictions, but we were concerned about an array of questions. The fuel in the external tanks would be used first so the tanks could be

Fig. 3. Dummy ramjet outfitted with the 20-degree cone and the local flow experiment on the 90-degree X-15A-2 ventral fin prior to the August 21, 1967 flight. (E67-17493)

jettisoned and then the fuel in the internal tanks would be used for the rest of the flight. However, the external tanks were quite heavy. The modified number two X-15 was also heavier than the other X-15s because it was longer and had added structure to take care of the tanks and the ramjet experiment (Fig. 3). We were concerned about how the additional weight of the fuel and the tanks would affect the vehicle and what would happen when those heavy tanks were jettisoned. Some changes had also been made to the wing for additional experiments and we didn't know how these changes would affect how it flew.

What concerned me most was that the sleek lines of the X-15 had been changed quite markedly by adding these sizable tanks on either side. The two tanks weighed different amounts, so the vehicle's lateral center of gravity (CG) was offset from the centerline. I really worried about that because the forces and moments caused by that offset lateral CG had to be trimmed out. This meant an appreciable change, with the tanks both empty and full, that we had to look at. We also had to look at the intermediate cases where, for some reason, like the engine shutting down, we needed to get rid of the tanks after only part of the fuel they contained had been used. We would be jettisoning partially full tanks in many combinations and the remaining fuel would be at one end of the tank. As a young engineer, I thought there were a lot of very exciting issues regarding how we'd work our way through this large change we'd made to the number two X-15. I never heard the pilots say much about any added risk or complexity, though. To me, the analysis we did left a lot of questions open as we got ready to fly the vehicle, with fuel tanks, in the full envelope flight, which, hopefully, would go up to Mach 8.

After we flew a couple of flights with the tanks, we started getting flight results that reduced the uncertainty in some of the other studies that we'd been doing. As that happened, we started to feel that things were in much better control. I'm sure the more experienced engineers knew that this was the way things were. We had the opportunity to fly this configuration in a very controlled fashion at low speed and get the first results back. After we had time to absorb those results and include them in our analysis, any additional risks would be reduced as we flew, first with the tanks empty and then with fuel in them, going to higher and higher speeds. Once we got above Mach 6, we would, of course, once again be in the uncharted part of the original X-15 envelope (Fig. 4).

The plan was to do an envelope expansion clearance from Mach 6 all the way up to Mach 8. Most of this would be flown at lower altitude, because the scramjet experiment would only work at high dynamic pressure and the scramjet and aerodynamic heating at high dynamic pressures were the primary reasons for the X-15A-2 configuration, after all. The aerodynamic heating would be vastly different at Mach 8 than it had been at Mach 6. That required a lot of thermal analysis be done. This was when I did most of my work to see if the same parameter identification techniques could be used on the heat flow and the heating loads as the X-15 went up to Mach 8. I didn't get much of the data I expected, because the vehicle, for a number of reasons, only went to Mach 6.7.

Fig. 4. X-15A-2 with the tanks and full ablative coating a few minutes before reaching its record setting Mach 6.7 speed on October 3, 1967. Pete Knight's flight resulted in shock impingement damage and the loss of the dummy ramjet experiment. (EC68-1889)

The data that I did get didn't have high enough sample rates to capture the transient I wanted to analyze with the parameter identification technique. Predicting the steady-state rate of heat transfer was well in hand with the tools available at the time. In fact, we had added an analog model that predicted the heating at various places to the X-15 simulator. This was the same heating model that I later used for the Shuttle and for various proposed vehicles including the X-33.

The ramjet, I think, was always a future program and we never really did get to examine the actual scramjet propulsion, in which I was most interested. I never did get to see the actual scramjet fly. A dummy ramjet was flown on the X-15A-2, carried along so we could do the envelope expansion and evaluate the effect of the dummy ramjet on the vehicle. The presence of the dummy ramjet was the primary cause of the burn-through on Pete Knight's flight to Mach 6.7, the last flight of the number two X-15 (Fig. 5).

From all the analysis and work that had gone into it, I learned a great deal in terms of the flight-derived aerodynamics, stability and control, and handling qualities. As we got up into the hypersonic region, I started looking into the aerothermodynamics, heat transfer, and heat loads on the X-15. As a result, I became familiar with a lot of the tools that I would eventually use on the Shuttle. I was still trying to analyze the heat transients using the same techniques that I'd used before. Nevertheless, my work on the X-15 involved the aerodynamics, heat transfer, stability and control, flight loads, flight dynamics, and air-breathing and rocket propulsion, a good list of things for a young engineer who wants to work on any vehicle that's going to fly in and out of the atmosphere to know.

We also studied advanced X-15 experiments and configurations. Primary

Fig. 5. X-15A-2 after the October 3, 1967 flight showing the scorching of the ablative coating and the absence of the dummy ramjet experiment. (ECN 1834)

among those was the delta-wing X-15. It was a 75-degree-sweep delta wing that would replace the original X-15 wings. The structure would have been changed and it would, I believe, have been a somewhat longer airplane. It was to be capable of Mach 8 flight. In those days, Mach 8 was, we thought, the limit of our materials, so the basic X-15 used an ablative heat shield to get from Mach 6 to Mach 8. This ablative heat shield wasn't very satisfactory because the material was messy to work with and some of it burned off the airplane, so that it had to be refurbished between flights. The delta wing airplane was a Mach 8 program (without an ablative heat shield) and it had been studied since 1963. I'm sure there were studies before that, although I wasn't involved in them, of course.

I did a lot of work on the project, but we didn't have any real wind-tunnel data on the delta-wing X-15, because the configuration wasn't really defined. We did a lot of estimation of the control and performance characteristics from some similar shapes that had been tested in wind tunnels. At that time most of us thought that a real hypersonic airplane probably would be a delta wing.

The ablative heat shield was necessary because the original X-15 wing was going to have heating problems at the higher speeds. The original wing did a nice job of putting the aerodynamic center near the center of gravity, which is always a good idea on an airplane. With the delta wing, the location of the aerodynamic center was one of the issues we were examining, since the center of pressure on the delta-wing X-15 would be quite a ways aft, meaning more weight could be put at the back of the airplane. However, we were working on other issues involving air-breathers and the delta wing. Evidently that work was even less official than the work on the delta wing itself, for I can't find very much that was written or published on it (Fig. 6) [32].

Fig. 6. One of the conceptual designs of a greatly modified delta-wing X-15 powered by an air-breathing ramjet. Adapted from reference 32.

I remember working with a couple of people during several visits to Marquardt Engineering in the Los Angeles area. Marquardt was the ramjet expert of the day. They had developed the ramjet for the Bomarc surface-to-air missile in the 1950s. This ramjet was later used to power the Mach 3.3 D-21 reconnaissance drone. Even in the 1960s they were quite experienced with ramjets and with what the ramjet could and couldn't do.

The delta-wing ramjet vehicle would have an inlet and be capable of sustained flight using the ramjet. [9] Marquardt had several designs, in terms of what shape, how the compression in the inlet and the expansion of the plume were going to be done, how much drag this would add to the vehicle, and where the rocket would and wouldn't be used. One example is shown in Fig. 3, chapter 1. Ejector technology was known then, so the rocket could be used to bring air in through the inlet to start the ramjet and to sustain it at conditions that weren't particularly well suited to a ramjet. I found it a very interesting area. The ramjet experiment on the number two X-15 were exciting, because in my opinion that would be the next step for going higher and faster with a more conventional aircraft. By "more conventional" aircraft, I don't mean X-planes such as the X-15 or the X-20 Dyna-Soar but production-type aircraft.

I spent a lot of time working on various configurations, trying to come up with some form of aerodynamic analysis. I could make assumptions about Newtonian flow and come out with some ballpark numbers and there were other people at Dryden better able to deal with the forces and moments that would be generated by the delta wing. I don't think that the effort ever got very far, because it was, I think, a Mach 4 or 5 airplane. Some people looked at that as a step backwards, but I looked at it as a step forward, because it was going to be an airplane going Mach 5 with a sustained air-breather.

I didn't know anything about the SR-71 and its characteristics in the 1960s. The SR-71 was publicly announced in 1964. We didn't know much about it except that it probably flew faster than Mach 3. We didn't know what kind of propulsion it used, either. There was always a question in my mind about whether the reluctance to go forward on the X-15 air-breather had something to do with black programs that I knew nothing about. Now, over 30 years later, I don't believe that was the case, but at the time I did suspect that some of these problems might have been solved on other aircraft.

I was greatly disappointed to find out that not only would neither the delta-wing nor scramjet experiment fly, but the X-15 itself would be terminated in 1968. I think there were a lot of reasons for that. The error in predicting the heating effect from shock impingement and interference of the dummy ramjet which caused the burn-through was part of it (Fig. 7) [33]. Right after that had happened, Mike Adams was killed in the number three X-15. I think people in

Fig. 7. Close-up photo showing severe thermal damage to the local flow experiment and the nearby X-15A-2 structure. Compare to the local flow experiment shown prior to the damage in Fig. 3 (E67-17526)

positions of influence backed away from supporting further X-15 flight because it seemed dangerous and the program appeared to have lost its way and had no clear goal. However, the program had accomplished all of its objectives and then some. So, at the end of 1968, they decided to quit flying the X-15. Even over 30 years later, the X-15 is still the only fully reusable spacecraft to have ever been flown and the only spacecraft to have ever been flown manually, by a pilot, into space.

Back to School

While I was working on these projects, I was also getting my master's degree at the University of Southern California (USC), partly on campus and partly at Dryden. Some of the time I had to commute south to the campus in Los Angeles for the courses. They were offered once or twice a week, so that meant trips to the Los Angeles basin from Edwards. Traffic wasn't nearly as bad then as it is now, so the wear and tear from the drive wasn't nearly as great as it would be now. I essentially finished the course work in 1966, testing out of one deficiency in 1967, when I received my master's degree in mechanical engineering. Part of the reason I chose mechanical engineering was that I wanted to understand more of that part of the airplane, because I felt was I was weak in that area.

Once it was clear that the X-15 was through flying, I decided to go to the University of California at Los Angeles (UCLA), to take more mathematics and advanced course work to help me with some of the techniques that I was developing. I've always been interested in aircraft configurations, but I'm even more interested in things that can be generalized in terms of the aerodynamics or the analysis of flight data. They all go together in my mind as one big thing. I thought I was not as competent in technique development and some of the mathematics required to support it as I needed to be.

In some of the parameter identification work that I was doing I had found that the math I'd had as an undergraduate at Iowa State was not sufficient. I hadn't taken any additional math at USC. When I started going to UCLA full-time, I did so by actually resigning from Dryden. When they found out that I was serious about leaving, though, they suggested that I take an unpaid leave of absence. In the spring of 1968, I returned to UCLA part-time, using my annual leave and commuting from Lancaster. In the fall of '68 I switched to full time, living on campus.

At UCLA, I took a lot of courses in math. I could have finished a master's degree in math, but UCLA wouldn't grant a second master's at the time. I did a lot of course work in what was then called systems science. It was a very scientific approach to engineering disciplines that eventually ended up part of the electrical engineering department. I took electrical engineering courses as well as math courses.

I could have gotten my doctorate in either math or electrical engineering. Since I didn't see myself as being a full-time professor, getting a Ph.D. in math didn't seem like it would be much better than what I already had, so I chose electrical engineering. Although my degree is a Ph.D. in electrical engineering, the chairman of my dissertation committee was Professor A.V. Balakrishnan. He is an engineer, but he's far beyond me and almost any other engineer I know, in mathematics. He's really a mathematician, but won't admit it.

I briefly considered a career change and, since biotechnology was a big buzz word in those days, I audited some courses in biotechnology. In the process, I met Willard Libby, who was very supportive of me. Libby had gotten the Nobel Prize for discovering and demonstrating carbon dating for determining the ages of sediments and fossils. The carbon-dating analogy has, of course, been extended to other kinds of dating so we can accurately date mammal, fish, and bacterial remains, and can now date the whole solar system. I was impressed with his approach of combining physics, chemistry, and biology. Although I eventually chose not to go into biotechnology, he did agree to serve on my dissertation committee and he's one of the people who signed my dissertation. It's unusual for an engineer to have a Nobel Laureate on his or her committee, because an engineer can't get the Nobel prize-only scientists and mathematicians can.

Besides becoming a much better theoretician for aerodynamic analysis and parameter identification, my doctoral work at UCLA also gave me a broader background in controls, which had become my specialty. My doctorate degree was in controls, and my dissertation was on identifying stability-and-control derivatives and turbulence simultaneously from flight data. The main application is ride smoothing for photography, military applications, and passenger comfort in airliners. I had finished my course work and the bulk of my dissertation by 1971, and it took me another two years to get it written up, printed, and signed off. I returned to Dryden full-time in early 1971, finishing my dissertation after I came back to work. I did have some new tools to hone and I was waiting to see what the next step would be.

An interesting aside to my days at UCLA is that on the day that the first packet of computer information was sent from a computer at UCLA to a computer at Stanford, I was sitting two offices away on the third floor of Boelter Hall at UCLA. I remember the computer science guys who worked for Professor Leonard Kleinrock at the time, primarily graduate students, were very excited about what had just occurred. To me, it seemed like a very peculiar thing to do, breaking up what a computer could do very well by itself into little packets, shipping them to other computers, and having other computers reassemble the packets and do the work. Of course, I didn't recognize the real significance of that event, because that event was the birth of the internet and, consequently, of the World Wide Web, which has changed the world as we know it now from both a commercial and an intellectual point of view. [34] While I had been sitting in one office, probably trying to derive something theoretical in the mathematical world, something very practical and very important had occurred two doors down from me. I've always remembered that event as something I probably should have given a little more attention.

There's always some confusion as to what research engineers are. My particular point of view is that research engineers working at Dryden are basically scientists. We're doing experiments and we're trying to understand phenomena. By getting specialized data with specialized instrumentation, we apply the same principles that scientists use in trying to tease apart their phenomena. As a result, research engineers—particularly in aerodynamics, aerothermodynamics, materials and structures-are really scientists and not engineers, even though their background may be in engineering. Most engineers are not interested in scientific phenomena except as the means to improve their design or to make a better one, so there's always some confusion when people talk about engineering results. When we're talking about research engineering results, we're really talking about scientific results. It's very similar to what materials scientists or fluids scientists would be doing, how they look at data and run experiments in situ to obtain the very best information and, therefore, the best understanding of an unknown phenomenon.

The reason that research engineers aren't recognized as scientists is that research engineers are trying to understand the overall aircraft characteristics as well as the local phenomenon. Therefore research engineers are involved in all aspects of studying the phenomenon, such as flight-planning and safety of flight. Research engineers have peer reviews for their publications; engineers are judged by the adequacy and usefulness of their research. The results will be used immediately by engineers in aircraft manufacturing or design. The proof of the validity of any research engineer's results comes only when the rubber meets the road in the flight of future designs and vehicles. There is no better test of any scientific result than this.

Although I had been going to UCLA full-time starting in the fall of 1968, I was back on the payroll at Dryden in the summers. There were any number of assignments that I could do in the summer. Most of what I remember doing was trying to sort out some of the more difficult parameter identification maneuvers from the lifting bodies (Fig. 8) and the XB-70, trying to hone the techniques, sometimes by using what I had learned at school and sometimes by using the experience I had acquired before I left for school. The summers of 1969 and 1970 were quite busy in the lifting-body world, with lots of new configurations, but not a lot of big surprises other than rocket-engine problems. There were lots of little uncertainties being discovered as the flight envelopes were expanded and the rocket-powered phases of those programs were extended.

On a less formal basis, a number of times I came to Dryden, on my own time, for a few days during the school term. I usually came up to help with the analysis of data, the advocacy of maneuver, or some issue on envelope expansion. By no means was I important to any of those projects, but I was more experienced in some analysis techniques than were some of the others who were trying to analyze the data.

In the interim, also while I was at UCLA, in the early summer of 1969, I was tasked with the evaluations of the F-15 configuration. The Air Force had study

Fig. 8. X-24A after launch from the B-52 on an August 1970 flight. (E70 2511)

contracts out for proposals from three different manufacturers for configurations for this new air superiority fighter, the F-15. I was at Wright-Patterson AFB for several weeks, evaluating the various proposals. I don't remember many of the details any more, but I thought the McDonnell Douglas configuration was by far superior, based on the data packages that were submitted. There were many other disciplines being analyzed simultaneously. Among the analyses I did was one using the computer techniques that I had developed to come up with pilot ratings for all flight conditions on all three of the manufacturers' proposals. [23] The McDonnell Douglas proposal was considered the best overall by the Air Force and they were selected to build it as the prime contractor.

From the Earth to the Moon

During a break in the F-15 evaluation, I had the opportunity to go to Cape Canaveral to watch the phantasmagorical transmogrification of the Saturn V into the surreal vehicle that I got to see heading for the Moon with three astronauts aboard! I had a tremendous opportunity to get a broader view of NASA's technology and engineering in the few days that I spent at Kennedy for the launch of Apollo 11, because it was a very different environment from Dryden's. Kennedy was very hardware-oriented, as that was the key thing that they did. They took these highly complex rockets and did all of the certification, assembly, checkout, and launches. This meant Kennedy was a much bigger operation than I'd seen at Dryden on the X-15. I remember being very impressed by that.

I had the opportunity to tour all the larger facilities, including going out to the Saturn V sitting on the launch pad, not yet fueled. I got to go right up to the vehicle, with proper escorts at all times. I don't know how much of that Fred Haise arranged and how much was just their courtesy to a fellow NASA employee. I was just overwhelmed by the size of the engines and the vehicle. I remember feeling tiny next to the vehicle and being a ways away from it and seeing it towering up into the sky, a very slender vehicle. Its size still impresses me more than the size of the Shuttle in its stacked condition, ready for launch, does. Of course, the Saturn was a great deal taller than the Shuttle, but it wasn't as wide.

During the few days that I spent at Kennedy, I attended the various parties that they had at night. I think the parties were probably by invitation only, but with my NASA badge, in the company of some of the other people that I'd met there, I was able to attend some of them. Between the parties and the launch activities, I did manage to meet most of the 24 astronauts who were to go to the Moon and the 12 who were to walk on the Moon. I met a lot of other people, too, and over the years I've worked with quite a few of those astronauts on projects unrelated to Apollo. My visit to Kennedy did give me the chance to meet people who were going to be in the news for many years, through the Apollo program, Skylab, and Apollo Soyuz Test Project, and into the early part of the Shuttle program.

The opportunity to go to Kennedy was, of course, because of Fred Haise, with whom I'd had the good fortune to work at Dryden. Fred arrived at Dryden in 1963 and left in 1966. It seems as if I worked with him a lot longer than that, but I think one of the things that you discover as you get older is that time doesn't pass as quickly when you're young as it does when you're older. I remember when Fred was trying to decide whether to apply for the astronaut program. He was obviously attracted to it, but he loved being a research test pilot. I'm not sure what actually made him decide to become an astronaut. I know I was in favor of him doing it because I thought he would have a good chance to go to the Moon, even though it was not clear whether the original seven and the succeeding nine astronauts would get the bulk of the flights to the Moon. As all of us notice in our lifetime, if you speculate on who or what is going to happen, even in the near future, you're usually wrong. People leave, have various kinds of medical conditions, and change what they want to do with their lives.

I know Fred was vacillating on whether to apply or not because he felt there was a chance that he might be an astronaut in name only. The worst outcome for an astronaut, of course, would be to do all the work, sit around, and never get a flight. As I recall, a fate only a little worse than that was, in Fred's mind, going up in a capsule, going around and around the Earth, and really not being able to do much. I remember him discussing these possibilities at the time. Much to my delight, Fred did apply and was selected. I'm sure he was very high on the list of those selected. After he was notified, he started winding up his work assignments at Dryden before going off to Houston to be trained as an astronaut. I have had occasion to work with Fred a little since he made the arrangement for me to go to the Cape for the launch and it has always been a real pleasure.

I didn't actually spend much time with Fred when I was at the Cape for the Apollo 11 launch because he was busy. Fred was the backup to Buzz Aldrin, and the backup crews were very important and very busy. First of all, they had to be ready continually to step in for the primary crew. Secondly, there were a lot of duties that they would sit in for the primary crew. He also had many preflight activities to ensure everything was as it should be. Fred was the last person to leave both the Apollo 8 crew and the Apollo 11 crew on top of the Saturn V, so he was the guy who said goodbye to them and closed the hatch. The three people were alone from when Fred left until the mission was completed.

In the three years he was at Dryden, Fred established a reputation with the engineers as an excellent research test pilot. I'm never sure what pilots think of other pilots. Sometimes they tell you and sometimes they don't. Fred was very serious about flying research vehicles as well as very much interested in doing any task, piloting or otherwise, on a lot of the engineering endeavors that I was working on. This was unusual, as the other pilots, at least at that time, didn't offer me that kind of help. Fred and I had wide-ranging discussions on engineering and he taught me a lot about how a pilot thinks about an airplane versus how an engineer thinks about an airplane, even though we both understand the same things about the hardware and the interface between the pilot and the hardware. A pilot's view is always different from an engineer's view because the pilot needs not only to understand the aircraft but to have that understanding be instinctive in case of an emergency. Engineers make most of their discoveries after all the data has been collected and analyzed. We can spend some time poking through the details, finding the real events and reaching whatever conclusions come from analyzing the flight data.

Not long before the Apollo 11 launch I asked Fred if he could get Lowell Greenfield and me passes to view the Apollo 11 launch. Because I'd asked so late, by the time we received the official paper work, badges, and everything, we ended up reserving the last available hotel room in Orlando. Orlando was a much smaller city in 1969 than it is today. I remember making many phone calls, looking for anybody who had an available room. Of course, Orlando isn't all that close to Cape Canaveral, but all the hotels, even there, were full and they didn't even want to discuss the subject of rooms. At the last minute, a cancellation appeared at one of the poorer motels in Orlando, one that was old and only barely air-conditioned.

We also managed to get the last rental car from a no-name company. All of the big ones had brought in cars from all over the eastern U.S. for the launch and were out of cars. Somehow, I stumbled on the one car for rent. The car was about five or six years old, and it didn't run very well. Every time Lowell Greenfield and I went back and forth between Orlando and Cape Canaveral, we wondered whether we were going to make it or not. Partly because of the unreliability of the car, we decided to spend the night before the launch in the car, on the premises of Kennedy Space Center. The reason for this all-nighter at Kennedy was that we knew there had been some monstrous traffic jams going to the Kennedy Space Center for previous Apollo launches. We figured that, whatever the size of the traffic jams for the previous launches, the traffic jams for Apollo 11 would be even bigger. We were dubious about how our car would do in a traffic jam.

The night before the launch was a very up time at Cape Canaveral. All of the people who had worked so hard on the program—the contractors, the politicians, and the other people who had invitations—were milling about in the general area, both celebrating and hoping that everything would go okay on the mission. Of course, there was a great deal of uncertainty in terms of reliability and no one was sure how the mission would go. This was going to be the first time that the Lunar Module was taken down to the lunar surface. It had worked well on Apollo 10, but there were a lot of issues that we didn't understand regarding landing on the lunar surface. We also weren't sure how the Lunar Module's takeoff from the Moon would go. As a result, there was some uncertainty underlying the celebrations.

Lowell and I attended some of these celebrations. Then, as I recall, about midnight, we took the car to a parking area near the VIP section on that July night and tried to sleep inside the car, with the windows open. It seemed to me to be a very hot and humid night. I don't think I got much sleep and I remember the mosquitoes were incredible. But spending the night in the car guaranteed that we were where we needed to be the next morning, without the uncertainties of the car's performance, the traffic jams, and wrong lanes or wrong turns. We really wanted to see the launch, so losing a night's sleep wasn't a big concern. We were able to go to the VIP area and get in place quite early, around six, as I recall.

We were there quite a while before the launch. All through the night we could see the bright white object that was the Apollo 11 spacecraft atop the Saturn V vehicle, because Kennedy had flood lights on it. Florida is flat, so wherever I was, if I looked in that direction, I would see the lights. This added to the surreal atmosphere. I will never forget the entire experience.

I'd ended up going to Kennedy between evaluations on the F-15. After returning from the Cape, I did the final evaluation on the F-15, probably extending from late July into mid-August of 1969. I returned to school after that, coming back to Dryden the summer of 1970, and finally returned to Dryden with my PhD and my wife, Mary Shafer, in the spring of 1971. I had met Mary at UCLA while she was getting her master's degree in engineering. Mary joined Dryden several years later as a flying qualities engineer.

While I was at UCLA, another major event in my career took place at Dryden. Larry Taylor, who as I've mentioned before had a very strong personality, got into a dispute with management over his giving a paper in Europe. As a result of that dispute, Larry transferred to Langley Research Center in 1969 and never again worked with me on any of the programs or techniques we had worked on together. When I returned from UCLA, I no longer had my leader and my mentor there, although I could always, of course, talk with him on the phone. I saw Larry frequently, because I made a lot of trips to Langley for various reasons. We also saw each other at conferences from time to time. However, the easy camaraderie I had with Larry at Dryden earlier was gone forever when I got back from UCLA.

I thought it was a disaster for Dryden when Larry left, because he was the quickest engineer that I've ever known at coming up with good innovative ideas for studying a problem or doing analysis. I also think it was an unfortunate change for Larry. Being so talented, Larry had been a big frog in a small pond at Dryden. Langley is much more formal and much more structured place than Dryden, without Dryden's focus on flight.

I continued to see Larry over the years and we had dinner with him and his wife a month or so before he died suddenly in 1993. He had been perfectly healthy, having just passed his private pilot's flight physical, and he succumbed to Sudden Cardiac Death.

The Lifting-Body Program 1968-1975

I continued to support the lifting-body program wherever I could, for anybody who asked my opinion, but I wasn't involved in the day-to-day work on the lifting bodies after Gentry's flight in November 1966. Alex "Skip" Sim, who had started as a co-op at Dryden in 1966 and continued working at Dryden after graduation, replaced me for the most part, doing some parameter identification derivative extraction on all of the lifting bodies except the X-24A. [35, 36, 37] I augmented this advisory work to hone some of the newer parameter techniques used on all of the lifting bodies.

Once the HL-10 flew again, beginning in March 1968, it lived up to all of its previous billing. Once the curve of the tip fins' leading edge was changed, the HL-10 had very good handling qualities. It had a little less L/D max than predicted, but that was not a problem for the vehicle. The pilots had watched us struggle in getting the M2-F2 ready to fly, only to have three PIOs in 16 flights, so they were happy with the HL-10 and its excellent characteristics. The vehicle was as good as it had been on the simulator in 1964, thanks to the effort of refining the fins. A lot of hard work had gone into getting the HL-10 back into flight. I don't mean to minimize that work, but the gliding characteristics of the airplane were much as Langley had said they would be once the fins were fixed.

The HL-10 began powered flights in October 1968, after a series of ten glide flights (Fig. 9). It had the little problems with the engine and other little difficulties that always seem to occur on complex one-of-a-kind flight vehicles with things that can fail. We always had trouble with the XLR-11 engine. It was a really good engine, but it had been designed back in the 1940s—and it showed. It took a lot of work for people to keep the engine working right. When it worked,

Fig. 9. HL-10 under power with shock diamonds evident during Bill Dana's Mach 1.59 flight of November 17, 1969. (EC69 2347)

we got what we'd set out to do. My memory of the lifting-body program is that an awful lot of the things that set us back or discouraged us had to do with the operation of the XLR-11 engine.

At any rate, the HL-10 did make the highest and the fastest lifting-body flights. Pete Hoag reached Mach 1.8 on February 18, 1970. Then Bill Dana set the lifting-body altitude record of 90,303 feet on February 27. Those records, although they were set fairly early, held throughout the program for various reasons. Of course, there never had been a goal to set a speed record or an altitude record with the lifting body. The primary reasons for flying the lifting bodies were to establish them as having credible terminal phase aerodynamic characteristics as reentry vehicles and to sort out and spot some of the problems, providing a fix for each of them as they were found.

The HL-10 was still flying when the X-24A (Appendix C) showed up (Fig. 10). It was built by Martin rather than by Northrop, as the M2-F2 and HL-10 had been. I think the people who worked on the vehicle spent a lot of time trying to either understand the system or turn it into a system they already understood, so it took quite a while to get the X-24A into flight. The X-24A had the highest volumetric efficiency of any of the lifting bodies. That had been one of the selling points when we were looking at it in 1964.

Fig. 10. T-38 chasing a February 1971 X-24A flight. (EC71-2593)

The X-24A had a very complex control system, with both upper and lower body flaps. Because of some problems with the yaw due to the aileron, it also had an interconnect. The interconnect gain, scheduled with angle of attack, had been hard-wired into the control system. It may have had a pilot override on it, but I didn't like the idea of the fixed interconnect gain schedule. What a huge mistake it would have been had we done that on the M2-F2 and guessed wrong on the 0.4 or 0.5 interconnect gain. As it turned out, a hard-wired gain would have worked. We would have avoided the Milt's PIO on the first flight had he not had the . interconnect gain wheel to set.

At any rate, some of the flying qualities of the X-24A were not too satisfactory on the first flight. Jerry Gentry flew this mission on April 17, 1969. The automatic interconnect gain scheduler stuck at too high a gain, so the airplane had a tendency to PIO at low angle of attack and the high speed needed for landing. As a result, Jerry intelligently used the landing rocket. Also, the split lower body flap control surface was rate limited, requiring a highly proficient pilot to fly it. These problems were understood and fixed by the third flight, so they were not a significant problem on subsequent flights. Jerry had also flown the M2-F2 and the HL-10 by then.

We had seen significant problems that were not predicted by wind tunnels or seen on the flight simulators on the first flights of vehicles flying beyond the state-of-the-art configurations. The X-15 had significant handling qualities problems on the first flight by Scott Crossfield, the M2-F1 was unflyable during the first attempts to ground tow it, the M2-F2 had a serious PIO problem on its first flight, the HL-10 had severe controllability problems caused by flow separation that could easily have resulted in a PIO on its first flight, and now the X-24A had significant controllability problems that could have easily resulted in a PIO with loss of control. As I will discuss later, the Space Shuttle had poorly-predicted control characteristics that could have resulted in loss of the vehicle on STS-1. These six examples of serious first flight problems demonstrate why high quality prediction, simulation, and study preceding a flight program don't always prevent the real problems that must be addressed and solved. These real problems are discovered when the rubber meets the road during flight test, perhaps during the first flight.

Although the X-24A had the highest L/D max of the three original heavy-weight lifting bodies (HL-10, M2-F2, and X-24A) and the highest volumetric efficiency, flying qualities, ease of flying, and control effectiveness suffered because of the compromises. Of course, the X-24A was intended, as a reentry vehicle, to have a high L/D max and a high volumetric efficiency, so it had adequate handling qualities. They just weren't as good as the HL-10's had been right out of the box.

The X-24A went into its rocket-powered flight program on March 19, 1970. The X-24A team had the same hiccups that the HL-10 had in trying to get the XLR-11 engine to work. I'm not bad-mouthing the engine. It was an old engine with many things that could go wrong and they seemed to go wrong one after another as the program proceeded.

The X-24A had a few interesting characteristics that were discovered during the envelope expansion. I think Martin had foreseen most of them in the wind-tunnel data prior to finalizing the design before the construction of the vehicle. The roll control was fairly limited and the X-24A had a fairly narrow transonic envelope. There were good reasons for looking at the subsonic characteristics in a flight program that would get into the transonic region, in order to see what the characteristics really were.

I think all the transonic problems were addressed and either understood or improved enough that the vehicle became a decent transonic aircraft, considering its shape. It's not a shape that would be designed for sustained flight supersonically or transonically—and probably not subsonically, for that matter. But it had one peculiar characteristic that had been noted both on the D-558-II and the Saturn V. [38] When the X-24A was under thrust, there was a change in the trim and stability characteristics of the vehicle. The characteristics changed markedly when the engine was on because of how the pressure field changed upstream of the engine. The problems this caused were addressed and understood. They developed flight plans and maneuvers that took into account the difference between the vehicle under thrust and gliding.

Another interesting characteristic could be observed looking at the rear of the

X-24A. As a stability-and-control engineer, I noticed an apparent roll stability-and-control maneuver, with a large aileron deflection in one direction and then the other before returning to the neutral position. What amazed me was that the vehicle showed no reaction at all in roll when these large control deflections were made. The roll control never dropped to zero, but because the motions on the vehicle were small, the view from the rear looked to me as if those roll controls weren't controlling roll. We had that footage in the project film we showed to visitors to Dryden, but I'm not sure that they knew what they were seeing.

One of the impressions I had very early in the SV-5 program and later on watching the X-24A program was that the X-24A did show the maturity of design and the care devoted to the entire flight envelope, the volumetric efficiency, the placement of the fins, and the shape of the body. Even the forebody was quite sensitive aerodynamically to small changes in shape. Martin had done an outstanding job of optimizing this configuration in the wind tunnels and with the other design tools that they used. As a result, the way it flew fulfilled the dreams of the designers. It may not have flown as elegantly as the HL-10, but it had some significant advantages over the HL-10 as well. Part of its design achievement was the result of design technology as well as the resources that Martin had put into its design.

The other thing that was significant to me—although I didn't become aware of this until early 1970 when I was working on another program—was that the X-23A was also an SV-5 configuration. The X-24A was the SV-5P. The X-23A had successfully flown three suborbital missions, showing it could maneuver at Mach numbers as high as 20. Once it slowed to Mach 2, it deployed a series of parachutes and a C-130 tried to catch it in midair. The C-130 was able to catch the third vehicle in midair and bring it aboard. The X-23A is now in the Air Force Museum with the X-24B and an SV-5J painted to look like the X-24A (the actual X-24A was rebuilt as the X-24B).

I didn't work directly on the X-23A program. I did get some of the flight data to look at the hypersonic data with maneuvers. It was very poor data, not anywhere near the quality that I was used to working with on the flights of our research vehicles. I don't think I got much out of the data, but I was impressed when I finally figured out that the SV-5 configuration had been successfully flown from above Mach 20 (X-23A) all the way down to landing (X-24A). It looked to me as if that was the first full proof of a lifting reentry shape capable of horizontal landing that had flight data from nearly orbital conditions on down.

I also believe, however, that you can't change much on that vehicle and still make that statement. I remember looking, in the 1960s, at some of the SV-5 data we'd kept in the safe. Martin had a lot of different configurations with slightly different forebodies on the top of the vehicle right behind the nose. I don't remember now what the nuances of the changes were, but there were quite substantial changes in the characteristics with just small profile changes on the front top of the vehicle.

To me, the most significant achievement of the X-24A program was that it demonstrated that all of the good work that Martin and the Air Force had done on the X-24A and the SV-5 was borne out by a flight program. Being a flight guy, I don't really think something's happened until it has been flown and the data from the flight has been analyzed. I think a very credible case was made for the SV-5.

The M2-F2 was eventually modified and rebuilt, with the center fin on, as the M2-F3 (Fig. 11). (Appendix C) A lot of negotiations and conspiracies went on to make that happen and in the process it was quickly decided that it would have a center fin. I claim it was because the center fin data had been available to us for

Fig. 11. M2-F3 under power on Bill Dana's July 25, 1972 flight to Mach 0.989 for the initial test of the CAS system. (ECN 3134)

two years. It wasn't the same center fin that flew on the M2-F1, but it had almost zero adverse yaw. This fin was larger than the M2-F1 fin and therefore had an even larger effect of decreasing adverse yaw. That was the main configuration change to the vehicle. They did control system changes as well. They had a triplex system and the vehicle tested some fancy control schemes, both in terms of the automatic control system—the SAS or the CAS—and in terms of the cockpit design. They evaluated a side-stick controller and a reaction control system (RCS) on the vehicle on the M2-F3. I think they had mixed results on all of these things, but for the most part they were able to conclude whether or not the key things they set out to demonstrate were a good idea.

Bill Dana made the first glide flight in the M2-F3 on June 2, 1970. After only three glide flights, it began a powered phase on November 25 (Fig. 12). When the M2-F2 was rebuilt as the M2-F3, the vehicle already had the engine in it. It was a very successful program. At least lateral-directionally it flew very well with the center fin and the control system designed for it, which was not greatly different from the control scheme on the M2-F2, other than the interconnect. I think the

Fig. 12. M2-F3 flown by John Manke on his first flight of the vehicle on October 19, 1972. (ECN 3270)

control authorities were increased on the M2-F3 because the M2-F2 had shown that more control authority on the roll and the pitch was desirable. Most pilots felt the M2-F2 and M2-F3 were at the lower limit of acceptable subsonic L/D max for landing. Some thought it was a little too low at an L/D max of just over 3.

The HL-10, at the very end of its program, made two powered landing tests. LLRV rockets were fitted to the HL-10 to demonstrate a powered lifting body landing at shallow glide angles. They did this because everybody who saw the exceedingly steep dive angles of the lifting bodies thought them impractical for bringing people back from space. The flights were made on June 11 and July 17, 1970 by Pete Hoag. Since the landings had a shallower glide, the pilot found it harder to land. Diving at the ground at those speeds, a pilot can see the touchdown point better at steep angles than at shallow angles.

The final shape of the heavyweight lifting-body series was the X-24B (Fig. 13). [Appendix C] It was a reconstructed version of the X-24A, but it didn't look very much like it. It wasn't as ugly as the X-24A, but in a group of any regular airplanes, the X-24B is still pretty ugly. Almost all of the lifting bodies are ugly. Even though I worked on them for a large chunk of my career, I didn't do it because they were pretty. I did it because they were interesting and filled a need.

The X-24B shape was scabbed onto the X-24A fuselage. It was quite a bit longer and wider at the back end. It had a very good set of aerodynamic characteristics. The design had gotten away from the emphasis on volumetric efficiency and gone more in the direction of high hypersonic L/D max, giving more cross range, something that any vehicle that's going to serve the needs of a spacefaring nation needs to have. This cross range allows the vehicle to come down from many different orbits to a particular place to land.

One of the primary goals with the X-24B was to examine a vehicle with a

Fig. 13. X-24B after B-52 launch on May 22, 1975. The pilot, John Manke, reached a speed of Mach 1.6 and an altitude of 74,100 feet, the highest altitude attained by the vehicle. (EC75 4638)

hypersonic L/D max of about 2.5. The other lifting bodies we were looking at had hypersonic L/D max values of 1.1 or 1.2. The Shuttle's hypersonic L/D max is a little over 1.8. The Shuttle, however, comes in at such a high angle of attack for so long that it doesn't have a cross range any better than that of a vehicle with an L/D max of 1.0 or 1.1. When we talk about L/D requirements, we're usually talking about the L/D max. If you're not flying at or near the hypersonic L/D max you won't get the potential cross-range benefits. The X-24B was able to use a good portion of that higher hypersonic L/D max and it also had very good flying qualities. Because it was bigger, it didn't have all of the constraints of the earlier lifting bodies, with all control surfaces stuck on the back end to make it fly properly. The HL-10 did the best job of any of the earlier lifting bodies in having the control surfaces on the back end and having them do what they wanted, but its volumetric efficiency was poor. From the side the X-24B was slender, compared to the M2-F2 or the X-24A.

The X-24B was, I believe, delivered with the propulsion system already installed (Fig. 14). It did a few glide flights and then went on to very successful low supersonic flight testing, mostly in the transonic region. It got up to Mach 1.76. The X-24B didn't go any faster because it was quite a bit heavier than the other lifting bodies because of the larger size and heavy material it had to get it to accommodate having the X-24A inside. Also, going faster wasn't a program goal. That was true with each of the lifting bodies, where we only needed to get to a

Mach number of about 1.5. Then we could look at the critical characteristics in the transonic region. The X-24B also had reduced directional stability when the engine was on, similar to that of the X-24A. [39,40]

Looking Back at the Lifting Bodies

At the end of the lifting-body program, I was already working on other vehicles, particularly the Space Shuttle. Many times I was asked by others working on the Shuttle program what I thought of lifting bodies versus winged Shuttles and what all of our lifting bodies had taught me. I think what they taught me was that we'd flown five significantly different lifting-body shapes and each had its pros and cons. All the same, all of them, with the exception of perhaps the M2-F2 and M2-F3, certainly had the ability to do a high-speed pushover maneuver into the flare and land almost effortlessly. That was true of all of the lifting bodies from a performance point of view and true of all but the M2-F2 from the controllability point of view, although the M2-F2 could probably have been fixed. Putting the center fin on wasn't a big deal, and that put it in the same category as the HL-10 and the X-24A.

Fig. 14. X-24B being chased by an F-104 (whose gear is down to increase drag and thus matching the glide slope of the lifting body). This was Dick Scobee's first lifting body glide flight on October 21, 1975. (ECN 4913)

The lifting bodies showed that this class of vehicle could have the character-istics that we needed for high volumetric efficiency and still have a high enough L/D max and a relatively dense configuration ensuring a high landing speed. The lifting bodies gave us a great deal of confidence that we weren't going out on a limb when we started to look at some of the Shuttle configurations. The shape and the way the aerodynamics blend together are quite different on a lifting body and on a winged body configuration like the Space Shuttle. However, for the pilot, inside the cockpit, or the automated system, what it looks like on the outside doesn't make much difference in how it flies. All of the lifting bodies had lower subsonic L/D max than the Space Shuttle. The X-24B approached the Space Shuttle subsonic L/D max. When we were working on the Shuttle configu-ration, we were really looking at an airplane that would be easier to handle in approach and landing than the lifting-bodies had been.

The characteristic I think we didn't fully evaluate on the lifting bodies (and on the Space Shuttle) is vehicle handling in crosswinds. That was the concern I heard from Milt from day one when we'd talk about flying the M2-F1. (It contin-ued through all of Milt's flights and through those of the other lifting-body pilots. They always worried about crosswinds. We never did fly the vehicles in a high crosswind (over 10 knots), but I don't think they would have done very well if we had.)

Fig. 15. F-104N chasing the M2-F3 near runway 23 (appendix E) on Edwards Air Force Base just prior to landing. The ascent portion of this flight was shown in Fig. 11. (ECN 3142)

The M2-F2 was purported to have landed in a crosswind of about 10 knots.[41] However, the wind was measured several miles from the landing location, and the pilot chose to land the vehicle into the wind instead of down the lakebed runway. The X-24A landed once in a crosswind of 5 knots and once in a crosswind of 10 knots. Once again, the crosswind measurement was several miles away from the touchdown point. [42] The pilot had some difficulty in these landing, and on one of them, the vehicle was forced well off the lakebed runway during the rollout and braking. The X-24B landed once in a 10-knot crosswind with little comment from the pilot, which to some extent confirmed the speculation that the X-24B was not extremely sensitive to crosswinds and turbulence. [43] There may be an automated system that could be put in a vehicle to help alleviate the problem. Nevertheless, I think that a strong crosswind, even at the high landing speeds of the Shuttle, would probably make for a bad landing—not necessarily a crash, but certainly an uncomfortable landing.

Something else we saw on all of the lifting bodies, from the M2-F2 through the X-24A, was the tendency for turbulence to make the pilot uncomfortable. That's always going to be true of a vehicle that has a low roll inertia and virtually no damping in roll. Lifting bodies have no wings, so they have a low roll moment of inertia, low rolling moment from the ailerons (small moment arm), and virtually no damping in roll. Any time a pilot is put into that sort of situation in turbulence, it's going to be uncomfortable. In addition, the response of a lifting body in turbulence can make the pilot feel the vehicle is unstable, when it really isn't.

I think the M2-F3 (Fig. 15) exhibited even stranger characteristics in turbulence and they tended to fly it on calmer days than they did the other lifting bodies. One advantage of a research environment like the one we have at Dryden is that you can pick and choose the days you want to fly. That might not always be true for a space-faring nation's vehicle coming back from orbit.

I don't actually remember the X-24B having the same problem in turbulence. That vehicle flew so differently than the others, I don't remember anybody being unhappy with it in turbulence. In fact, the comment I usually heard from pilots was that it felt more like flying a real airplane. If it felt more like flying a normal airplane, I suspect it didn't have an attention-getting tendency to roll when it encountered turbulence. The X-24B was the icing on the cake showing that the lifting bodies could land on the concrete runway at Edwards and come close to the intended touchdown point. I think everybody knew then—at least the people at Dryden did—that there was no special problem with landing any of these lifting bodies unpowered and landing them more or less where you wanted to land.

VI

Origins of the Space Shuttle

Some of the perspective that I have on the runway-to-orbit concept comes from my earliest days at Dryden in 1962. Shortly after I arrived I was told of a concept that Jake Drake and Robert Carman at Dryden had come up with of a two-stage hypersonic research aircraft. This was part of a research program that would eventually put the second stage into orbit. It was proposed in 1953, well before the X-15 program began. I recall that particular runway-to-orbit concept as being vague. It may not have been vague to Jake Drake or to the other people who worked on it, but it was to me. I always mentally pictured the aircraft in the concept as somehow going down the runway with the first-stage rocket, going up into a high-altitude/high-Mach number and then launching the upper stage.

Both stages were airplane-like in that they had wings and tails and both were meant to fly back to Earth to land. At some high Mach number, the second-stage rocket would be lit, and then it would go to very high Mach number or perhaps into orbit. It would also come back from orbit and land. When I started working on the X-15 and the lifting bodies in 1962, the concept still made sense, at least from a conceptual point of view.

While I was at UCLA between 1968 and 1971 period, returning to Dryden for the summers and doing occasional tasks for Dryden during other times, I was reading Aviation Week and other magazines, journals, and technical papers, well aware that the Space Shuttle was becoming the next thing as Apollo wound down. I don't remember being particularly disappointed in that, because I thought the important problem to solve was getting a vehicle from the ground up into orbit and then having that vehicle return to Earth and land horizontally. I had seen a lot of combinations of airplanes, rockets, air-breathers, and rocket stages permeating the literature regarding what NASA's next program would be.

The Shuttle was going to be, in turn, the vehicle that would go up from Earth to a space station to deliver supplies and people, then return bringing down products made in space, as well as the garbage. This had always been the role projected in concepts of a Shuttle vehicle, as far back as the 1950s. Construction of a space station was seen as vital. I don't know why we thought it was vital back then, for we still seem to have some trouble figuring out exactly what we need it for. However, I'm a big supporter of the space station. I would like to have seen a smaller version so we could first learn what it is we can do in a space station, prior to the present station. But since it seems we can only make big things happen, I did understand why they were talking about a space station. That was the concept that we were doing in the Skylab program, so we knew a little bit about space stations, and it was clear then that we didn't quite know what to do with a space station. It was a lot of fun to watch the people inside Skylab in zero gravity. We also learned a lot about long-duration missions and their effect on human physiology. There were always people talking about manufacturing techniques or observatories.

People also talked about using a space station as a staging area for assembling more complex vehicles that would go to the Moon and Mars. I think we always wanted to go beyond Mars, but since getting to Mars was obviously going to be hard, that was the next logical step to those of us who were big supporters of manned space efforts. We already had some idea of what was going on in the planets from the early robotic unmanned missions. The Viking program was being talked about as the next logical step that we would go through in the unmanned space program.

But nobody had committed to sending humans to Mars yet. What most space enthusiasts thought was important was going to Mars and finding out more about its history and geology as well as the possibility of finding evidence of life either as we knew it, or in some different form. The space station was all things to all people then, as it continued to be through 1990.

People who didn't really have a good feeling for orbital mechanics always threw the space station in as an intermediate step to anything else that we wanted to do, such as going back to the Moon to get out of the gravity well of Earth. But it's really a poor place to be for routinely going to the Moon. From an energy point of view, we really have to have the orbit of the space station in the same orbital plane as the one from which we want to go to the Moon. We can, of course, change the orbital planes, but that takes a tremendous amount of energy, and it's wasted if we could have done it either directly from Earth or from some other position that had the momentum lined up with the direction that we wanted to go.

There was a lot of talk about how a space station was going to be the way that we were going to be able to make more frequent and easier trips to the Moon, because we wouldn't have the gigantic problem of getting all the equipment out of the gravity well to the space station's orbit, where we could do assembly before going on to the Moon. However, since the Moon and the Earth-orbiting space station are not lined up most of the time, and we are worried about night landings on the Moon, how we would use a space station as an all-purpose way for going to the Moon gets fairly complicated. It could be useful. I'm not saying that we would never want to have it involved, but it does complicate things; as with a launch from the Earth to a given orbit at a given time, we would have only a very small launch window for matching a space station's orbit to a lunar mission. Whatever the uses of the space station would be, they were beyond anything that we had planned in detail.

Nevertheless, I really believe in the space station and humans in space, because every time researchers are offered a new environment, new and unanticipated capabilities are developed. I envision the space station to be a new environment.

Initial Shuttle Concepts

NASA thought that the Shuttle should be reusable, because that would make it cheaper than using expendable rockets, and also we would not be continually leaving debris in space that would have to be contended with later or that would reenter Earth's atmosphere and perhaps cause some issues. By 1970, the idea of taking a reusable vehicle to orbit and returning had been around for at least 20 years.

Of course, everybody's favorite choice was a single-stage-to-orbit vehicle. With the technology that we had then, there were concepts that people claimed would be fully reusable and would go to orbit and return. These proposals in the late 1960s were well thought out in terms of the concept, but not in terms of the reality from an engineering point of view of how we would make all of the pieces needed, how much this thing would weigh, and how we would deal with the weight and the structures of a vehicle that size.

I did see some things that were in the Saturn V class in terms of weight that people were proposing, but with the technology and the materials that we had then, they would have ended up being much larger. They would have been the largest vehicles that had ever been launched, flown, or designed to move. Those concepts included one for an HL-10 (Fig. 1) single-stage-to-orbit reusable vehicle. It was, as I recall, a demonstrator. The payload and the payload shape weren't considered at that time. It was simply a concept for how we could use the

Fig. 1. HL-10 accelerating to Mach 1.59 on November 17, 1969. This photo was taken immediately after the photo in Fig. 9 of chapter 5. (E 21090)

HL-10 configuration up to orbital speeds. As a concept, from an aerodynamic and structural point of view, it was doable, but I don't remember it ever being on anybody's official list of favorites. There were other concepts and shapes that also were single-stage-to-orbit. Some of them even launched horizontally, which seems like it would take a lot more energy than the vertical launch if we were going to use a rocket.

But as people went through the physics and the analysis and breaking these down into doable parts, it became clear that with the technology and, more importantly, the budgets that were available at that time, the only way we would get this done would be a two-stage-to-orbit vehicle. The concepts that we saw were various two-stage vehicles: two vehicles of similar shape and sometimes similar size, or two of the same shape, but dissimilar size, or still other concepts that had dissimilar shapes and sizes for the first and second stages.

We went through a period in the late 1960s and early 1970s of trying to figure out what the Shuttle would look like, which was really a national effort, because all of the major aerospace companies at the time were putting quite a bit of effort into it. They had some of their better people working on turning the Shuttle into a viable concept that would meet the still uncertain requirements, and be able to do so within a budget that Congress would be willing to approve.

There are many perspectives on how we do a large task such as building a Space Shuttle, but it ends up involving two different approaches-the engineering and the political. The engineering approach has fairly narrow specifications and is determined by what is within the realm of possibility—in other words, being just a small step from things that we'd done before, using the "design a little, test a little concept" mentioned before. In setting out to do a Shuttle from that point of view, the idea would be to do something that was not terribly ambitious. We would simply demonstrate how we would do it and have enough margin in the design that, as we went through the design of individual components and the fabrication, we would be able to accommodate changes and still end up with a vehicle that would go from Earth launch to orbit and back again.

How we do something such as build a Shuttle is quite different from a political point of view. The way programs have been run in the United States throughout my career is that the politicians lay out certain constraints or require-ments that the engineers are then forced to accommodate in their design. I'm not saying that Congress, which is accountable to the taxpayer, should not have that task or that authority, but I am saying that Congress needs to interact with the basic engineering concept to find out where the sweet spot is in the design before it sets things in concrete that this thing is going to be all things to all people.

My first encounter with that sort of thing, even though it didn't affect me directly, was with the F-111. Secretary of Defense Robert McNamara was trying to make a really advanced world-beater airplane that satisfied both the Navy and the Air Force. However, the Navy and Air Force's requirements were different. So, by the time we had accommodated the needs for each service and combined them into one vehicle, we had made the vehicle far less capable than it could have been had we designed two vehicles, one for each of the armed services. The F-111 should have taught us about that, but it didn't. It was somewhat of a heroic effort getting the F-111 to do as well as it did, given the problems involved in trying to satisfy too broad a set of criteria. The result was not a very elegant, efficient, or successful design.

This is my perspective. It may not be the official perspective on the Shuttle, but at the same time that NASA was looking at the Shuttle, the Air Force was also becoming interested in advanced space vehicles. The Air Force, of course, had always been a strong advocates for having a presence in space. And well they should have been, because military history shows us that occupying the high ground is usually a big advantage to the military. Space is the high ground from many points of view—for observation, gathering information, and for launching an attack. Consequently, our military has always had a strong focus on things dealing with space since Sputnik.

The military had advocates within Congress and other agencies arguing that the military's needs in cross range and payload for their missions be included in any Shuttle design. NASA had a more nebulous idea of what it wanted for the Shuttle in terms of something to carry payload up to the space station and return to Earth with used or manufactured goods or trash. To do that job, any engineer would say, "I don't want this vehicle to get too big, because that continues to make all of these problems grow." We would want to start out with our first useful operational vehicle having a modest payload requirement. My opinion on that would have been several thousand pounds and a 5-by-5-by-10 foot payload bay, or something in that neighborhood, because, other than with the Saturn V, we had not been able to put up large vehicles and large payloads. Even though ultimately we would want to put a large vehicle with a large payload into orbit, that might not be what we would want to do with our very first attempt at a fully reusable orbital vehicle. It was my opinion then, and still is, that the smartest thing we could have done—and it wouldn't have been the fastest, and it wouldn't have been politically the most salable thing—was to have a pair of vehicles that were in some sense reusable, then demonstrate that the second stage could manage to get to orbit and return back to Earth.

We had the technological basis for such a project. The X-15 had probed many of the areas of reentry flight, hypersonic flight, aerothermodynamics and the problems that we encountered. The lifting bodies had provided ideas in terms of what shapes we could use for such a vehicle coming from above Mach 28 down to Earth. Finally, the Saturn V had demonstrated the rocket capability that existed within the country. So, even though it would have been a tremendously ambitious program, even if the payload had been just a couple of astronauts to go up and perhaps do a little maneuvering in space, it wouldn't have been intended to be an operational program. Instead, it would have been more like the traditional X programs that the Department of Defense (DoD) and NASA had done over the years. Its real purpose would have been the single requirement of going into orbit and returning in a reusable vehicle.

Because of the political situation and the way big programs are sold, Congress said that the Air Force could not have a separate space program and that they had to get consensus with NASA. But the DoD had requirements in terms of payloads and missions that were well beyond any demonstrated capability by either NASA or the DoD at the time. Both NASA and DoD were looking at the Shuttle as being the only game in town, and they needed to get their specifications in so that it would satisfy their needs for the 1970s and perhaps the 1980s. The DoD's requirement was that the Shuttle be able to put 50,000 pounds (sometimes 60,000 pounds had been stated as the requirement) into low Earth orbit and accommodate a payload of 15-by-60 feet (that is, a cylinder 15 feet in diameter and 60 feet long), roughly the size of a railroad boxcar.

Considering what we'd managed on Apollo, we probably were a little overconfident in our ability to solve really difficult engineering problems. But with Apollo, we also had a demonstrated reliability. We knew enough about rocketry. We knew enough about orbital mechanics and physics to figure out how to get people to the Moon and back by using huge vehicles. For Apollo, reusability had not been an issue, nor had atmospheric flight. But atmospheric flight is really the tough nut to crack in terms of a reusable space vehicle, in my opinion. The objective of the Apollo program was really to increase the reliability and the capability of our systems in order to do things where we understood all of the physics.

The reusable space vehicle had all of those requirements in terms of reliability. However, it also had the requirement of returning to Earth using true atmospheric flight, which is very dependent on the aerodynamics, aerothermodynamics, and thermal and structural properties of what we were flying. Consequently, it's got a lot of uncertainty, even with our X-15 and lifting-body experience. There are also uncertainties in the atmosphere in terms of density and molecular constituents, pressure, temperature, turbulence, density gradients—all kinds of things that occur in the upper atmosphere. That became a very hard problem. I don't think the managers or the politicians of that time ever really considered that doing the Shuttle program where everything was reusable was in the same category as going to the Moon and returning. Perhaps this was because Shuttle missions were done in Earth orbit, so it wasn't nearly as scary as the risks and consequences of stranding people on the Moon or in lunar orbit, either of which could have happened with Apollo.

In summary, this is why, to me and many others, it seemed as if the next step should be an X-vehicle to demonstrate all of these capabilities. These were smaller steps because the X-15 demonstrated the atmospheric flight and reentry of a lifting vehicle. Even though the X-15 had been at Mach 6 instead of at Mach 28, it had demonstrated the materials and the aerodynamics and how we could handle them for something that having no mission other than to explore that region as a research vehicle.

We also knew that the lifting bodies were predicted to be able to go from orbit to landing in a safe manner in terms of the aerodynamics and the atmospheric uncertainties that we knew, but those were only predictions. We had flown lifting bodies from Mach 1.8 down to landing speeds, while the X-23A vehicle had demonstrated a lifting-body concept that went from above Mach 20 down to the altitude where they put the drogue chutes out, which was supersonic, but they demonstrated that it would work in both of those areas. Those had the goal of really exploring the flight characteristics of these shapes. As a result, we were confident that a lifting body, particularly the X-24A and the X-23A, could somehow be combined to be able to do that.

The idea was to demonstrate a two-stage vehicle, with both stages flyable and reusable. We would launch both of them. At some point, based on the design, we would run out of fuel on the first stage. We would stage the second vehicle and return and land the first vehicle either with no power or with an auxiliary power system to fly it back to the launch point. The second stage would continue on with its own rocket power and go into orbit.

We could do some orbital demonstrations—perhaps small orbital inclination or altitude changes, or rendezvousing with an existing satellite to show the usefulness of having this large of a vehicle as a ferry vehicle and its ability to visit other orbiting craft, in particular the space station, which hadn't been defined or built yet. So it seemed like a logical step to do an X-airplane to show how this could be done. Once we had done that, we would have learned enough that when we did the first operational vehicle, it could then be larger, and have more payload capability. It would have been a demonstration in terms of technology as opposed to concepts, so there might have been two X-vehicles instead of one to get there.

However, because Apollo had no successor, the X-15 program had been cancelled and the space station had not been defined yet, I think NASA suddenly was put in the position in the late 1960s of wondering, "what next?" The only thing that we could define completely at that point, using the momentum of the Apollo program to help us, was the Space Shuttle. NASA could have chosen, at that time, to do an X-airplane. I know there were advocates for all kinds of experimental ways of demonstrating these things, but it was going to cost quite a bit of money to do that, and we needed something to continue our space effort. As a result, NASA decided that we would build a Shuttle that would ultimately be an operational vehicle. We would skip the X-plane demonstrator, and we would skip the smaller vehicle concept because the payload was too small. As a result, we would go directly to what the military and NASA thought was the next requirement for orbiting vehicles, and that would be done, in this case, with a reusable vehicle.

Between the DoD and NASA having to come up with a common program and NASA only being able to sell something that was a significant step to Congress, it was agreed that we would do a Space Shuttle program that met the perceived military as well as NASA requirements. Consequently, the 15-by-60 foot, and 50,000 or 60,000 pound payload became a requirement. I'm not exactly sure when this large payload weight and size capability became an absolute requirement, but it was early. It was fairly clear by 1970, and certainly clear by 1971, that this was a requirement, even though many of the proposals coming in had payloads at a half or a quarter that size. I think everybody who was involved officially probably recognized that this requirement that the DoD had was an absolute requirement and that if we didn't meet that requirement, the political climate would be such that there would be no Shuttle whatsoever.

Whatever the reasons were, whether I agreed with them or not is of no significance. Nevertheless, this is how we got into taking, in my opinion, too big a step. Ultimately the vehicle that we ended up with barely made it from almost every point of view. It's an engineering miracle that it works at all! Because it pushed so many technologies of the 1970s to their limit, it took a great deal of scrutiny, inspection, and refurbishment that wouldn't have been required had we been able to take the steps in between.

Over the years, there have been many stories and claims that the Shuttle was poorly done because it requires so much inspection and refurbishment as well as 2,000 people to take care of the vehicle, or that the design was poor, or that the contractors, NASA, and the DoD did a bad job in building it. In my opinion, none of that is true. In fact, quite the opposite is true. Every time the Shuttle makes it into orbit and back, I think of it as another stellar accomplishment. I wish we could have done intermediate steps so the current Shuttle would not be as complex and ungainly as it is. However, the way it was done was the only way we could solve the problems as mandated by Congress. And I believe that the final design was very close to the only design that would meet the mandated requirements.

Picking a Configuration

When I returned to Dryden in 1971, the X-24A was just completing its flight program, and the M2-F3 (Fig. 2) program was into envelope expansion in the powered phase. Among my other assignments, I assisted with the analysis of some of that lifting-body data, but I was not assigned to any lifting-body project. The 3/8th-scale F-15 remote piloted research vehicle (RPRV) program [51] was one of my main assignments. It was just coming to fruition as a program based on having the computers for the control system on the ground as opposed to in the aircraft. The vehicle was an exact 3/8-scale replica of the full-scale F-15 aircraft, which of course I'd been involved with during the selection in 1969. The program was designed primarily to look at some of the more extreme subsonic flight conditions of the F-15, so there was work to be done on the F-15 itself as well as on the 3/8th-scale F-15 RPRV.

In addition, the SR-71 and the YF-12s had started flying at Dryden. These were highly classified, under the code name Senior Crown. Since I wasn't working on the program, I did not obtain a Senior Crown clearance. I did work on some of the data that they were having trouble analyzing, but they put it in a form where I could use it and not recognize what I was seeing. It turns out that

Fig. 2. December 1971 flight of the M2-F3 with Bill Dana setting up for a landing on Rogers Dry Lake (foreground). (ECN 2923)

the YF-12 and the SR-71 used what was essentially a ramjet engine above Mach 1.6 or 1.8. The turbine of the turbojet was essentially bypassed, but I wasn't working on it, so I didn't know that at the time. The D-21, which was launched from the M-21 (a modified A-12 which was an earlier version of the SR-71) in the mid-1960s and a B-52 later on, had a Marquardt ramjet that had been used on the Bomarc missile several years earlier. The D-21 had successfully flown in the neighborhood of Mach 3.3 and made reconnaissance missions over China.

At the time I returned, I was aware of no whispers about any kind of a ramjet or scramjet program. That sort of work had died when the X-15 program was cancelled. The program had been fairly well defined before the X-15 was cancelled. It was really a scramjet, because it was going to have supersonic combustion. Langley was in charge of the actual development, which was running a little late, but the dummy ramjet (scramjet) that had been on the X-15 was the same external configuration that would have flown with an active scramjet engine. Langley continued to work on the scramjet concepts, as did several other organizations, including some for the DoD, but there were no planned flight programs to work on in that area.

At the same time, the Space Shuttle was being put through the various stages of reconfiguration in the attempt to select the final design concept. When I returned to Dryden, Milt Thompson asked if I would be interested in contributing to it. Dryden was not much involved in the Shuttle at that time, but Milt was very interested in the Space Shuttle, the first airplane with wings that would fly at speeds faster than the X-15 did. He'd probably been following the Shuttle since the late 1960s, when it first was being seriously discussed. I know he was a strong proponent of getting the Shuttle to land dead stick with no engines, which became a tremendous savings for the program. I'm not sure that the program would have been successful had it retained the requirement for landing engines on the vehicle. Landing engines sound good to managers, but I don't think they sound very good to people who fly. If they're required for landing, the pilot hates to have to start them just before he lands in order to accomplish the landing, because if they don't start, he is in worse shape than he would be if he had no engines at all.

Milt was actively involved in trying to convince others that there was no need for landing engines as we had shown in the X-15 and the lifting-body programs. The engines would have been turbojets in this case, not ramjets. I started looking at some of the Shuttle concepts, making comments primarily through Milt, although I'd started to get contacts at Langley at that time, too.

When I got back to NASA in 1971, the two-stage vehicle was still the one being considered. By the spring, however, it was apparent that there would not be enough funding to build the vehicle as originally designed. As a result, NASA and the possible contractors began looking at fall-back positions of having an expendable first stage and a reusable second stage. There were a lot of proposals where a lifting body would be on top of a stack of expendable rocket stages to put it into orbit. Besides the lower cost, there were other advantages from the engineering point of view. If we had two fully reusable vehicles, they both had to be able to fly throughout their flight regimes with all of the potential failures that we might have in either stage, a very complex thing to do.

I initially did some Shuttle work with Milt after I came back to Dryden in 1971. We had developed programs in the late 1960s and developed more while I was away at school because of the dissertation topics that I had investigated. We had developed some trajectory optimization programs [44] even before I went to UCLA that I could use to figure out what was real—that is, what could we really get as a first cut between some of the proposals that were around, either officially or unofficially. Could we actually get something into orbit? How big would it be, and how much would it weigh? It really was nowhere near a design, but it was a reasonability check in terms of how much thrust, fuel, and weight the first two stages together had to have to get the vehicle to some staging condition. Then, based on that staging condition—which would consist of the velocity, the momentum vector, the altitude, and the angle of attack—could we stage the two vehicles so that the first stage would return to Earth, and the second stage would continue on into orbit? That was mostly what I remember doing for Milt, because I already had those programs, and I was very interested in that problem. I don't think I made any major contributions to the field of trajectory optimization, but the programs worked, and they showed Milt and me fairly early on, as I recall, which concepts were nonstarters because they wouldn't make orbit and which ones were at least likely to have that possibility.

With everything we looked at, we assumed the vehicle would come back from orbit and land without engines. We didn't consider the vehicle's stability and control characteristics. We would give it performance characteristics, and those would basically be the lift, drag, and thrust as a function of angle of attack and Mach number and the rocket performance as a function of altitude and Mach number.

The techniques that I was looking at were fairly basic, but digital computers and general software were not available, at least to me in those days, to do anything beyond that. I didn't have a database of the stability and control characteristics, so there wasn't anything I could put in for them. I did have lifting-body characteristics from orbital speeds down to landing that I analyzed, and I did look at the same kind of characteristics I'd looked at for our transonic lifting bodies a few years earlier. We did look at that, but it wasn't a matter of looking at other people's configurations and proposals as to what those might look like. I know I felt that a two-stage, fully reusable vehicle was a viable concept and that we could do it, but that the huge payload size and weight made these vehicles really big. And because they were so big, they were going to be too expensive and have too much technical risk, so that they probably wouldn't happen.

I think Milt's feeling at that point was that Dryden as a center was interested in the pieces that went to orbit and came back. If there was an aerodynamic shape for the first stage, we would be interested in that, but it was a moving target as to what that might look like. Some of the first-stage configurations would be recovered by parachutes, or perhaps only the engines would be recovered. What we were looking at mostly were two similar vehicles. The first and second stages were the same shape, more or less, as much as could be, and we would try to integrate the two together. Therefore, if we looked at the aerodynamics of one, we were looking at the aerodynamics of the other as well.

I remember doing this work over a couple years, so I probably was doing it beyond the point when the actual Shuttle configuration was selected. The final vehicle was the stage-and-a-half configuration. It had a delta-wing Orbiter, an external fuel tank, and two solid rocket boosters (SRBs). The Orbiter and SRBs can be refurbished and reused, while the external tank is lost and destroyed on each flight. I know why those decisions were made, and there were really no others that could have been made at that time. But probably there remained a little wishful thinking on my part and Milt's, for we continued to look at some of the lifting-body concepts even after a winged design had been selected.

We also were following the official program as much as we were permitted. I was not directly involved with anybody at either Marshall or Johnson on the Space Shuttle at that time. Everything I did was through Milt, who interfaced with these other people. I think there was at least a gentlemen's agreement somewhere along the line that Dryden, since we had extensive flight-test and flight-research capability and experience, would be a major player in the aerodynamic entry aspects of this vehicle, and not very much in the rocketry or the design. Although we weren't struggling to have a part in it, I think Dryden's upper management had a lot of struggles in that regard, aware that the NASA budget was going to be spent largely on the Space Shuttle. Consequently, if you didn't have fairly significant contributions, your likelihood of surviving as a center was diminished. I know they were looking at that.

At just about the time I returned to Dryden, Paul Bikle retired as Dryden's director. I don't think he ever told me directly what his reasons were, but the general consensus throughout the years has been that he was willing to make one trip a year to Headquarters, but not frequent trips. However, due to the complexity of the space program, all center directors were expected to be much more visible at Headquarters than Bikle was willing to be. When Bikle retired, many of us thought that someone else at Dryden would become the director.

However, Headquarters appointed Lee Scherer, someone I had never heard of, as the new director of Dryden. I'd been at Dryden long enough by that time that I was uncomfortable with that decision. We'd always been a large family, and Bikle had been the head of that family. We knew what we could and couldn't do, and we had a good working relationship with the Air Force at Edwards. But our perspective beyond that, from my point of view, was somewhat limited. It was evident that Scherer had come from the space side of NASA and was probably well thought of at Headquarters. When he came out to Dryden, it was really the first time that somebody from beyond the point of view of airplane flight had been the director of Dryden. Walt Williams had certainly had that perspective, and Bikle had been well familiar with Edwards and flight-testing aircraft when he took over a few years before I came to Dryden. The working relationship became much more formal with Lee Scherer than it had ever been with Bikle. Bikle had spent a lot of time in the halls, on the hangar floor, and in the shops. He knew everybody. He knew who was good and who wasn't so good.

Scherer, we soon found, was a nice guy and easy to talk to—at least for me— but he had no interest in activities beyond management that we could detect. He had strong relationships with Headquarters and started building fairly strong relationships with the upper management at Dryden, the division chiefs at the time. But he was purely a manager. He made decisions based on the information given to him, not based on experience in flight tests or flight research.

However, with the Shuttle coming along as the big thing that NASA was going to do for the next few years, Scherer's appointment probably wasn't a totally bad decision from the point of view of what was best for Dryden. It just seemed a little bit strange and risky for those of us who were familiar with Dryden as being more of an entity on its own, rather than being a fully integrated entity within a space agency that had some aeronautics capability. I think that the little work on the Shuttle that Dryden was doing in 1971 when Scherer came probably set the stage for what Dryden would be getting as far as its funding for the next eight or ten years. We did have a good relationship with Johnson Space Center at the time, because a number of Dryden employees had gone either to Headquarters or to Johnson (some via Langley) by that time.

The space side of NASA, to me, was where the excitement was. While it was not my first choice of how to continue on in the runway-to-orbit arena, it was a choice, and it was a good one. It wasn't where I would have wanted the focus to be, but there were lots of problems that needed to be understood and solved. I saw it as a fairly rich area, and I was a little frustrated that Dryden was not a bigger player in trying to come up with what this concept would be. It was, I think, largely driven by Marshall and Johnson. How these things happened, how they were orchestrated, and how to make an input, were also completely outside my experience.

About 1973 or 1974, it became clear how the Shuttle was going to be done between the contractors and the civil servants within NASA and the DoD and where Dryden fit in. There was a period from 1971 to probably late 1973 when I was on the outside looking in, doing some analysis, and, I think, giving Milt some ammunition for use in his struggle to get rid of any concept that included landing engines for the Shuttle.

As I recall, still in 1971 the number one-option was to still have turbojets on the returning vehicle from orbit to the landing. Milt's and Dryden's experience had been that dead-stick landings of vehicles with a subsonic L/D max of three or greater were really easier without having to worry about landing engines. This was not only because of the complexity of the engine in terms of another system to fail, but also due to the extra weight and payload penalties of the engines. The engines would have to be protected during reentry, then deployed somehow to land. Fuel would also be need for them so that they could be used as landing engines. This represented a weight penalty of as much as 30,000 pounds (to assist only in the landing, with no cross-country capability), if we were looking at a mature design of at least two engines with fuel tanks, and probably a JP fuel of some kind, and all of the backup systems and thermal protection needed for those engines. The Shuttle design without engines barely made it as it was, and I believe if they'd tried to do the Shuttle with the landing engines, either they would have had to significantly reduce the size of the payload or they would not have succeeded with a design that closed at the time of manufacturing.

Through advocacy at the Air Force's Flight Dynamics Laboratory (AFFDL), which Dryden worked fairly closely with in the 1960s and 1970s, a proposal for a "stage-and-a-half" Shuttle was developed. That's ultimately what the Shuttle ended up being; rather than two full stages with two flyable vehicles, the Shuttle became one flyable stage with recoverable SRBs and a discarded external tank. I think that as others went through the same kinds of studies that Milt and I were doing from a fairly simplistic point of view, they could see that trying to make the first stage come back and land was probably not going to happen, because the total vehicle would be huge. It would also probably cost far more than Congress was willing to give us.

For quite some time after I returned, proposals were still being considered by engineers (although I don't think by management) as to how we might still do the Shuttle with a reusable first stage. As we went to the two-stage and then to the stage-and-a-half concept for the Shuttle, there were really only three candidates that got much effort put into them. The first one was the lifting body, which in the early 1970s I figured was by far the best concept because of its flight-demonstrated aerodynamic characteristics. The other two were the delta-wing and straight-wing concepts.

Advocates at that time for a wing body proposed a delta-wing planform with a cylindrical fuselage. This would be something on the order of a B-58 or an XB-70 (Fig. 3) shape. Among the advocates for a straight-wing vehicle was Max Faget. I've never been sure why that concept was pushed for so long, because I didn't know how they were going to solve the heating problems on that straight wing and on the tail. I'm sure they had a way to do that, but I didn't understand

Fig. 3. The XB-70A number one aircraft on an August 1968 flight to define its structural dynamics characteristics. (ECN 2127)

what that was. They also were violating another Air Force requirement that it have cross range of 1,200 to 1,500 miles to allow the vehicle to go to orbit, then deorbit immediately and land at the launch site. A cross range of over 1,200 miles is needed to be able to do that from Vandenberg, taking into account the Earth's rotation in the 90 minutes it takes to make one orbit. Faget's straight wing concept was advertising a very small cross range of two or three hundred miles. I never understood why that was thought to be a viable concept, even without regard to the DoD's requirement for the cross range. The DoD needed to launch from Vandenberg and land at Vandenberg once around, due to military missions. It had to be brought back to Vandenberg because, located as Vandenberg was on the Pacific coastline, there wasn't anywhere 1,000 or 1,500 miles out to sea where the Shuttle could land. We could bring the Shuttle down in Alaska, but then we'd have to figure out how, in addition to addressing weather issues, to get it from Alaska back to Vandenberg. So the DoD had this cross-range requirement that went with its large payload requirement. I agreed with the cross-range requirement, because from an operations point of view, it seemed right that the Shuttle ultimately should be able to land at its launch site with both its first and second stages, if it were a two-stage vehicle, or with just the orbiting stage as it is in the present Shuttle.

There's a perception that the straight-wing vehicle never got serious consideration because this concepts inherently had low cross range, but that's not true. It was true, however, for the way they proposed to fly the straight-wing vehicle, which was at a very high angle of attack where it had low cross range. The X-15, for instance, was a "straight winged" vehicle with a flight demonstrated L/D max of 2.3 at Mach 5, which would far exceed the 1,200 mile cross range, putting it into the same category as the X-24B and the Hyper III configurations of the Air Force Flight Dynamics Laboratory at Wright-Patterson AFB. Most of the high L/D max capability comes from the side farings on the X-15, which were put there to allow all of the hydraulics, electronics, and wiring to go around the fuel tanks that are the center fuselage. It is these side farings that give the X-15 an increase in hypersonic L/D max. I don't know what the hypersonic L/D max for the Faget straight-wing configuration was, but I suspect that if we could have figured out how to fly it at a low angles of attack, it probably would have had close to adequate cross range.

Although the straight-wing Shuttle concept was eliminated because it couldn't meet the cross-range requirement, I think many people were concerned that we'd have a very hot structure, and it was difficult to figure out how to do that with the technologies of the early 1970s. The lifting-body shapes hung in there for a while, but the huge size of the payload bay eliminated the lifting body for what, in my opinion, was the most unusual reason. The large payload bay took away the lifting body's volumetric efficiency. Now, total volume wasn't important. A particular volume was important, and that volume was 15 feet-by-60 feet. This volume also had to go near the center of pressure of the lifting body, which is traditionally in the neighborhood of 60 to 70 percent of the length of the vehicle. Putting the center of gravity of that large payload bay that far back on the vehicle meant that the boat-tailing of the lifting bodies and the control surfaces on the rear end of the lifting bodies were no longer compatible unless we made the lifting body really large. Consequently, it was the need to accommodate that very large payload shape that eliminated the lifting body as a Shuttle concept. The cross range of most lifting-body shapes was adequate for the DoD requirement.

If you look at the Shuttle, it's the optimum design for carrying something that's 15-by-60 feet. Once we determined the payload it was to carry, and where that payload must go over the center of pressure of the fuselage and delta-wing combination, we had the general shape, more the cartoon shape of the vehicle. Then, we tried to enhance its characteristics in the subsonic and hypersonic regimes. The basic problem was to develop a hypersonic reentry vehicle with the requirement that it land horizontally.

This was done by changing the cylindrical fuselage to one with flat sides.

With the vehicle reentering at a high angle of attack, the vertical tail and rudder are blanked out. The flat sides give the Shuttle stability and lift under these conditions. The size of the wing is determined by the weight it needs to support for reentry and landing, while the shape of the airfoil of the Shuttle wing is defined by the subsonic lift. The double-delta wing was determined to be—from a design, manufacturing, and aerodynamic point of view—the best trade off. The forward highly swept section was to accommodate hypersonic flow at the high angle of attack, while the forebody and its thermal protection was determined by the hypersonic, supersonic, and subsonic requirements of the Shuttle. The lesser sweep of the outer portions of the wings was for the supersonic and subsonic regimes.

Partly by understanding the various flow regimes, which wind tunnels were capable of doing in the late 1960s and early 1970s, little by little the shape of the current Shuttle evolved for aerodynamic or aerothermodynamic considerations in the hypersonic, supersonic, transonic, and subsonic flight regimes. Its internal structure, and to a certain extent, its shape was also due to the structural requirements of the ascent loads that occur on the Shuttle because of it being offset from the external tank and the solid rocket boosters. Finally, there were the structural loads on the Shuttle from up to 40 degrees high angle of attack above Mach 10, down to more or less horizontal flight subsonically with or without the payload in the vehicle.

There's really not a lot of designs that would accommodate all of those things. I by no means believe that the Shuttle shape is fully optimized, but with the uncertainties of materials and the aerodynamics and the atmosphere, to have enough robustness in the design, it was going to look something like the current Shuttle. The Soviets' Buran vehicle looks a lot like the Shuttle without the Orbital Maneuvering Systems (OMS) pods. The OMS pods probably shouldn't have been on the U.S. Shuttle in the first place, so the Soviet Buran is probably a better design. They benefited from our having demonstrated the flyability of the Shuttle. From two points of view, it isn't too surprising that the Buran and the Shuttle looked as much alike as they do. One is that we had already flown it, so it was an obvious place for them to start looking in terms of configurations. Secondly, almost everything on the Shuttle is there for a requirement of the aerodynamics or the structure, or in terms of the canopy, for the pilot to be able to fly the vehicle. When you put all those things together, it's going to look like the Shuttle. It's my opinion that the general Shuttle configuration is the only design in 1974 that would meet the payload, the cross-range, and the reentry requirements.

As a result, the changes in the Shuttle shape between the time when it was selected in mid-1972 until it had the final mold lines two years later, was due to thousands of people looking at the wind-tunnel data and the structural design, then refining the shape because it was not going to meet one of the technical requirements. As the Shuttle shape was evolving, Milt and I started looking at other issues.

We started focusing primarily at that time on the entry. There were others at Dryden who were helping Milt. I don't really remember who they were, but I knew I wasn't the only one that he was talking to, and it wasn't a big secret that he was concentrating in this area. Rather than being a team, we were really a group of consultants to Milt, and he knew who understood hypersonic aerodynamics, structural loads, and heating from the X-15 program as well as other things that we'd done. I'm sure he picked the brains of many people in terms of their experience and what they knew how to do.

After I had looked at the initial trajectory optimization and reentry profile of several concepts up into 1973, I started writing computer programs that would basically be what we call batch programs. I could try to fly different vehicles from orbit to landing by putting in predetermined flight conditions and control inputs. These were what angle of attack and Mach number combination got us from some point near orbit at Mach 28 down to a landing on the dry lakebed at Edwards. Looking back at it now, I believe Milt probably had conversations with people at Johnson, Marshall, and Langley as to what Dryden's contributions and responsibilities would be for the Shuttle Orbiter.

We were also talking with the Air Force, because I think all of us knew that any vehicle like this was going to end up landing at Edwards. However, it may not have been quite as clear as to what we might be doing other than watch it land. There were advocacies by Milt and others, and I am sure Lee Scherer came into play, about how we might carve out a piece of the pie for us to do under the direction of others and about which pieces of the pie that we might be lead center on. At that point, we didn't have, at least as far as I knew, an official Shuttle project. It was still a loose collection of engineers interested in working on various aspects of the Space Shuttle Orbiter.

Writing the Aerodynamic Data Book

This process became a formalized official procedure for determining the baseline data, the actual predicted value, and the tolerances and the variations about the predicted values. That process was the first time that I'd been involved in a democratic organization with respect to engineering data. We had people with experience in various kinds of flight vehicles, various kinds of wind tunnels, and various techniques for obtaining wind-tunnel data. We then formed groups to look at all of the huge amount of predicted data. It was always summarized in some form before we got it. Then we would all go over it. Sometimes this was at our own locations, and sometimes we would meet at a common location. It was usually done at Dryden, at Rockwell in Downey, California, or at the Johnson Space Center in Houston, Texas.

The process started up in 1974, and was fairly formalized by the time we flew the Shuttle Approach-and-Landing Tests (ALT) in 1977. I said earlier that the process was democratic because we would each have our own take on wind-tunnel data and what wind tunnel it was run in. What size was the model? What was the Reynolds number? How good was the data in that kind of tunnel? And finally, what kind of flight data did we have that we could say was or wasn't well predicted by those tunnels? After much discussion, we would vote on what we thought the mean values should be, the predicted value that went in the data book, and what the tolerances and the variations should be for that predicted value. It was a good process because we had varying experience levels in both wind tunnels and flight test. They decided that some data should be quite different than I thought it should be. Since it was a democratic process, each of us had to accept the vote of the committee. It was a very credible group of people, and a lot of detailed work had gone into coming up with the presentation of the data on which we were voting.

Joe Weil and Bruce Powers made efforts throughout Dryden to take every piece of flight data—specially at supersonic speeds with a delta wing where we had extracted aerodynamic information—and compared that data with the wind-tunnel predictions available for that flight. [45] Then, we tried to come up with what we thought the expected, reasonable error might be. These determined what were called the variations, based on past flight to wind-tunnel comparisons. A similar activity determined what were known as the tolerances, one of many such activities that went on among the wind-tunnel people who were familiar with a wide range of wind tunnels and tunnel-testing techniques. The variations are usually larger than the tolerances. [46]

It was an interesting process because it wasn't something that an engineer normally would vote on. Instead, an engineer would do an analysis and perhaps an error analysis to accompany it, and that would be all. But in this case, we were flying a delta wing at Mach 28 down to Mach 3. Delta wings had not flown anywhere above Mach 3, so that was unknown territory. The only actual reasonable flight data that we had above Mach 3 was for the X-15, but its configuration was quite different. As a result, lots of engineering spin was put on the analysis and predictions to come up with something useful for us to put into a formalized aerodynamic data book. Everybody who worked on the project used this book for design of the vehicle, the systems, the loads, the heating, and the control system. All of it was revised several times. The data book started with trying to get a large group of people to collaborate on flying a type of vehicle that had never been flown before and that we were eventually going to be flying in regimes where no one had ever flown any vehicle before.

We had official Shuttle technical panels, groups of people from various NASA centers who sat in on weekly—sometimes more often than weekly—telecons. The telecons included people at Ames, Langley, Lewis, Dryden, Johnson, and Marshall, as well as people from various parts of Rockwell. These flight panels included people in various specialties. There were also ones that tried to specify the instrumentation needed to obtain the data so we would be able to explain the flight behavior. There were other panels for interpreting the aerodynamic data from flight or from wind-tunnel predictions. As the vehicle got more refined, we kept getting more and more wind-tunnel data. As we approached the ALT program's first flight [47, 48] we had a flight-test panel as well that coordinated all flight-related items.

I don't know how many total technical panels there were on the Shuttle, but I participated in four of them over the years. We used that process to create an official traceable set of data on which all design would be based. Most aircraft designers do what we did but do it more or less formally. I don't know that they do it as completely as it was done on the Shuttle, because building an airplane usually does not involve building something that is drastically different from what has been flown before. Consequently, there is a trail that can be followed. Aircraft designers have official data books as well, because every design, bolt, piece of metal, and control actuator has to be traceable to the official configuration-controlled set of data. In this case, it was system, structural, aerodynamic, and aerothermodynamic data.

VII

Getting Ready to Fly

Once it was determined that the Shuttle Orbiter would be a delta-wing-body combination, the specific issues of how to build the entire vehicle, get it into orbit, and bring it back were somewhat defined. Since there weren't any liquid-fuel rocket motors available for use on the Shuttle, a completely new engine, called the Space Shuttle Main Engine (SSME), that pushed the state of the art was to be designed and built by Rocketdyne. This engine had to be designed, built, and tested to be sure it wouldn't fail, because it would be making multiple flights. Other than the X-1 through X-15 engines and the RL-10 engine, liquid-fuel rockets were used on expendable vehicles and they weren't reused. The requirement of reusability determined that a new design was needed for the Shuttle.

Another technology that needed to be studied and matured was the thermal protection system (TPS). Was the Shuttle going to have a hot structure? Was it going to have ablatives? Was it going to be all metal? Those of us on the aircraft side of the house really wanted the Shuttle to be built with generally standard aircraft construction techniques. It was to be reusable, making it more like an airplane than a spacecraft. It had to carry aerodynamic loads and be flyable, too. We had a lot of experience with internal aluminum structures. The X-20 DynaSoar was to have had a hot structure all-metal vehicle design, but there were too many problems with this to solve at the time. Fairly adequate ceramic-coated materials that would accept the heat and radiate it away were also available. The ceramics could be bonded to the aluminum fuselage structure. It was fortuitous that the ceramic technology was just maturing about when they needed to decide. They did run into many problems as they manufactured and tested the thermal protection system, but they made the decision to use the ceramic technology because that would make the Shuttle much lighter than would anything else available.

These were the key decisions required for the Shuttle to be a stage-and-a-half reusable vehicle. The Shuttle weight and exact shape were fairly well defined by 1974. I don't know of any major mold line changes made after then. During the development process, NASA decided that the vehicle would be built and tested subsonically in the Approach-and-Landing-Test (ALT) program. [47, 48, 49]

Justification of the Shuttle Approach-and-Landing Tests (ALT) Program

In the ALT program, a prototype Shuttle Orbiter would be air-launched at 0.7

Mach and fly down to a landing. This decision was made, I think, because there was time in the schedule to do it and because it would allow us to test the subsonic aerodynamics of the vehicle, its subsystems, and some of the software in-flight. It also provided the opportunity to calibrate the air data probes on the side of the Shuttle with a flight nose boom probe. You can't fly a boom through a reentry because it would burn off because carbon-carbon nose cap, where a boom would be located, is the hottest part of the vehicle. The nose boom would give us the air data, which is very important in all flight regimes, but is especially important during the approach-to-landing phase. This is because the vehicle is maneuvering and the flight conditions, such as angle of attack, altitude, and velocity, must be measured. Measuring those really starts with the air data system, which measures dynamic pressure, velocity, air density, and angle of attack. It would've been nice to have had angle of sideslip, but because the probes were on the side of the vehicle, that wasn't available easily. The ALT flights were flown to demonstrate various maneuvers and landing [47, 48, 49]; obtain subsonic stability-and-control and lift-and-drag information; and test the structural design, the loads, the actuators, and some of the software.

There was yet another reason for the ALT program, of course, which was that the Shuttle was to be launched from Cape Canaveral in Florida and landed at Edwards Air Force Base (AFB) in California. We had to have return the Shuttle back to Kennedy to fly it again, so we had to develop a way to move the airplane. Getting it to the Pacific Ocean, about 80 miles from Edwards, to move it by sea was considered. To do so, it would be put on a barge and taken through the Panama Canal to Cape Canaveral. However, there are mountains between Edwards and the Pacific, as well as a lot of people, increasing the difficulty of doing this.

Because it was too difficult to get it from Edwards to the ocean, it was quickly decided that we needed a way to fly it there, to mount the Shuttle on a large aircraft and fly it to the Cape. There were a number of proposals for doing so. Most of the early proposals involved two fuselages with a common wing between them, where the Shuttle was mounted. I think there were several reasons that seemed to be the natural way to do it, at least from my point of view. In particular, the air launches would be done with the Shuttle under the mothership, making it easy to avoid recontact on the launch.

I do know they looked at two C-5s with a special wing between the two fuselages. I think they looked at two 747s, too, as well as other pairs of aircraft, trying to optimize the carrying ability while minimizing size and cost. Somewhere along the line somebody decided we could launch it from the top of a 747 or a C-5. To those of us who worked at Dryden, that seemed awfully risky, and we asked why we should do it that way when we knew how to do it by dropping it from the mothership so well.

I believe Boeing came up with a proposal showing how they could mount a Shuttle on top of the 747 with minimum modification of the airliner. The proposal also showed that the basic 747 had enough lift and propulsion to be able to ferry the Shuttle from Edwards to Kennedy. In the end, the 747 was selected.

The ALT program was implemented using the same 747 Shuttle Carrier Aircraft (SCA) that would ferry the vehicle from Edwards to Kennedy. The Shuttle would be mounted at a higher angle of attack than the 747 SCA to help the two vehicles separate at launch. This meant that there were three tasks on which Dryden and Edwards AFB could make major contributions. First, they would oversee Boeing's modifications of the 747. Second, they would be involved in the mating of the two vehicles, putting the Shuttle on top of the 747. Third, they would oversee flight-testing the Boeing-modified 747 without and then with the Shuttle attached. The flights without the Shuttle attached were to be flown in Washington state and the flights with it were to be flown at Dryden

It was first necessary to flight test the SCA to make sure that it satisfied the design requirements. Next we would test the two vehicles mated together, first in the ferry configuration and then in the ALT configuration. The differences between the two configurations were the angles of attack or incidence and the length of the forward attach point structure. The attach points on the SCA were matched to the Shuttle external tank attach points.

In addition, the ALT program also would launch the ALT test vehicle Enterprise, obtain aerodynamic data, and demonstrate the Shuttle horizontal landing capability. This last was a fairly massive exercise in itself. Putting the Shuttle on the 747 was also a challenge. Once we decided we would carry the Shuttle on the 747 SCA, we had to lift the Shuttle Orbiter and gently set it down on the 747 attach points. The structure we designed for this task was called the mate-demate device or facility.

The engineering tasks of modifying and testing the 747 SCA, performing the flight tests of the mated pair, and flying the ALT flights were the elements of a major flight operation for both Dryden and Air Force Flight Test Center (AFFTC). There were other organizations involved, of course. Nothing went on with the 747 SCA or Shuttle that Johnson or Marshall wasn't in charge of, even though Dryden's facilities, technicians, engineers, and shops were used throughout the ALT program. Working on this program became my focus from 1975 to 1977.

These were also the primary tasks Dryden had signed up to do at the time. I, like many other people at Dryden, was a member of various technical panels set up to communicate among the various partners involved in building and flying the Shuttle. Because Dryden is a fairly small place, the ALT program was a major endeavor for us, requiring a great deal of cooperation with Marshall and Johnson, Edwards, and Rockwell. Several of us also ended up with tasks related to the ascent of the Shuttle. Al Carter at Dryden took the lead in the structural analysis of ascent issues. I was looking at the flight dynamics, to determine the aerodynamic characteristics and how they might differ from predictions, particularly for the "stack" of the two SRBs and the Orbiter attached to the external tank. For the ascent I used the same techniques that I'd used for earlier parameter identification with the X-15 and the lifting bodies. The Dryden contribution to the ALT program was extremely large, second, I think, only to the X-15 in terms of manpower.

The interplay between the airplane and spacecraft aspects of the Shuttle affected the types of people and their experience working on the Shuttle. It ended up being difficult, in many ways, for the aircraft-oriented people to work with the expendable-rocket people and the spacecraft people. The environments and the perspectives are far more different than might be expected from thinking only of the similarities. Each of us was concentrating on an individual aspect of these various programs and all of us had to be aware of what was going on with the Shuttle Carrier Aircraft, the Enterprise, the mate-demate facility, and the Orbiter itself.

For many of us the Shuttle work very quickly went from a small part-time effort to a very large effort, because it had a very fast schedule. We had to develop techniques for large vehicles because we were accustomed to small vehicles, such as the X-15, M2-F2, and HL-10. Some people had to be retrained. Some people were reassigned briefly to other facilities to work with people at those facilities. Fairly early on I started working with Rockwell at Downey, California, and discovered a former Dryden employee, Joe Baumbach, at Rockwell working on the Shuttle program. The ALT team was spread over three NASA centers (Marshall, Johnson, and Kennedy), three NASA research centers (Ames, Langley, and Dryden), Rockwell, and a number of groups involved in Shuttle construction and subsystem components. As a result, the Shuttle had a lot more people than the three or four hundred that had worked on the X-15, spread from coast to coast. It was fortuitous for me was that the Shuttle was being designed at Rockwell in Downey and built at Plant 42 in Palmdale, both fairly close to Dryden.

Because Rockwell got the contract, I could again watch how a company did the design and construction of a new vehicle. My experiences of this, of course, were with Northrop, on the M2-F2 and the HL-10, in more of a skunk works environment. This was going to be a very large project, involving a hundred thousand people in total, and the vehicle was going to be designed about 80 miles from where I lived, so I could participate in some design reviews and discussions there. I was impressed with the overall leadership at Rockwell and with the people I got to watch and work with, including Don Schlosser in aerodynamics and Guy Bayle in controls. Some of the people involved with the design, of course, were at Dryden and many more were at Johnson and Marshall, but the actual primary detailed design of the Orbiter itself took place at Rockwell. I got to work closely with those people through the construction of the Enterprise in Palmdale at Plant 42, in Palmdale, California, and then through the greater effort of designing and constructing Columbia. This was quite a contrast with what I'd done with the skunk works kind of lifting bodies in the 1960s, and something I'd get to do again in the late 1990s for the X-33, which was designed and built at Plant 42. In the case of the X-33, I got to see how yet another company, Lockheed Skunk Works, did the design and fabrication of a vehicle. The X-33 didn't fly, but the vehicle was built, except for the liquid hydrogen tank.

I felt fortunate to be able to watch these three independent approaches: the truly skunk works environment that we had on the lifting bodies with Northrop, with a small team and collaborative effort; the totally non-skunk works environment on the Shuttle, with tens of thousands of people and lots of paper floating around; and then back to the real Skunk Works environment at Lockheed. On the X-33, which was once again a project distributed all over the country, the work was more like what I'd observed at Rockwell doing in the 1970s than what we had done with Northrop in the 1960s. I found these to be interesting contrasts.

Getting the Data

As with any new vehicle in flight-test, one of the first requirements is to specify the types of sensors needed to make measurements, the location of those sensors, the dynamic characteristics of the sensors, and the accuracy of the sensors, as well as how the information from the sensors gets to the user. In other words, how does the information get put into engineering units so that the person can see the control, performance, or loads data? The requirements are primarily formulated by each discipline. Some sensors are shared by many disciplines, but different disciplines often have different accuracy or sample-rate requirements. An example of this is control surface positions, used by the controls discipline and by the stability-and-control estimation specialists.

On the Shuttle, there were groups who already had specified things such as control surface positions. They wanted, first of all, to get information about what position a control surface was in. Because the airplane had a feedback control system, the control group was going to use the same onboard data for control surface positions to drive the control system that they were using as measurements. This meant that they were having that data sampled frequently. I got the control positions from them for my stability-and-control estimation because our requirements were the same. The air data—angle of attack, angle of sideslip, dynamic pressure, and velocity or Mach number—were specified by people working primarily on measuring lift-and-drag characteristics. Their requirement was primarily pseudo-steady-state information, so they requested lower sample rates than I wanted. I didn't get to specify additional, independent sensors, so I had to live with the low rates. Other data that were specified by me, or shared from others' specification, were the rate gyros, accelerometers, various intermediate control movements, and the attitude of the vehicle. The altitude came both from radar and static pressure, used to derive the density and pressure altitudes.

My first problem was that the information I needed was coming from three or four different sources. Those different sources had different qualities of information, different accuracies, and different sample rates. Moreover, they weren't all timed to the same master clock. This meant there were time skews from one system to another to account for. We didn't have much control over the instrumentation system that we had specified, but we knew we had to live with it, and we did get a pretty good analysis from it.

Because we really wanted to do some research as well as get flight-test data out of this, we advocated a separate instrumentation package that had very sensitive, very accurate three-axis accelerometers and three-axis gyros. This was called the Aerodynamic Coefficient Identification Package (ACIP) system. Alex Sim from Dryden was detailed to the Johnson Space Center for one year in 1975 to help meld Dryden and Johnson in some of the flight-test areas. Alex had done much of the parameter identification analysis on the lifting bodies. He alerted Johnson to the lack of flight dynamic measurements and insufficient accuracies, especially longitudinal acceleration (AX). It was through his efforts that Johnson and Dryden defined the ACIP package and lobbied for it to be put on Enterprise.

The ACIP package was not quite as good as we asked for in the ALT program, but that was still where we got our best dynamic data for the analysis that we were to do after the ALT flights. We could've gotten better data with a better instrumentation system, but that's the lament of every researcher. We can never have too many sensors or too much data. Most engineers know as soon as they see a researcher that he is going to want to put 10,000 pounds of instruments into their vehicle that's already overweight. There's a lot of truth in that stereotype. I've never had a vehicle that was adequately instrumented for what I wanted to do, and I doubt if any other researcher has either.

With the asynchronous Shuttle instrumentation systems that we were working with, one of the things we had to figure out was how to merge the data. In the parameter identification that we do in the aircraft world, we need to know, within a millisecond or so, when a measurement was made—not when it was recorded, but when the measurement actually occurred. Because the techniques are very sensitive to time skews, a time skew of 5 to 10 milliseconds will start to mess up our analysis. In addition to needing to keep track of when within each data frame the sample was made, we had to merge together data running asynchronously at different sample rates with different clock settings. Rich Maine took on that task, developing techniques used all the way through the Shuttle program, techniques that are still used on all of Dryden's current flight programs where we need to do any sort of data massaging.

We knew we weren't going to get very many maneuvers from five to eight flights, only two or three of them with the tail-cone off, so we needed to get good data and get it in a hurry. The software that Rich developed is what everybody analyzing ALT data ended up using. Because it was such a good clean code, it also ended up being the official system on the first nine or ten orbital flights. The official tape from Johnson that was generated for the orbital flight had significant errors in it. They had to go back and rework their software to get it into agreement with reality. We used Rich's software there, and it really saved our bacon during the first Space Transportation System (STS) flight.

In defining the maneuvers that we wanted flown on the ALT flights, we ended up working with a group led by Rick Barton at Johnson that still exists. One of their young engineers at the time, Doug Cooke, was detailed to Dryden for, I believe, about six months to work with us to learn the art of parameter identification so that the Johnson group could assess this independently and learn how to analyze and interpret the data. Because Johnson was primarily a spacecraft center, people who worked at Johnson really didn't have experience working in lift-and-drag and stability-and-control flight data. They may have had the information and understood the process generally, but they didn't have the flight experience.

So Doug Cooke came to Dryden, and he worked with us to learn the tools and work on data extraction. I think he actually was assigned to the F-8 digital fly-by-wire program. He supported their work and got to know everybody at Dryden. He really helped with building the team, because when he went back to Johnson, he knew the people at Johnson we needed to talk to, and he knew the people at Dryden that the Johnson people needed to talk to. It worked out very well. He analyzed the data on the ALT program with us and continued on with us, doing analysis, through quite a few of the early orbital flights on the Shuttle before he went into Johnson management. He became the head of the manned Mars program at Johnson.

With the ALT program, we were starting to create the model of how we would work on the much more complex orbital flight test (OFT) program. The OFT was the first four orbital flights (OFT-1 through OFT-4), also called STS-1 to STS-4, which stood for Space Transportation System. The ALT program was very good experience because we had people analyzing the same data at Dryden, at Johnson, at AFFTC, and at Rockwell in Downey. Consequently, we really had four groups analyzing the data independently and Dryden was also working with Langley on the lift-and-drag analysis. We were all using the same data, and most of us were using the identification techniques that I had developed back in the 1960s. I was doing the analysis at Dryden as did Rich Maine, on the ALT and the first few orbital flights. I think the Air Force was doing a little independent work using analog matching and trying to understand some of the nuances of the data, and they also used the Maximum Likelihood Estimator to some extent.

The same four groups each had one or more representatives who worked on the Shuttle on the flight-test panel, so we had people from other centers participating in the weekly telecons. The other panels that I was on, the instrumentation and the aerodynamic panels, also had frequent telecons. It was nice to be able to do that work over the telephone. If we had charts to go over during a telecon, they would be mailed to each person in advance or faxed if they were hot off the presses. That was before we could send charts via the Internet. Sometimes we knew each other only through the panel telecons, but after working with people, hearing their comments, their reactions, their approaches to problems, their reactions to solutions, we got to know everybody and had them fairly well calibrated during the ALT program. The only way to do the work that we were doing was with a well-focused and dedicated team where members really trusted each other and had some way of reaching consensus on things. Some of the things that we worked on were really flight-safety issues that we needed to fix or change before the next flight, while other things we were working on could be put on the back burner until there was more time for them. That was how we worked the flight-test conditions we wanted into the program, the types of maneuvers we wanted, the kind we required, and the minimum number of maneuvers that we needed to certify the values of stability-and-control derivatives or the lift-and-drag numbers.

I think all of the panels had at least one, if not several, members from Johnson, who coordinated with other people at Johnson and elsewhere to get the information flowing to everyone who needed it. One of their tasks, once we started building our flight-test requests in terms of maneuvers, was to get our inputs to the people who did the flight-planning on the Shuttle simulators at Johnson. They worked the maneuvers into the training of the two ALT crews. The crews then flew the maneuvers manually.

One of the other really good things about the program is that Fred Haise and Gordon Fullerton had been assigned to be the prime crew on the ALT program, Joe Engle and Dick Truly were assigned as the secondary crew, and the two crews (Fig 1) would switch back and forth for each flight. So I had the opportunity, 10 or 11 years after Fred had left Dryden and after he had been on Apollo 13, to once again work with him in a completely different capacity. He was the pilot of the program that we were going to fly. It was a first flight, a dangerous

Fig. 1. The four astronauts who flew the ALT missions posed in front of the Shuttle Enterprise after the vehicle rollout. Pictured from left to right are Gordon Fullerton, Fred Haise, Joe Engle, and Dick Truly. (EC76 5893)

thing to do, an exciting thing to be involved in, and, therefore, a big deal. Even though I didn't work with Fred one-on-one, as I had back in the lifting-body days, he always was very hands-on, very involved in understanding all of the subsystems of the Shuttle and all of the eventualities and contingencies. All training for the flights was done on the simulator at Johnson, not Dryden. What Fred did at Dryden in terms of training was in the Shuttle Training Aircraft, which was a business jet configured by modifying its electronics system to approximate the flight characteristics predicted for the Shuttle, in order to give the pilot practice diving at the runway to a simulated dead-stick landing.

We got to work not only with Fred once again but also with Gordon Fullerton, the pilot who flew with Fred on three ALT flights. Gordon would go on to become a Dryden research pilot after making two orbital Shuttle flights. Both Fred and Gordon were well-known to Dryden. Joe Engle had, of course, been an X-15 pilot, and we knew him from that program. A lot of Joe's old cronies were still around and he, being very extroverted and outgoing, made new cronies, too. Dick Truly, the pilot who flew with Engle, eventually became the Administrator of NASA between 1989 and 1992. Overall, it was a stellar crew, one that I had a great deal of confidence in, both for their piloting skills and for their ability to gather the data that we wanted to analyze.

The Enterprise had the same mold lines as the orbital Shuttle but had fake tiles that approximated the actual configurations. Actual tiles were also used at several locations on the vehicle, being flown on the ALT to get information on tile installation, durability, and response to flight loads on a vehicle. The actual tiles were on three different parts of the vehicle: up near the nose, near the leading edge of the wing, and along the bottom. The Enterprise, used for the ALT program, didn't need the thermal protection system because it wasn't going to get hot coming down from 20,000 feet to the runway at subsonic speeds.

One of the advantages of the ALT program was that we got to see how the vehicle performed, in a controlled manner from 20,000 feet down to landing, but we also got to check some of the subsystems and software and resolve some of the operational issues of how to handle things, such as who was going to be responsible for what and how we would communicate. These were details developed and enhanced by the ALT program.

Doing all of this on the Shuttle reminded me of what I'd done on the lifting bodies and the X-15. We had data requirements and a way of inputting those requirements. They were then prioritized and passed on to the people who did the flight-planning. The pilots and the flight planners practiced them together and had various go/no-go situations where they would or would not do the maneuver, depending on other things that might happen. Then, when the flight was flown, we got the data from various sources and combined it into a single set of information that each particular discipline could analyze. I think it was done this way because this was the Dryden way and it was being flown at Dryden. Most of the people at Rockwell hadn't been involved in a program like this before, at least the people I worked with. Some of the more senior people, obviously, had been on other programs, probably with other companies. The people at Johnson, for the most part, didn't have flight-test experience either, so they accepted the Dryden's method sometimes modified.

As Rich Maine and I prepared for the ALT flights, others at Dryden had tasks similar to ours on the 747 SCA aircraft and on the mated pair of the 747 SCA and the Shuttle. We helped them, too, because we were, obviously, interested in their part of the program. However, the two of us would have been spread too thin if we tried to try to adequately cover the SCA program as well. I was still working on the 3/8th-scale F-15 at that time as well. The people Dryden had doing the parameter identification work on the SCA were not as experienced with it as Rich and I were, so we helped them with their analysis and their advocacy of maneuvers. However, they had a different schedule than did those of us involved with ALT program, because they had to finish their job before we could do ours.

I think one of the reasons the ALT program went so well was the relationship that developed between Deke Slayton and Milt Thompson. From Johnson, Deke was in charge of the whole ALT program. He was a good pilot, a good communicator, and an excellent manager. He was very much like Milt in that regard, for Milt was an outstanding pilot and a good engineer. When Milt got into management, he got the right people to work on the problems. I saw the same characteristic in Deke.

I gained a lot of respect for Deke's capabilities watching him manage the complex issues from center to center and from contractor to center while keeping things pretty much on schedule. Deke was hard to budge unless one had a really good reason but he would support something with a good reason. Everything worked out well on the program. I think this was largely due to his management skills and, from Dryden's point of view, the mutual respect that Milt and Deke had for each other. They certainly didn't agree on everything. I remember quite a few discussions where they were obviously on opposite sides. I don't know if either of them ever changed his mind, but on the important things, they managed to agree.

That was when I discovered that the best way for me to make an engineering input into the program was to talk to Milt. If Milt agreed, or at least thought it was worthy of further consideration, he'd talk to Deke. That meant it got brought up at that level without my having to go through the official system of panels and reviews, discussions, and meetings. This wasn't really going around the system, and I only used it for things that I thought were being missed elsewhere because the experience levels were so different. With Milt and Deke involved, the program addressed issues that I thought needed further consideration so they could be changed early on if necessary.

Another change occurred at Dryden during the buildup for the ALT program. After Lee Scherer was reassigned as the director of Kennedy Space Center and Dave Scott was appointed director of Dryden in his place. They were both good guys to have at Dryden as we learned to work with the space side of NASA. Scott, in particular, had a reputation because he had walked on the Moon. He was different to work with than either Bikle or Scherer had been, but he was a very intelligent guy, quick to see a point. He didn't always change his mind, but he did understand your point. He was very interested in seeing the ALT program through. He resigned right after the ALT program was completed and went into private industry. I worked with him as a consultant for a while after he left Dryden.

Flow Fields and Separation

As we were preparing to fly the captive-carry tests, Milt and I talked about the air-launch. Neither of us had experience with launching off the top of an airplane instead of off the bottom. We were concerned whether the people doing the analysis had really taken into account everything that might happen with two differently shaped vehicles in such close proximity. We were worried that the uncertainties in how the lift of the Enterprise, when in the flow field of the 747, were accounted for under the worst possible conditions. We began asking questions to be sure that the Shuttle would clear the tail and not recontact the 747. We weren't satisfied with the answers we got, so I took an independent look at the problem, which I thought was really important because any incident definitely would put a big delay in the Shuttle program.

I went through the analysis, repeating some of it independently, and concluded that the air-launch from the top of the 747 was not going to be a problem. Still, never having seen it done before, I was very nervous right up to the first ALT flight. I worried about whether those vehicles really would separate the way all of our analyses had shown, or if we'd all missed something and they would recontact for some unanticipated phenomenon. However, the photo coverage showed that the vehicles separated very smartly, especially with the tailcone off, and the Shuttle was clear of the 747's tail within a second, and well ahead of it. As it turned out, Milt and I had no real reason for concern, as the various analyses had been done competently, but we did worry about it.

Doing the wind-tunnel testing of the two vehicles in proximity to each other was almost like making a claymation movie. They mounted the two vehicles the equivalent of a foot or so, in scale, apart and then measured the flow fore and aft at that distance, looking for interference effects. When they did that, we ended up with good ol' wind-tunnel data. With the uncertainties added on top of the flow characteristics of the 747 and the Shuttle, determined in these tests, there seemed to be no way that the two would collide in midair.

For years after that, when Milt and I were discussing something new that somebody was going to do, we'd get to chuckling about how unsure we felt about the physics of the launch because we weren't directly involved in the work. We both had to learn to trust people to do the analysis, to trust that they would come out with the right answers. We were both a little embarrassed that we'd spent the time and doubted people. It probably irritated them that we questioned their analysis. It certainly would have irritated me if somebody was questioning mine. Most of us who are involved in the flight business realize that any challenge from any credible person really needs to be examined, because we all can get complacent and sometimes overlook the obvious, or the not so obvious. Although it was proper for us to do so, later we were a little embarrassed that we'd been concerned about a close call when it wasn't anything to be concerned about.

The approach to getting the two vehicles to separate was quite different from the approach we used on the B-52. The B-52 keeps flying along, basically, while the airplane that's dropping away is at less than 1g, so there's no way they can recontact. We just had to convince ourselves that inside the flow field where they're flying, there's no tendency at all for the X-15 to recontact the B-52. If you watch movies of the X-15 or a lifting body dropping away from the B-52, you can see that the separation is very positive and very much downward, easy to have confidence in.

However, with the Shuttle ALT launch technique, we wanted to put a lot of lift on the Shuttle to get it out of the way of the mothership. Through various analysis techniques, we decided the best way to make sure that everything worked out was to have the mated pair climb to a higher altitude and then, just prior to separation, push it over into a slight dive, increase the air speed, throttle back the 747, and release the Shuttle. The release call was made by the Shuttle pilot, who pressed a button that fired explosive bolts to separate the Shuttle at the three attach points. Once that happens, we don't want the Shuttle to stall because then it would roll too close to the 747. We wanted a lot of lift, but not so much lift that we got into a stall. If the increase in lift (not the lift itself, the increase) as the angle of attack increased was reduced, or became negative, the Shuttle would stall. When this happened the flow around the vehicle would make it roll in one direction or the other (the direction can't be predicted) and the Shuttle might contact the 747 SCA.

For the Shuttle to separate cleanly from the 747 SCA, the Shuttle has to be at a significant angle of attack, or incidence, relative to the 747. For the ALT flights, the nose mounting post was longer than the two posts at the back, giving it the higher angle of attack. I think the angle was about eight degrees. In the ferry condition it's about three degrees. That eight degrees gave the Shuttle a good amount of lift. The launch was at a fairly high dynamic pressure, with favorable flow above the 747. The flight went pretty much as predicted. The two vehicles separated, one vehicle rolling one way, the other rolling the other way, providing a lateral separation in addition to the vertical separation.

That was the technique we worked on and honed in the simulator and the pilots practiced in the captive-carry phase. The dive let us check the loads on the attach points and certain characteristics of the Shuttle in terms of vibration or any unsteadiness in the flow. We got flight test data, to add to the predictive data, from the captive-carry test by putting the vehicle in a slight dive and doing everything that we were going to do in the real air-launch except the launch itself. After the landing, we looked at the data and, if it was what we expected, we could go on to the actual release on the next flight. This build-up process was very similar to the one we used on the B-52 drops, using the captive test in a way similar to the taxi test with conventional powered aircraft. That the Shuttle was launching off the top of the mother ship didn't affect the procedure we used to verify that it was under the expected loads. The loads measured on those attach points tell us what the lift of the Shuttle is at the instant it is released. If there are no big changes in lift between initial and full separation, the separation will go as planned.

Obviously the flight tests were performed conservatively, but some risks must be taken or no progress will ever be made. The concern with the separation was concern for the 747 crew, because if the two vehicles did recontact they would have no way to escape. If the Shuttle took off the vertical tail of the 747, for instance, the 747 would go into a spin, making escape very difficult. A big hole, with a hatch, was made in belly of the 747, so that, if the 747 crew needed to escape from the vehicle, they could leave their seats, run back to the hole, open the hatch, and parachute out of the 747. I don't think any of the 747 pilots were terribly confident that this was going to work. If they ever really needed to bail out, the 747 would probably be going through such gyrations that they would be unable to reach the hatch. Nonetheless, the hatched exit was put in the airplane, despite the expense and probability that it wouldn't work, because we have do as much as we can to make a vehicle safe and survivable. The crew comes first.

For the ALT program, the Shuttle had two ejection seats: one for each of the pilots. If, despite the predictions, the Shuttle did smash into the 747 or have a failure on its own, the pilots could eject. This is the standard technique of first making the flight as safe as possible while still flying it and providing as many ways to have the crew survive, even if the vehicle is lost, as possible. Practicing the separation maneuver with the two vehicles mated, so that we could get data and look at it before we actually did the separation, was a part of making the flight as safe as possible.

I remember watching that last captive-carry flight carefully, to be sure that during the simulated launch the loads on those three mounts showed that the Shuttle was not going to pitch up or roll over in the actual launch. [48] My examination of the captive-carry data gave me confidence in the actual launch. If everyone did the actual maneuver as they had practiced on the final captive-carry flight, the Shuttle launch would be safe.

Of course, that's only the first issue of flying a first flight on a vehicle of an unusual configuration with no engines. The 747 crew was as brave as the Shuttle crew, doing this first separation. Once the risky separation was completed and the two vehicles were clear of each other, the flight plan proceeded as it had on the first M2-F2 flight, with the crew doing the practice flare at altitude, putting in some little maneuvers to make sure the vehicle was moving around as it should, and then flying the energy management approach and flaring over the runway to touchdown.

VIII

The Approach-and-Landing Tests

The first big event for getting ready for the Approach-and-Landing Tests (ALT) flights was bringing the Enterprise from Plant 42 in Palmdale to Edwards AFB, then to Dryden. Because the Enterprise is a big aircraft, in preparation for the move, the roads were widened and telephone and power poles moved off to the side. A lot of planning and careful work went into making sure that this was done with minimum impact on the residents of Lancaster and Palmdale. Tenth Street East (now called Challenger Way) in Palmdale and Lancaster ran from Plant 42 north to the Edwards boundary. This was the road used to move the Enterprise, which was the ALT aircraft, to Dryden. A road was then constructed on Edwards property, which ran from the end of Tenth Street East to Rosamond Boulevard. This was much easier in that we didn't have to worry about the impact on any residents other than the desert tortoises, jackrabbits, and sidewinders. The road bed was extended, widened, and made as smooth as possible. Then a gravel coating was added to the roadbed all the way up to Rosamond Boulevard, the main road that comes onto Edwards from Rosamond, and the same road on which Milt Thompson used to challenge me to race in earlier days. Once it got to Rosamond Boulevard, the Shuttle could be brought onto the heart of Edwards. Once it was on the taxiway, the Enterprise then could be brought to Dryden.

Those of us who had been working on the program were very much looking forward to the arrival of the Shuttle. The mold lines were essentially the same as those of the Orbiter that was going into space. The vehicle had fake tiles on it, and fake nozzles and reaction control system (RCS) jets, and some fake venting areas, but it looked much like the Shuttle, at least from a distance. Only by getting up close to the vehicle would anyone notice that these things were dummies. The fake items were not used for crowd appeal in any sense. They were used to simulate the actual aerodynamic shape and roughness (surface texture) of the vehicle that would be flying. Consequently, if there were nozzles on the Orbiter vehicle, we needed to have nozzles on the ALT vehicle to get a proper assessment of the basic lift and drag, stability and control, and the various flow effects due to protuberances.

I hadn't expected the huge crowds that gathered all the way from Plant 42 to Edwards. A lot of people knew the Shuttle was being built and knew it would be the next big space program for NASA. A lot of them came up from Los Angeles and much further away to get a glimpse of this thing as it passed them on one of the world's slowest parades. In its own stately manner, it made the transit from Plant 42 to Edwards.

The first time I saw the vehicle was as it came up the taxiway toward Dryden. I had other duties that day, but there was video coverage displayed close by, so I knew when to go outside to see it. The vehicle was exceedingly slow in arriving from the time when we could first glimpse it some distance down the taxiway. It came by me and pulled behind the main building at Dryden. There were large crowds all along the flight line made up of personnel from the Air Force, NASA, and various contractors at Edwards. Most of us were engineers, technicians, or people who spent a good deal of time doing administrative tasks. I think everybody at Edwards felt a part of this vehicle. It was quite exciting to watch it go by and see how big it was.

Shuttle/SCA Captive-Carry Tests

Since I was not primarily involved in the captive-carry tests of the SCA and the Orbiter [47, 48], I had the opportunity to watch some of the operations that I normally wouldn't see on a program where I was required to be in the control room, meetings, and briefings. It was interesting seeing this huge Shuttle sitting on top of an even larger 747. It could be seen in the mate/demate facility as they lifted it up and drove the 747 under it and did all kinds of fit testing and checking. It spent quite some time in that mate/demate facility. Even though things were built according to specifications, there was still some tweaking that needed to be done to get ready to do the final mating. For sentimental reasons, I went up to watch the mating of the Endeavour to the 747 after STS-111 in July 2002. I was escorted by Joe D'Agostino, who had been involved with the Shuttle operations at Dryden since 1976. He has been in charge of Dryden's Shuttle operations since the 1980s and is one of the few at Dryden that I can still talk to about the initial mating in 1977.

I remember well the first day that the 747 was towed from the mate/demate facility with the Shuttle on top, no longer having any of the superstructure of the mate/demate in the way. The Shuttle was sitting on top of the 747, attached only at three points. Some of us who had access to that area were somewhat in awe of the size and the depth of the mated pair. I remember my impression then was much the same as it had been when Milt and I had first started worrying about the separation of these two vehicles. It just didn't look like a very good idea because it didn't look like anything else that I'd ever seen done with an aircraft.

During the takeoff, the mated pair didn't climb very fast [47] or didn't maneuver very much until well above the ground. The mated pair was pretty easy to see due to being very large in terms of things that fly. I'd never seen any airplane taller than that mated pair fly. It was easy to see the white Shuttle mated to the silver-colored 747 anywhere in the pattern during its envelope expansion. The captive flights were to verify the integrity of the mated pair. This required the moderate amount of maneuvering [47, 48] needed for any ferry flight and for the ALT flights. There were also maneuvers required to get the stability-and-control parameters and lift-and-drag from the mated pair. [47, 48] As a very interested observer, I understood the things that were going on, directed by many people.

Ike Gillam had joined Dryden just prior to this program, replacing David Scott as Dryden's Shuttle director. Gillam, who became Dryden center director (again replacing Scott) after the ALT program, was in charge of the program from Dryden's point of view. It was very apparent at that time that this was a big program and an important one. We had lots of help from Boeing on the Shuttle carrier aircraft and from Rockwell on the Orbiter. We also had a lot of help from elsewhere in NASA. And, of course, Johnson and Marshall were the primary organizations in charge of the overall Shuttle program. There were lots of people that were on temporary duty at Dryden until we completed the ALT tests, so Dryden's size grew quite a bit during that time.

It was clear to me at that point that the things I had been noticing in terms of the design—the number of meetings, people, charts, memos to be read—were huge compared to anything that I'd worked on before. I could see that this had another aspect. There were many people who had tasks that I wasn't used to other people having. It was partly to train them, to get them ready to do things that were very new to them, because they were mostly from the rocket world and not the airplane world. The Shuttle was then, and in my opinion still is, the most extreme example of the merging of two very different flight perspectives and very different approaches to problem-solving.

These differences sometimes still confuse me in regard to why specialists in vertically launched vehicles take such a strange approach to problems that have a relatively straightforward solution from the aircraft point of view. That's not to say that there's anything bad about either group. It's just clear that they view problems differently. They view risk differently. They view physics and applying it very differently than the way aircraft people do. Some of their ideas I've adopted because they made more sense to me after I became used to them. But in regard to other things I am still confused about why traditional rocket-launch specialists don't do them in much easier and much more straightforward ways.

Meeting the Press

I was also surprised by the amount of press coverage once the Shuttle was brought to Edwards. At that point, it was clear that Edwards had been a very high visibility place during the 30 years prior to the ALT program when the X-1, X-2, D-558-II, X-3, X-4, X-5, X-15, XB-70, LLRV, and the lifting bodies were being flown at Edwards. People who had been at Dryden before I arrived told me that Dryden was a popular place with reporters and other dignitaries at the beginning of the X-15 program because it was somewhat open to them. I've heard it mentioned that Walter Cronkite was a big fan of the X-15 program. By the time I got to Dryden, the Mercury program was underway, and almost all of the press had moved on to Cape Canaveral.

However, not even in the 1950s and 1960s had we the masses of reporters at Dryden that were around for the ALT program. There were many hundreds of people representing the media—the press, radio, and television. It was a major effort, I'm sure for Edwards in general as I know it was for Dryden in particular, to have all of these media people here to observe. There were long periods in the program when nothing happened, but some of the press stayed on. They needed to be kept occupied in some manner because they couldn't be given free run of the engineering facilities and hangars. The ALT was not our only program. We needed to maintain the work environment on all vehicles for the engineers and technicians so that things were done on schedule and done properly and safely, with all normal checks and balances and inspections.

Center director Dave Scott, who'd been through the Gemini and the Apollo programs, had a lot of experience with the press. He felt that we needed to come up with presentations by various managers and engineers in case the press got bored enough to listen to more technical information. There was also broader information in terms of programs or history, films on things we'd done before, and other aerospace endeavors around the country. Quite a bit of that was used.

I made two presentations for the press. The one I did on some of the past work including the lifting bodies, the 3/8th-scale F-15 RPRV, and the X-15 was very well received, and I felt the reporters who were there at the time at least got something out of the briefing. I also gave them a presentation on my version of parameter identification techniques and extraction of aerodynamic characteristics from flight data, trying to give them perspective on the differences between airplanes and the Shuttle. Not a lot of people turned out for that one, and there were a lot fewer people when I was done than when I started. I think there may have been a few technical people visiting Dryden at the time who may have gotten something out of my presentation. I always felt that Scott's actual goal was to use this opportunity to gain publicity for Dryden and for NASA generally as well as to convey information to the general public. I think the media events were fairly successful in some areas.

We were seeing media people who I don't think had very much interest in aviation at all. In talking to them, I found they were always amazed by all sorts of obvious things that the more aviation-oriented press knew all about. We had to be a little more careful what we said in the cafeteria or the hallways because we didn't know the other people around us. If they happened to be reporters, they might pick up something and try to make a story out of it, when it was just somebody trying to say something humorous or venting a little frustration, as all of us sometimes did.

Normally, the reporters were identified, but it was a new environment for me, and I had to be more careful about what I said. I was used to that feeling only for classified or proprietary programs. You just don't talk about them with anybody that doesn't need to know, not even staff engineers working on different projects. There were other programs that were not either classified or proprietary that you could chat about a little more freely in the hallways with your friends who were assigned to other programs. They would also be able to tell you about what they were doing. That's just what engineers do. We're interested in other engineering things, so we tend to gossip about other programs and their progress, as well as the success of management and some of the bad things that some of us think that management does. We gossip about those, but we always get back to work on our program, no matter how bad a decision we think it was, and try to do the very best job that we can. With the reporters having fairly free reign in the places where our gossip took place, we had to be a little bit more careful about who might be within listening distances. I found myself doing a lot more of my gossiping in either my office or the offices of other engineers.

Prelude to ALT

In preparing for the first ALT free flight, I had many things to do profession-ally, but there was also an emotional attachment that I think is not always obvious to people who don't do this kind of work. The minute that engineers are assigned to a program, they probably have a feeling about whether it's going to be a very interesting program or whether it's a particularly good idea, but once they are assigned to it and start to work on it, they tend to become very emotionally attached to it. It becomes a friend. It makes no difference how ugly or inappropriate the configuration is; it always happens. I wasn't directly involved on the X-24A, which to me at that time was the ugliest airplane that I had ever seen. Even though I was only an occasional project support person, I developed an attachment to that vehicle and would have felt very bad if it had not been able to fly or if it had had an accident. So engineers develop an emotional attachment to a vehicle, and if any-thing happens, they go through the same kinds of emotional readjustment that people do when a friend is injured, dies, or has a fatal disease. Anything we work on, we very much want to see succeed.

As we got close to launching the first ALT free flight, I was starting to feel as I had during the first M2-F1 flights or the first M2-F2 flights. We had a brand new configuration. We had a lot more engineering effort on the Shuttle than we'd had on either of those two programs, but even with the greatest care and the greatest expense, we could still miss something. That's what all engineers and all manage-ment of engineering projects worry about. That's why we have a lot of reviews, and why we tend to have a culture where if somebody has a really bad feeling about something, they're encouraged to bring it up so that everybody can take a look at it. Then, if everybody ultimately agrees that the real risks are understood and there are no hidden risks, we proceed. Sometimes the issues are things that are being done either poorly or inappropriately. Then the project goes through whatever it needs to go through to fix them, to minimize the risk, or to state the risk that's involved in the flight.

The Shuttle was in many ways a much easier vehicle to fly than the lifting bodies or the X-15. However, the Shuttle had never flown before, the launch altitude was lower, the launch concept off the top of the 747 was different, the weight and the size of the Shuttle were also in a class of their own from the other vehicles, and, of course, the Enterprise was also unpowered. Anytime something that huge is launched from a low altitude with no go-around capability adds to the intensity of it also being a first flight of a unique configuration. The Shuttle, as it was going to be launched from the 747, was 150,000 pounds. The X-15 with the external tanks full was a third that weight. I had seen drawings of the Shuttle. But in getting up next to it, as it was being built and now getting ready to fly, I wasn't used to seeing big fat wings like that. The base area was enormous, roughly equal to the floor area of a two-car garage. It really was a big blocky vehicle, and its dimensions in that blockiness were huge. The maximum thickness of the wing is about six feet, so large that it was out of my experience. Then, of course, there were the uncertainties. We had studied the predictions and done all kinds of variations and tolerance studies on the predicted aerodynamics from wind tunnels and other vehicles, so we were as confident as we could be in the aerodynamics. [46] Nevertheless, this was not like any other configuration that had ever flown.

The Shuttle was a totally fly-by-wire vehicle. It had four digital computers that did the computation of the control inputs, mixing the pilot inputs with the feedback from the control systems. Much work had been done studying the computer and software with their nearly identical computers in the F-8 digital-fly-by-wire aircraft. [50] The Shuttle had good hydraulic systems, but they'd been known to fail, too. All of the power for the vehicle was going to come from auxiliary power units (APUs), which had been a big problem on the X-15, delaying or aborting flights. As a result, there were lots of things that could go wrong that we could worry about, but the ones that usually worry us the most are the ones we don't know much about, the unknown unknowns.

The first three ALT flights were going to be with the tailcone on. [49] This smoothed out the airflow around the base of the Shuttle, and meant less stress and buffeting on the 747's vertical tail. Another result was to reduce the Shuttle's drag, and its subsonic L/D max was predicted to be well over seven, making it fairly docile in terms of what I considered low L/D vehicles. A pilot of a glider with an L/D max in the thirties would think that the Shuttle had a low L/D max. But to a pilot at Dryden, having flown the M2-F1 and the M2-F2 with an L/D max less than three and a little more than three respectively, over seven sounded really big. The X-15 has an L/D max in the four and a half range as well. So this was going to be a better glider, even though it was much heavier than the X-15 or the lifting bodies. The major concerns were if the maximum control surface rates were going to be fast enough for the pilot to be able to do what he needed to do. Was all of the software and hardware in the digital computers checked, or did either have bugs in it? There was a backup flight-control system that the vehicle could go to if those more complex algorithms were not correct. So there were off-ramps, and the Shuttle crew also had ejection seats in the Enterprise. If things went really bad, we might lose the vehicle but the crew should survive.

Nonetheless, those of us who put our money where our mouths are in terms of analysis really had butterflies on that first flight. Of course, I knew Fred Haise from many years earlier, I knew what an excellent pilot he was, and I wasn't really concerned about his piloting skills. However, we couldn't do normal envelope expansion with the vehicle. With the Shuttle, once it separates from the 747, the crew is committed to a landing with no go-around capability, so if anything isn't hooked up or wired up right, the crew would find that out shortly after separation.

I knew what the flight plan was, and there was nothing very racy in terms of what we were planning on doing. For the Shuttle's first flight, we were using runway 17 in the middle of the dry lakebed. (See Appendix E) With some rough spots in it, runway 17 is over seven miles long, located about two and a half miles from Dryden. We planned to come down somewhere in the middle of it; so if the energy was low, the Shuttle would end up on the north end of the runway. If the energy was high and the pilots were unable to correct or dissipate it, the Shuttle would land on the south end of that runway.

The first ALT flight was going to be very much like the M2-F2 flight from the point of view of the uncertainty in flying a new configuration. This was true even though it was from a lower altitude, the vehicle had a higher L/D max, and the glide slope would be much shallower than what we'd seen on the M2-F2. The 180-degree turn that the pilot does to change from a northward launch to a southward landing gives him the opportunity to measure the responsiveness of the vehicle to fairly modest control inputs. If the pilot is high on energy, he does the turn more slowly or later. However, if the pilot feels he's low on energy, he can do it more aggressively and sooner. The pilot gets a feeling for the energy with all of his practices in the simulator at Johnson. The pilots also practiced with the Shuttle Training Aircraft (STA), its systems configured to have more or less the predicted Shuttle characteristics. The pilots had also trained with the Shuttle motion-base simulator, so they had a good feel for what this should be like.

First Flight

The first ALT free flight was scheduled for the morning of August 12, 1977. Even though Johnson was in charge of the flight, Dryden had set up the control room in much the same way as we had for the X-15 and lifting body programs. This way, the Dryden engineers could monitor the flight in the same fashion as we had with those earlier programs. More than a thousand reporters and between 60,000 and 70,000 visitors were at Edwards. The Shuttle/SCA mated pair was backed out of the Mate/Demate facility, the 747 started its engines, and the combination began the taxi to the runway. It began the takeoff roll at the scheduled time of 8:00 a.m. The climb took a few minutes longer than expected, but the Shuttle/SCA and its attending T-38 chase planes were finally in position. Fred Haise, in the Shuttle, radioed that he was ready, then pushed the separation button.

I don't remember any comments on launch, but from my point of view the Shuttle cleared the 747's tail smartly. (Fig. 1) The Shuttle was many tail heights above the 747 by the time the base of the Shuttle was above the tail of the SCA. I believed that was going to happen, but I was sure glad to see that it really did happen.

Haise and Fullerton were heading north in the Shuttle. They did little inputs to see that they were getting responses in pitch, roll, and yaw. Pilots will do that whether they are asked to or not. It's how they know that everything is hooked up right, the computers and the hydraulic system are doing what they should, and the

Fig. 1. The first separation of the Enterprise from the SCA with the tailcone on.
(E77 8608)

vehicle is responding in each of the three axes due to the stick and the rudder inputs. Fred then did a 180-degree turn to line up with runway 17. Then, as had been done on the M2-F2, he did a practice flare at altitude to see that the vehicle was going to arrest its sink rate much as the simulator and the wind tunnels had predicted that it would. Then Fred came onto final approach and did his final energy management with the speed brakes, part of which was planned, I believe, and did a nice touchdown and rollout of close to two miles on the lakebed.

Getting ready for the first flight was exciting. It was certainly very visual to watch it separate and to do its 180-degree turn, then set up for the landing and complete the flare. Looking back, it was anticlimactic in a way, because the L/D max was so high that the glide angle (about 17 degrees) never got too steep, so it didn't evoke the feeling that we had with the lifting bodies and the X-15 of the vehicle diving at the ground. Visitors watching the vehicle's flight had been impressed, because it looked steep to them, but they hadn't seen the really steep ones that I had observed in the early 1960s. The first flight was a success. Although there were a number of gripes from the various disciplines, and some things had to be worked on quickly before the next flight, there was nothing major as I recall.

There were no real stability-and-control or performance pulses done on the first flight. But the pilot was still maneuvering the airplane, and those inputs and resulting vehicle responses could be analyzed to make spot comparisons for the lift and drag, and these were as predicted. So we got a little bit of a feel that things were about as advertised from the lift-and-drag point of view, and the same was

174

true from the stability-and-control point of view. It may not have been as predicted, but it was close enough that the pilot was probably not going to have great difficulty doing the lift-and-drag maneuvers and the stability-and-control maneuvers on the second flight.

We had the strip charts in the control room from the first flight, and we also had telemetered data from the Shuttle to Dryden. The primary things that the stability-and-control and performance engineers were interested in were on the aerodynamic coefficient identification package (ACIP). We also had other information on the operational instrumentation (OI) system, and the backup-flight control system (BFCS). Our first task, in terms of analysis of the data, was to merge those three data streams and account for any time skews from sensor to sensor on a given system and then the time skews between the clocks on the three different systems, so we got some practice doing that.

Rich Maine had developed the software for doing it. Some things weren't quite as they were specified in terms of official time skews, but we had some techniques that we could use for refinement. It was very important for all of the time histories from these sensors be within a few milliseconds of the time the measurement had been. We then interpolate between those samples so we can line them all up, as if they were all simultaneously sampled. That's a very important thing for us to do, and we learned how to deal with this problem on the Enterprise during the ALT program. We got a feel that the aileron, elevator, and rudders were all about as effective as predicted, which was within 10 or 20 percent. The control effectiveness wasn't going to cause any big problems with trying to get the maneuvers that we'd requested for subsequent flights. The same was true of the ACIP instrumentation for the lift-and-drag data from the maneuvers that we were going to perform starting with the second flight.

So my memory of the first flight is that it was exciting getting ready for it, the flight itself was exciting, and it was exciting getting our hands on the data the first time and seeing what it had to tell us. It's probably a difficult thing for other people to understand, but the nerdiness of an engineer with a new set of data to analyze is a lot like Christmas to a five-year-old. Everything's exciting—the package, the box and what's in it, and all the things to play with. There's definitely an overload at that time. It's very pleasant to an engineer who is going to analyze that data when things go well. When things don't go well, it's equally exciting, but if we're on the critical path, it can be overwhelming in terms of trying to understand what the data is telling us so that we can feed that to program management. If it's a serious problem, it either gets fixed or avoided. If it's not a serious problem, we can show that it is a problem that's not going to impact things.

Those were the kinds of things we were looking for with the data from the first flight. We had the normal crew debrief that we have after every flight. The Johnson Space Center was in charge of those flights from the time the mated pair started taxiing until the Shuttle was back on the ground, but it didn't really feel that way because our control room was very active. Deke Slayton was there. A lot of the Johnson people were there, even though control was being run out of Houston, so even though Dryden was not really controlling the flight, we had all the information that we needed.

At the crew debrief, we were very interested in hearing the reaction of the pilots to what they had just flown, along with responses from various people who were in charge of monitoring other things in terms of vehicle health—the hydraulic pressures, the APUs, and the problems that occurred. It was quite a long postflight debrief, since it was a first flight, but those are always very interesting. We always had those for all of the airplane programs that I've worked on. We usually have the postflights as soon as we can, while things are still fresh in the pilots' minds. There could be more extensive debriefing and questioning later, but listening at the crew debrief is how we capture the feeling of the crew.

We had the debrief, then we went into Lancaster and had a party. It was a different kind of party than we had for the X-15. It was more formal and, I think, less noisy and less enthusiastic than we had for Dryden or Edwards-driven programs such as the X-15 and the lifting bodies. But the atmosphere was the same. It was a celebration for those who were actively working on the program, the chance to let their hair down and get rid of some of the tension.

The next thing was to get back to the data. Initially we just had the strip charts, and over a period of a few days we started to get information from the other systems so that we could merge all of the data and interpret it. Once the time came to do our analysis, we did it exactly the same way as we'd always done it. The way we fed it back into the program was more formal and much different because we didn't just reveal our results or concerns to others working on the program or program management. There was a very formal structure, and within Dryden we had a formalized structure of how the various ALT engineers and technologists worked—that is, what we did with our data, what kind of review and oversight it had by our management, and how that was then fed up through the various disciplines and levels of management up to the top of the Shuttle program. That was different from the smaller programs that I'd worked on before, but the analysis and the excitement of the data was exactly the same as they had been with the other vehicles. I did have the feeling that I was a much smaller frog in a much larger pond than I'd been with the M2-F1, the M2-F2, and the X-15 to a lesser extent. It was the same kind of job, but it was in a bigger world than I'd been in before.

After I had looked at the time histories from the first flight, it was clear why the separation had been so positive. The Shuttle pulled about 1.8 g's as it separated from the 747. With the Shuttle launching at about 1.8 g's, and the 747 probably being a little less than 1 g, they separate at about 25 feet per second squared. A launch at 0 g from the B-52 separates the two by roughly 32 feet per second squared acceleration. The Shuttle was just going up as opposed to going down like the X-15. Once I saw that, separation never concerned me again. At launch on the first ALT flight, Fred Haise had rolled it off to the right as Fitz Fulton, who was flying the SCA, rolled to the left. The roll that Fred did was actually the very first Shuttle maneuver that I could analyze. It wasn't a large roll, but it was a very definite input on the aileron.

ALT-2 and -3

For the second ALT flight, we were scheduled to have different kinds of control inputs. One was called aero-stick inputs (ASI), and those are what pilots have always done for getting doublets on each control surface so that we could extract the stability-and-control derivatives. Another type of maneuver that they would give us were pushover/pullups (POPU). Those are a quasi steady-state maneuver that we do to get lift-and-drag information from the accelerometers. The pilot initially pushes forward on the stick so that the angle of attack goes to a lower value slowly. Then, once he reaches the planned lower angle of attack, the pilot slowly pulls back on the stick, which becomes the pullup part of the maneuver. Then, he goes back through the original position and continues on up to a higher angle of attack. When he reaches the pre-planned higher angle of attack, he pushes over slowly, back to the initial condition. The whole maneuver usually takes from 20 to 40 seconds to do, depending on the type of vehicle. Even on the Shuttle, with the tailcone on and its high L/D max, it was going to be a fairly rapid POPU. The angle-of-attack variation on the ALT flights wasn't going to be huge, but it was going to be larger than we'd be able to do hypersonically on the orbital flights many years later.

The third kind of maneuver was something that I had originally played with on the 3/8th-scale F-15 RPRV in 1973. [51] Since on it, everything went through the computer on the ground and then was radioed back up to the airplane, we had come up with a way of actually initiating the input with a switch. When the ground pilot turned the switch, the control surfaces went to a predetermined amplitude. Then, when the ground pilot turned the switch in the opposite direction, he'd get the other half of the doublet. When he returned the switch to the normal position, he would have completed the maneuver. A doublet is, ideally, one cycle of a square wave. When the pilot inputs it, he isn't quite as controlled as a square wave is, but he tries to approximate it. But between the various filtering and actuator characteristics, even with a computer, it looks more like a sawtooth wave at the control surface than it does a square wave.

We found the techniques that we developed with the 3/8th-scale F-15 RPRV to provide superior maneuvers, compared to asking the pilot to try to do it from the ground cockpit. So I extracted most of my stability-and-control derivatives on the F-15 RPRV through that technique. I also programmed pilot-initiated input sequences such as sinusoid sweeps or varying series of different frequency square waves. We had proposed that this same technique be used for the Space Shuttle, partly because of programmatic concerns that we were going to overload the pilot or that the pilot might do something inappropriate. These maneuvers were called programmed test input (PTI) maneuvers. The way we would do that, particularly for the Orbiter when it was coming in from a reentry, was to have the pilot initiate a preprogrammed sequence of control motions which would command doublets on each of the surfaces.

The timing and amplitude of these pulses was determined from simulator studies and doing various changes in the simulator characteristics [49] to reflect the variances and tolerances that had been determined for the Shuttle, so it was going to be a very formal procedure for determining these PTI maneuvers. They would be initiated by the pilot. If the vehicle was within certain constraints of angle of attack and air speed and had no system irregularities, the computer would accept this input and command the various control surfaces to move in the sequence designed in the simulator. We would then take the responses to those inputs, along with the inputs themselves, and we would extract the stability-and-control derivatives.

That was the first step in a formal process that would eventually give us literally hundreds of stability-and-control maneuvers throughout the first 75 or 80 orbital Shuttle flights. [52] Most of those were during the first 40 flights, but we continued to get a few as new questions occurred. So we would put in a request for a PTI, it would be designed in the simulator, and the pilots would use it in their training.

The next two ALT flights (ALT-2 and -3) were also planned to be with the tailcone on. They were going to have POPUs and stability-and-control maneuvers done by the pilot with the ASIs and also done by the computer with the PTIs. We were looking at our ALT-1 data, knowing that on the second flight, we were going to be getting specific maneuvers to analyze. We were able to get stability and control derivatives from the first flight because the pilot in the flight plan was supposed to bank in each direction. Each time he initiated the bank, we'd get a lateral-directional maneuver. Each time he recovered from that bank, we'd get a lateral-directional maneuver. They're not ideal for analysis but they're fairly good. So we did get some decent analysis from the first flight that gave us some confidence in going to the next flight. We also got pitch maneuvers on the first flight, both with the pilot's practice flare at altitude, because that is a longitudinal motion, and with his actual flare as he neared the ground. So we were pretty confident that we were going to get even better maneuvers in the second flight.

The POPU maneuvers were initiated from steady-state flight conditions, at about 10 degrees angle of attack. That means the pilot would slowly rotate the Shuttle from 10 degrees down to 5 degrees angle of attack. Next, the pilot would slowly rotate the vehicle up from 5 to 15 degrees angle of attack, then he would slowly rotate it back down to 10 degrees, and that would complete the pushover/pullup maneuver.

The stability-and-control pulses that we planned—both the PTIs and the ASIs—were small maneuvers for those of us used to working on high-performance airplanes. The Shuttle was a sluggish vehicle to fly, compared to most of the vehicles I had worked on, due to its large size and mass. The doublet amplitudes for the stability and control pulses would be about plus or minus 5 degrees on the aileron followed by plus or minus 5 degrees on the rudder. For example, the pilot would start the longitudinal maneuvers at a longitudinal trim of about 5 degrees angle of attack, then made a pulse that went down to zero degrees angle of attack, then back up to plus ten degrees, and back down to five degrees. These maneuvers were made in 3 or 4 seconds, so they are much more rapid maneuvers than

POPUS. These doublets were all done with the stability augmentation system engaged.

The second ALT flight was going to be flown by Joe Engle and Dick Truly. I was looking forward to that because on his first flight in the X-15, Joe had done a full 360-degree roll right after launch. I didn't think he'd try to do that with the Shuttle, but you never really knew with Joe. He had a lot of confidence and skill. The ALT-2 flight actually was going to be a better one to watch from the Dryden ramp. A thunderstorm had made runway 17 too damp to land on, but by then, we were confident we didn't need that long of a runway. We chose to land on runway 15, which is about six miles long, and it is less than a mile from Dryden. (Appendix E) To have roughly the same energy management scheme, the launch point was roughly the same, with the turn roughly at the same point. The landing would occur right next to the Dryden area.

ALT-2 was flown a month after the first flight, the delay caused by the wet lakebed. However, my experience with complicated airplanes is that the length of time between the first flight and the second flight is usually months, not weeks. The speed in scheduling showed the pace and the priority of the Shuttle program, the need for getting the ALT flights out of the way in order to get on with the orbital launch of the Shuttle at the Kennedy Space Center.

I remember some sort of informal discussions just before ALT-2 when Milt wondered if we'd really looked at everything that we should. I remember a fairly decisive response from Deke Slayton that we'd done everything we set out to do and we didn't have any real problems. His attitude was to press on to get the next flight, but the weather delayed it. If it hadn't been for the weather, I think we probably would have flown ALT-2 two weeks after ALT-1, instead of a month.

The ALT-2 flight looked to the outside observer much like the first flight. There was the same separation maneuver and the same playing with the controls a little bit to make sure they all worked. Then we started to get stability-and-control pulses and a POPU from the pilot. Joe gave us several ASI maneuvers at different angles of attack and, of course, slightly different Mach numbers. Mach number wasn't a big parameter, because it was probably going to vary from close to 0.4 up to nearly 0.7 Mach. It's a fairly benign Mach number range normally. So Joe set up three different angles of attack and did three longitudinal and lateral-directional ASIs and PTIs He did his practice flare at altitude, then landed north of the Dryden facility. As he came to a touchdown and the rollout, he was right abeam of Dryden. It was a nice thing for the crowd that was there, one of the best looks that you can have of the Shuttle up close and landing. The flight was also five minutes, while the rollout was almost a minute. It took the Shuttle awhile to slow down from its landing touchdown speed of about 170 to 180 knots.

The flight went very well. I know that on each of these flights many systems were being evaluated that were either prototypes or actual Shuttle hardware, along with the computers and some of the computer codes. Also, because this was an approach-and-landing test, there was quite a bit of activity by the pilots after it touched down. There was a lot of concern about braking, steering, and cross winds.

[49] On various flights, they used the rudder or the aileron for steering on the ground. They also had nose-wheel steering and differential brakes to evaluate.

These were also of some limited interest to the aircraft performance people because the distance the Shuttle rolled out was a function of vehicle lift on the ground and drag on the ground. The Shuttle lands first on its two back wheels, then does what they call a derotation down to the nose-wheel touching. All of that was looked at. We were also interested in ground effect, which is the change in the aerodynamics as we get within a vehicle span or two of the ground. The Shuttle span is in the neighborhood of 80 feet. We would try to see if there were differences in lift and drag as the Shuttle got within 150 feet of the runway. We also looked at the change in acceleration before and after derotation, which gives additional information in terms of the aerodynamic drag in the ground effect. Of course, we did have to account for the change in lift before and after derotation, because the force on the wheels was different with the nose up and with the nose down. There's more force on the wheels with the nose down than just the weight of the Shuttle, because the Shuttle has a negative lift coefficient when the nose is on the ground. So the initial force when the nose comes down on the runway is the weight of the vehicle, plus some negative lift pushing it into the lakebed. We had to account for all of those factors, but it gave everybody information in terms of braking and deceleration before and after derotation and what the various causes in terms of aerodynamics or dissipation forces were due to depressing the dirt on the lakebed.

After the ALT-2 flight, once again we had to pick out our maneuvers from the strip charts available in real time in our control room, then later refine those times when we got the final data and take care of all the data skews. It was a little more important this time because we had better maneuvers to analyze, and we got good analysis on all of them. They were never as good as we would like, partly because they were smaller than we were used to when working on high-performance vehicles and we were used to their larger motions and more rapid inputs to control surfaces. The rate limits on the Shuttle's control surfaces are lower, mostly due to their size, so we didn't get the change in the control positions as fast as we would on high-performance aircraft. We also got lower rotation rates. I think the highest roll rates we got to work with were about 10 degrees per second, and the pitch rates and yaw rates were in the range of 4 to 5 degrees per second. On a high-performance fighter, these maneuvers would have rates several times those on the Shuttle.

The analysis of the data showed, for the most part, that the lift and drag were fairly close to predicted. I think the L/D max actually was a little bit lower than had been predicted. It was predicted to be 7.6 and it was actually 7.5, and that could have been due to a lot of issues. So I would say, considering the limited amount of flight data that we had, the lift and drag had been predicted accurately by the wind tunnel. All of the refinements that went on after the wind-tunnel program in terms of the various reviews and voting on the data went into the official data book. The main thing I remember is that the elevon was less effective in pitch than predicted, and the vehicle was found to be somewhat more stable in pitch as well. The lateral-directional maneuvers weren't really as good, and there were a lot more unknowns in them.

The ALT-3 flight went very much like the first one, except this time Fred Haise and Gordon Fullerton gave us some specific stability-and-control pulses. Those were analyzed along with the ones we'd gotten from the first two flights. Once again, I don't remember anything being particularly different, other than the two longitudinal characteristics that I have already mentioned. I think the damping in pitch was also smaller than predicted. I do remember that we changed the aerodynamic data book when the flight numbers were different than predicted.

Tailcone Off and Pilot-Induced Oscillations

The next flight, ALT-4, was with the tailcone off. The program planned to fly two or three flights with the tailcone off, because that would have been absolute confirmation of the database in terms of the flight-determined aerodynamics of the Shuttle vehicle that would be launched into orbit. Any uncertainty removed prior to the first orbital flight would be very important. It also gave us an opportunity to check out more of the systems. The two sets of pilots alternated. Since Haise and Fullerton had flown the last flight, Joe Engle and Dick Truly would fly ALT-4. This was the first Shuttle flight in the normal Shuttle Orbiter configuration. There were some differences in directional stability and pitch stability predicted with the tailcone off, but those were well within the ability of the control system and the pilots to handle.

The biggest change was in the drag and L/D max. The L/D max for the Orbiter with the tailcone on had been shown to be over seven. But with the tailcone off it, was predicted to be about 4.7 or 4.8-that is, in the X-15 category, not in the lifting-body category. For those of us who had worked on the lifting bodies, this change didn't seem spectacular. However, it was apparent that because of the high base drag component, the flight would be fairly brief, so if there were differences in the airplane, the pilot would have a lot less time to sort it out. I don't think I considered the possibility that Joe Engle would roll the Shuttle on this flight. That probably would be too scary even for Joe to do.

The first three flights had used the horseshoe pattern of launching north with a 180-degree turn for energy management. For the ALT-4 flight without the tailcone, it was more of a direct flight to runway 17. With no turn in there, the launch point was moved just north of Highway 58, lined up fairly closely on the runway. I think there was an offset to give the pilot some task in terms of finding the runway and managing his own energy since he couldn't do it with a turn.

Engle and Truly launched it at 21,500 feet above sea level, or 19,300 above ground level, and again had the 1.8 g separation. (Fig. 2) The vehicle looked like it was diving much more steeply than it had for the first three ALT flights. I think this was even more apparent to those who were watching from the ground. There was still a fairly large crowd watching because this is the first vehicle that really looked like the Shuttle, and most people following the program knew that this flight was

Fig. 2. The Enterprise without the tailcone, separating from the SCA. (E77 8923)

going to be a little bit more interesting. I remember the crowd reaction to the launch and the approach angle of ALT-4 because it would have been much like the X-15—about minus 22 or 23 degrees or so. I think on the previous flights, with higher L/D max and the tailcone on, the approach was probably more like minus 17 or 18 degrees. (Fig. 3) Some of the lifting-body flights that we've done in the past have been done at more than minus 30 degrees, so for a big airplane with slow responses, that looks very daring. It was an exciting but short flight. We did get four stability-and-control maneuvers, and we got a POPU on that flight as well. [53] Then, Joe did his preflare a little more than a thousand feet above the ground and did a nice touchdown and landing rollout. The actual flight-determined L/D max of the vehicle was essentially the predicted value of 4.7.

Everything had gone as planned, and no particular issues were raised. So we decided to press on to the next flight with the tailcone off, and that was with pilots Fred Haise and Gordon Fullerton. The primary goal of the ALT-5 flight (Fig. 4) was to demonstrate landing on runway 04, the main concrete runway, which is 15,000 feet or nearly three miles long, with a 1,000-foot overrun onto the lakebed. By landing on runway 04 (see Appendix E), the Shuttle would be landing headed toward the northeast which is unusual. Normally, landings were made going to the southwest on the same runway, which is then designated as runway 22. This flight was going was going to have a designated touchdown point on runway 04 for the pilot to do a precision landing. It had been shown that the pilot could come very close to landing at the intended touchdown point with the X-24B, so we decided to attempt that with the Shuttle on the main concrete runway. I think they were going to demonstrate that the Shuttle could land on a 10,000-foot runway.

Fig. 3. The Enterprise on approach during the fifth ALT flight. (EC77 9058)

Fig. 4. The Enterprise mated to the SCA lifting off for the fifth ALT flight. (ECN 8905)

The flight plan was fairly simple. This was going to be Fred Haise's first flight with the tailcone off, making the flight different enough that it was going to be a difficult task to hit precisely a predetermined touchdown point, for the Shuttle was known to be susceptible to PIO under many circumstances. [49] The other problem was that having a specific touchdown point announced will get the pilot's gain up. When pilots practice in the simulator, [54] they won't find discrepancies or undesirable characteristics. They will find them in flight, even though the two were more or less the same, just because their gain would be low in the simulator. Once they are risking their own life, the lives of others, or the aircraft, their gain will be higher, which means that they are more apt to find any bad characteristics. It was known at that point that there was considerable lag between the pilot's command and the motion of the surfaces.

There's always some lag on all aircraft, as the surface can't move as fast as the stick because it's rate-limited. In addition, there's always an apparent additional lag due to the systems dynamics of filling the actuator with hydraulic fluid and having it move. The Shuttle, of course, had lower rate limits, and it flew more like a transport than a fighter. In addition, the Shuttle had lags due to its digital control system.

The digital lag comes from the time a state of the vehicle occurs to when the control surfaces react to that state. This is inherent in any digital flight system. Roughly, it is caused by an accumulation of delays. There is a delay from when the sensor detects the state (such as the angle of attack or roll rate) and the time it is written on the bus, which occurs once per cycle on the system. The computer CPU reads that signal on the next cycle of the bus. The CPU then computes control surface commands based on the last sensor signal that it has from the bus. When the CPU finishes computing the control surface command, it writes it on the bus when the cycle is finished. Then, the actuator reads that command off the bus on the next cycle and responds to that command. Therefore, the time from when the state is sensed until the command is implemented is several cycles, which on the Shuttle amounts to between 200 and 250 milliseconds.

This doesn't sound very long, and if the pilot's gain is fairly low, he'll wait for it. But if his gain is fairly high, and he doesn't see a response in a quarter of a second, he is going to think something's wrong.

There were also issues in the software in that it gives priority to pitch commands over roll commands. If the pilot put in a roll command when the pitch system was busy, the system would ignore it because it was busy moving the actuator. The Shuttle did not have enough hydraulic and computer power to handle both lateral-directional and longitudinal tasks simultaneously if they were both being worked near the rate limit.

Something else increasing the gain of the pilot was that Prince Charles was in attendance for that landing. He had visited Dryden earlier in the day. I had shook his hand, as had several hundred other people. (Fig. 5) He was also near the landing area on the main runway, along with a very large crowd and a lot of the press. If there ever was a situation designed to get the pilot's gain fairly high, that was it. All test pilots, particularly high-performance test pilots, have fairly large egos. I mean that in the nicest possible way. They work very hard at perfecting their skills so they can do precisely what needs to be done. A pilot does not want to look bad when faced with a new and difficult task. I'm unaware of any other situation where we put a pilot into that kind of an environment in front of a big crowd. I don't think very many pilots would have done very well under those circumstances.

Fred and Gordon had a normal launch from the SCA at 19,000 feet above sea level or 16,800 feet above ground level. That is 2,500 feet lower than the previous flight with the tailcone off. They appeared to be going faster than planned, which I think surprised them with the tailcone off. Fred tried to slow the vehicle with the speed brakes, but it was clear he was going to miss his aim point. He was trying very hard to get the vehicle to change its dive angle so that he would still hit the intended touchdown point. In the process, he started wiggling the stick in a pitch mode, which most pilots do prior to landing just to make sure the vehicle is responding correctly. I think in Fred's case, since most of his experience and training had been done in high-performance aircraft, he was probably putting in larger inputs than he normally would just to feel the response on a sluggish airplane. As he approached the touchdown point, he was going too fast.

Fig. 5. Dryden Director David Scott presenting Prince Charles with documents commemorating his visit prior to the fifth flight of the Enterprise. Fred Haise is seen over Scott's right shoulder. (ECN 8899)

When Fred started putting in these pitch motions, he got into a pilot-induced oscillation (PIO) in pitch. It was caused initially by the rate limits on the surfaces and Fred's inability to tell that the vehicle was doing what he asked it to. With a time delay of a fifth to a quarter of a second between his input and the command moving the surface, it's hard for a pilot to sort out the response to his command.

There's an additional problem with the Shuttle, and that's due to the pilot's position being fairly close to the center of rotation of the aircraft and its very large control surfaces. Even though the aircraft is responding at the back in the way he wants it to, he can't tell that because he can't feel the rotation as he could if he were not located near the center of rotation of the aircraft. In addition, with these really large surfaces, if he tries to pull the nose up by deflecting the elevons upward, his efforts actually decreases the lift before it rotates the aircraft. This is a normal characteristic of a delta-winged vehicle. This decrease in lift is felt by the pilot as if he is accelerating towards the ground, so when a pilot pulls back on the stick to try to arrest the sink rate, he actually feels the vehicle accelerate faster toward the ground. Then, as the aircraft rotates and the lift builds up as angle of attack changes, he gets the arrest in the sink rate.

Consequently, as Fred is trying to sort this airplane out, it's rate-limiting, which he's not aware of because he's at the center of rotation. He's putting in commands that the airplane either ignores or appears to be doing the opposite of what he's asking it to do. As a result of all this, the vehicle landed a little prematurely, with a higher sink rate than planned, and then it bounced back into the air. A lot of pilots have done that with other aircraft, but in this case, when the Shuttle bounced into the air, in addition to having the longitudinal problem, Fred now had a lateral problem because the vehicle didn't bounce quite straight up. It started to roll slightly. In addition to having the rate limiting on his elevator, the pitch priority in the software resulted in the vehicle ignoring his lateral inputs.

So now he had a lateral PIO initiated by the rate-limiting longitudinally and then bouncing him into the air. He finally figured out that he was PIOing it, that he was the problem, so he more or less just let go of the stick, as Milt had done earlier when he was in a PIO in both the X-15 and the M2-F2. That was the right thing to do, and once Fred let go of the stick, the vehicle settled down and was responsive to what he put in. The Shuttle came back down to the ground a 1,000 or 1,500 feet down the runway at about 140 knots.

It was, I'm sure, embarrassing to Fred and caused many people to worry about the safety of the Shuttle, since a pilot could PIO it with just the landing task. But it was a very hard task. Fred had only a minute or a minute and a half to get the Shuttle to the touchdown point. Then, he found out he was going too fast, everything he did to correct it didn't work, and everything was happening in a hurry. So he was set up, and it was not a fair thing to do to a pilot on a first flight of a new configuration. If he'd had several flights in it, I'm sure this wouldn't have happened. But from my point of view and in the view of the rest of the Shuttle program, it was very fortunate that it happened the way it did, because it gave us some time to try to resolve it.

It really took us months to sort out what I have just described in a few sentences. To expect any pilot to sort that out in real time was unrealistic, so we put in a suppression filter that was designed at Dryden [55, 56], tested in the F-8 Digital Fly-By-Wire aircraft, and sold to the Johnson and Rockwell people as something to postpone the PIO tendency. It's still the same vehicle. It's still got lags, and it's still got huge slab controls, so the problem is still there. However, a pilot now has to put in larger inputs for a longer period of time for it to appear.

We did many studies for the next few years on the PIO problem. We did some variable stability tests on the CalSpan TIFS airplane where several pilots and some of the astronauts evaluated the problems that Fred had seen. We discovered that we couldn't make Joe Engle and John Young PIO the airplane because they only pulsed the stick. They didn't really ever fly it the way most people see pilots fly airplanes with a stick. They'd just pulse it each direction, and by pulsing it they never got into this rate-limit situation.

The Shuttle pilots practice many of these things in the simulator now, and the Shuttle has the suppression filter in it. Between the training and the suppression filter, PIO is no longer a big problem on the Shuttle. Nonetheless, if we ever have an emergency landing of the Space Shuttle where the pilot's gain gets really high and he's got to control both axes at the same time, we may see the same PIO. The tendency is simply hidden. Any time something like that is hidden from a pilot, once he discovers it, it can have very substantial consequences.

The only issue on the ALT-5 flight was the PIO. There were many meetings, discussions, and phone conversations from coast to coast on the PIO. All the people who had been in these communications had a meeting to decide whether to have a sixth flight to investigate this PIO a little more. I went to the meeting, Milt having invited me. I've never liked meetings, and I usually avoid them like the plague. But in this particular case, partly because I was interested in the problem and partly because I thought I might have some knowledge that would help in this decision, I attended the meeting. However, it was discovered that I wasn't on the list of those allowed to be in the meeting, so I was actually asked to leave. So I'm even less inclined to go to meetings now than I was before. You really find out something about yourself when you get thrown out of a meeting when normally you don't even like meetings. It's very embarrassing.

That was the end of the ALT program. The decision was made not to have a sixth flight. After the analysis was done, we had hard data to update our official flight database. [57] We could reduce the uncertainty a great deal. The variances got quite small because we'd already flown there with a vehicle that was almost identical to the orbital vehicle. As flight guys will tell you, it doesn't make any difference what the prediction is, for the only real answer is the one you get in flight. But at least from 7/10th Mach number and 20,000 feet on down, we were fairly confident that we understood the aerodynamics of the Space Shuttle.

IX

Counting Down to Launch

When the Enterprise left Dryden at the end of the ALT flights, I was in the same position with the Shuttle that I had been with the M2-F2 in 1966 after its first few glide flights. With the M2-F2, I had a very bad feeling about its characteristics; however, with the Orbiter, I found that my confidence was increased in the subsonic portion of its flight envelope. Nevertheless, there were still many issues and many questions to get the entire vehicle to launch, perform on-orbit operations, then do the reentry, and come from Mach 28 down to 8/10th or 7/10th Mach number, which is where we'd been with the ALT flights.

The inaugural orbital flight was almost a full-envelope first flight. Most of the flight was going to be where we'd never been before, with a configuration that had never flown at those speeds before. So from the point of view of an engineer who had spent his entire career doing envelope expansion, this was going to be a new and daunting experience. For instance, on the X-15 we expanded the flight envelope in angle-of-attack in increments of two to five degrees and Mach number increments of about a half a Mach number. That way, we were never too far from where we'd flown before. With the Shuttle, we were going to start at Mach 28 and come all the way down to 7/10th Mach number on a single flight.

Working the Problem

The only way we could take those big steps was by doing a great deal of analysis and ground-testing, and by taking into consideration the experiences of other people who had worked in other flight regimes. We picked their brains as to what we needed to consider and what we needed to study. It's usually a big task to do even a normal incremental flight-envelope expansion. There's a lot of analysis that goes on in between the flights and a lot of comparison with other data so we are sure we understand something. If we don't understand it, we tend to go fly that point again, with special attention to detail, to get additional data. That's the way we build a little and test a little, as engineers like to do. On the Shuttle, we wouldn't be given that opportunity. We ended up with a requirement for a huge amount of wind-tunnel tests, and those wind-tunnel tests included aerodynamics and aerothermodynamics for the various configurations.

To start with, we had the large stack, which consisted of the two Solid Rocket Boosters (SRB) and the Shuttle attached to the external tank. We also had the aerodynamic predictions from the wind tunnel of the SRB separation which occurs at about Mach 4 and around 130,000 or 140,000 feet. Then, there were tests for the return-to-launch-site abort maneuver. The Orbiter would still be attached to the external tank and would have to rotate completely against its launch direction, then use the Shuttle's main engines to decelerate the vehicle until it was coming back toward the launch point, so that the pilot could land on the runway at Kennedy. A lot of effort went into trying to figure out how to do that return-to-launch-site maneuver. Being able to abort to the launch site was a requirement of the program. We never did get anywhere near enough data to be very confident that the maneuver would work. We didn't know that it wouldn't work, but we also didn't have a lot of high-quality data that said that it would work as advertised. However, it was studied a lot.

Then, there also was the matter of the Shuttle flying with the external tank attached. The separation of the tank from the Shuttle is more or less a nonaerodynamic separation because of the extremely low dynamic pressure. A very small amount of force is needed to guarantee the Shuttle separates from the external tank. In addition to that, we had all of the Orbiter-alone configurations that we tested for pressures, aerodynamic forces and moments, for aerothermodynamic heating where such tunnels existed. We also needed to extrapolate and understand how we might take the highest Mach number predictions that we had from wind tunnels (Mach 10) and use what then was the existing computational fluid dynamic techniques build a data table between Mach 10 and Mach 28 for the vehicle. Starting in 1974, it was a huge effort to define all of that.

Then, as holes in the data were found—or as analysis, simulation, or study by some of the disciplinary specialists indicated—we would go back into the wind tunnel and do yet more wind-tunnel testing, trying to refine all of the predictions. There were also tests being done on the thermal protection system in terms of how the tiles would behave under heating in an air load. So there was a lot of testing and a huge database that was being built between 1974 and 1977 that continued to be refined until the first orbital flight in 1981. After Columbia was transported to Kennedy on the 747 SCA, some of the tiles and gap fillers were found to have loosened under modest air loads. To understand and solve this problem, flight tests of the tiles were flown on Dryden's F-104 and F-15 aircraft with air loads to simulate Shuttle flight conditions. Two areas of the Orbiter's thermal protection system were redesigned as a result of these initial flight tests. [57A]

When I started working on the current Shuttle configuration in 1974, we needed to figure out how to have such a large group of people work toward a goal of having certified databases that were checked, understood, and accepted by all of the individuals who were doing the work. By accepted, I don't mean necessarily that they believed the data bases were correct, but rather that they believed they were as close to a consensus as could be achieved on such a vast program and with such a vast number of people working on the program.

All of the configurations somewhat similar to the Shuttle were used in helping build a database as to how good predictions had been compared to the flight-derived values for many aircraft, mostly high-performance aircraft including the XB-70. Nevertheless, there's still a huge difference between Mach 3 and Mach 28. The way that we traditionally break the flight regimes up is that the subsonic regime is roughly Mach 7/10th on down, the transonic is from 7/10th up through Mach 1.5 or so, and the supersonic is from Mach 1.5 up to Mach 5. In the hypersonic regime at Mach 5 and above, we start to have large thermal characteristics that change the chemistry of the airflow. There are a lot of issues that don't occur at speeds much below Mach 5. In addition to the hypersonic effects, of course, because the Shuttle is flying in the exceedingly rare atmosphere at the orbiting altitude, we had different predictions in terms of the type of free-molecular characteristics at the very high altitudes and as we get into the transition region and down into the continuum region. So we had other physics different from the aerodynamics most of us work on that occurred at the very highest altitudes down to where there is a substantial atmosphere-say, 250,000 feet—where things start to behave more like the rest of the atmosphere for hypersonic flow.

Although we had all of those regions to work in, we had virtually no experience in them. Of course, there were sounding rockets that had been in those regions that gave some data. We had all of the capsule reentries, both manned and unmanned, to help build the atmospheric model. Consequently, we did learn something about the density or the pressures, but only a very small amount on each of those. There was a very limited amount of information as to the characteristics at the altitudes up to 400,000 feet, which is what we call "entry interface." Some people take it as some literal change that occurs there, but there is, of course, no such thing. It's where there start to be just enough molecules in the atmosphere that the first tiny measurements can be made of the upper atmosphere and its effects on the Shuttle. For those of us who work in flight dynamics and stability and control, the actual reentry interface for a vehicle going in the neighborhood of Mach 27 to 28 is at an altitude of about 250,000 to 300,000 feet. That's where there is enough aerodynamic pressure that the Shuttle can start flying with aerodynamic controls. I think the Shuttle gets up to about 1/10th pound-per-square-foot dynamic pressure at about 350,000 feet. This is, of course, near the altitude record for the X-15.

The flow regimes usually get a broad-brush treatment, that's multidisciplinary: sources of heating, sources of the change in chemistry, and finally the catalytic effects are very important above Mach 10 or 12, especially for metal airplanes or for any contaminated tiles on the Shuttle. So there are flow regimes where there is catalytic heating along with boundary layer heating and shock impingement heating. At the higher altitudes, there is laminar boundary layer heating and shock impingement heating. After the boundary layer transitions, then there is the much greater turbulent boundary layer heating.

In addition to the primary heating that occurs on the windward side of the vehicle, there is also leeside heating on the back that is substantial. It's nowhere near as high as on the windward side of the vehicle, but it's high enough that the leeside has to be protected. The hottest portion of the Shuttle is the nosecap, which is around 2,800 degrees F, and the leading edges of the wings. We needed a lot of predicted heating data before we could do the final layout of the tiles on the Shuttle. That's how we ended up with some 30,000 individual tiles on the windward side, plus the thermal blankets that go on the leeside of the vehicle, and the carbon-carbon nosecap and wing leading edges. It's quite complicated and very interrelated. How the tiles and the gaps between them are laid out on the vehicle affects the heating and the boundary layer state of the vehicle, which also affects the heating, so it's an iterative tradeoff. As previously noted, we also were working in areas where there was very little real data.

There were some ground-test data that, with the proper adjustment and extrapolation, allowed us to make some good educated guesses as to what the heating environment would be, whether we would have much in the way of catalytic effects, what the state of the boundary layer was, and at what locations and flight regimes the boundary layer transition occurred that will effect the total heat generated by the Shuttle coming in. If the boundary layer is laminar, there will be one level of heating, and if it's turbulent, it will be a very different level of heating.

The Importance of the X-15 Data

All of those issues needed to be answered, but the only really high-quality research data available at the time of the Shuttle was from the X-15. [15] However, the X-15 didn't go fast enough for catalytic effects to be an issue. Actually, some of the things learned with the X-15 were that there was very little laminar boundary layer, and that hypersonically the boundary layer was almost all turbulent. Before the X-15 flew, there were various theories about whether we would have all laminar, a mixture of laminar and turbulent, or all turbulent flow. There were also the heating issues between various components of the X-15. Much was learned about boundary layer heating, and modifications were made to the X-15 to reduce that kind of heating so that the skin-buckling that occurred on some of the earlier flights would not be repeated.

There also was the shock impingement heating that occurred so dramatically on Pete Knight's record-setting Mach 6.7 flight. It had burned through several portions of the aft fuselage of the X-15 and burned the dummy ramjet experiment off the X-15's lower fin. Shock impingement heating would also occur on the Shuttle at some point, and we needed to predict where that would be. We knew that the bow shock that comes off the nose of the Shuttle at the higher Mach numbers was going to impinge on the wing. The wing, of course, would be generating its own set of shocks at that point, so there was the possibility of shock impingement heating.

The information that we learned from the X-15 flight that terminated our ramjet studies made us a lot smarter in terms of how to deal with such heating issues when we got to the Shuttle. The X-15 had given data on shock impinge-ment with the interference of two shocks intersecting and those impinging on the surface. Even though it melted pieces of the X-15, we had a good idea at what temperatures those things would melt. Since we had thermocouples on parts of the vehicle and knew the melting temperatures of the metal components, we were able to extrapolate a good set of information that showed the actual temperatures. [33]

We could use the answers in the X-15 cases as benchmarks for the relatively simple codes that people were then using to calculate the heating rates for the shock impingement and the boundary layer heating. We had very good instru-mentation on the X-15 that told us about the boundary layer heating that could be used to calibrate the codes that we were going to use to predict the Shuttle's heating rates. Of course, there were large uncertainties put on all of those num-bers. We didn't really have enough data to say with absolute certainty what the heating condition would be, if the boundary layer state would be laminar or turbulent, precisely where the shock might intersect, and impinge on the wing.

Early in the Shuttle's development, we had technical panels from various disciplines and with varying levels of experience. We would discuss the source of heating, how we were going to predict it, and how we were going to bracket it in terms of the best case and the worst case for various parts of the Shuttle. These panel discussions led to the initial design of the thermal protection system—the tiles, the carbon-carbon, and the thermal blankets that would be used on the Shuttle and where they would be placed. Of course, everyone on the panels learned a lot in working together.

In a broad sense, there were really two major issues at the time. Where normally we would be talking only about the aerodynamics of the aircraft, we were talking about aerothermodynamics and aerodynamics. In some cases, they're very closely related, and in other cases, although they use the same physics, the focus is quite different in how to deal with them. Once we had preliminary data for the aerodynamics and aerothermodynamics and where various kinds of tiles were going to go, we got into the analysis of the structural heating and the structural loading of the Shuttle. There was a basic design in terms of the internal aluminum structure that would carry the aerodynamic loads. Then there would be some heating of the internal aluminum structures. The design limit for internal temperatures was around 200 degrees F, so there would also be some thermal loads, due to the structure getting hotter and the expansion that would occur throughout the vehicle, in addition to the aerodynamic loads. Taken all together, this was the total thermal load that would go into the vehicle.

All of these things would not occur at the same time. The highest surface heating rates occurred earliest. However, since the surface stayed hot for a long time, the heat would flow into the vehicle and start heating the internal part of the structure. Because this takes time, the actual highest temperature for some of the internal parts of the Shuttle actually occurs after the Shuttle has landed and has been sitting on the runway for a little while.

We had all of those issues of the internal design of the vehicle. They all had to be compatible with all of the systems on the Shuttle—the hydraulic system, the actuators, the avionics, and the RCS jets. Some of those things generated significant heat on their own, so there was a huge effort in analyzing the internal heat and external heat flowing inside and making sure that no part of the structure and no component came anywhere near its temperature limits.

Dryden and the Shuttle Technical Panels

As the various technical panels were formed and began operating, early on in the program, quite a few Dryden people were active on them and made important contributions. However, for whatever reason, the closer we got to the definition of the actual Shuttle and getting into the missions and the definitions of sensors, sensor locations, sample rates, and accuracies, a lot of the Dryden people lost interest. We were used to being in charge of the instrumentation, and that was not going to be true in this particular vehicle, because, first of all, everybody could not have what they wanted. The Shuttle would not be able to lift off if all of the sensors that each researcher wanted were put on the vehicle. The equipment needed for the sensors, the wires and signal conditioning, and the sampling and the recording of all of that data would have weighed tens of thousands of pounds.

So there was a very intense effort to wade through all of the measurements that each researcher wanted from the vehicle. Each researcher had to fight for his data, have justification, and have reviews on his own technical panel. After he convinced the people on his own panel, his proposal for sensors would be presented to other panels. As the proposal got more and more support, it gradually worked its way up through the official part of the program. Ultimately, sensors that involved the flight-testing of the vehicle and the safety of the vehicle were the ones were included on the flight. For the most part, all of the ones wanted merely for research so we could be smarter for future Shuttle improvements were denied. This effort is called scrubbing the instrumentation list. I never saw a complete list, but it had to be tens, if not hundreds, of thousands of requests in terms of sensors. Sometimes it was the same sensor with different filtering or different sampling rates.

On the first orbital flight, I believe there were in the neighborhood of 10,000 to 12,000 sensors aboard the Shuttle. I guarantee that no researchers on the Shuttle program, including myself, were anywhere near happy with the sensors that we were allowed. The problem faced by Shuttle management was trying to keep off measurements or sensors that were on a wish list as opposed to on a requirements list. I think they made some poor decisions, but for the most part, considering the total weight of the instrumentation, the recording and the telemetry requirements, I think they did a good job because we did get a large amount of research data from the vehicle. Other research data programs were advocated throughout the first 50 or so Shuttle flights where we did get to fly some experiments to get additional data, both to help the Shuttle and perhaps to help the code development for computational fluid dynamics (CFD), thermal prediction, or various subjects of interest to people working on the Shuttle or a follow-on to the Shuttle program.

Since Dryden had significant flight experience with hypersonic and delta-wing vehicles, we got to play a fairly major role early on in contributing to the various technical panels. We had varying levels of experience and knowledge of the various disciplines we were tasked with, but as we worked together over a few years, it soon became apparent who, in addition to Dryden, would take the lead in certain areas, who would really have the largest vote as to whether we did or didn't advocate something, or how we advocated that it be done. It was a bigger deal than any program I'd been involved in before, but it was also very much a situation where we could get down to arguing the physics and the engineering with other people familiar with physics and engineering and get all of the issues down to a fine point. Even if I didn't agree with the decision, I at least understood what the issues were. Once the decision was made, I would live with that decision and do the best I could.

I recall that every panel was led by a person from Johnson. Some of the key Johnson people leading the flight-test panel over the years were Rick Barton, Paul Romere, Doug Cooke, Dave Knipe, Rob Meyerson, Jimmy Underwood, and Steve Labbe. The fact that Johnson personnel led each group didn't mean they were autocratic in any sense. It just meant that they would be coordinating the project and feeding the information from the technical panel into the rest of the program. Furthermore, most of the decision-makers in management were located at Johnson. However, a lot of the people with experience in wind tunnels, CFD and flight-test did not work at Johnson. They worked for other NASA centers, the Air Force, Rockwell, and the other contractors. Consequently, even though people from Johnson chaired the panels, they did a very good job of making sure everybody understood the requirement and that each person was listened to as we went through defining the problem and coming up with potential solutions. Then they would record those solutions and present them to those who were impacted by each solution working its way up through the system. The process worked fairly well. I don't ever remember being unhappy due to others not listening to me or feeling I wasn't making a significant contribution to the panels.

However, as we got into 1976, it became clear that a lot of the decisions of our panels were being denied. This started to become very frustrating. We thought we'd come up with—if not the right answer—a good answer in many areas, only to be told that it wasn't going to be done this way. In short, "thanks but no thanks." This became very frustrating for the people at Dryden, myself included, because we were used to having a fairly big say in everything that we worked on, because everything we worked on we flew.

That wasn't true, of course, on the Shuttle. I think for that reason a lot of the really experienced people from Dryden little by little dropped out of, first, the panels, then the program, the advocacy, and finally any responsibility on the Shuttle. I think that was very unfortunate. I think management for various reasons, particularly of the research group at Dryden, felt that that was okay. We had plenty of important programs to work on. To waste our time on something where we would get a very tiny accomplishment for a very large amount of work was not justified. Little by little, it soon got down to what I felt was just a handful of Dryden people contributing to the panels, staying abreast, and being available for briefings.

I was among those people who chose to and fought to stay on the program, and I wasn't really discouraged by any of our management from doing that, even though I had other duties at the time. I was a primary research person by 1975 on the 3/8th-scale F-15 RPRV, and I became the primary engineer and researcher working on it through 1981. I was also trying to do parameter identification on any adequately instrumented aircraft configuration being flown at Dryden, by industry, or at other locations. This is when another facet of my career began, when I became very interested in the high angle-of-attack flight regime, aerodynamics, stability and control, and flight dynamics. I was probably working about half of the time on the Shuttle and half on my other assignments.

While Dryden's participation in the Shuttle continued to decline, I actually started participating in more of the panels than I had before, because of my experience and my proximity to other people who had experience. I was included in groups that were working on the Shuttle lift-and-drag characteristics, aerodynamics, aerothermodynamics, flight dynamics, and stability-and-control characteristics. My major thrust was in the parameter identification, because I did have far more experience in it than anybody else working on the program. I'd done parameter identification of flight dynamic stability and control on the M2-F1, the M2-F2 primarily, the other lifting bodies as a secondary contribution, and on the X-15 and the XB-70, so for the various flight vehicles that were most applicable to the Space Shuttle, I had significant experience.

Dryden's Shuttle Simulator

Little by little, my main focus was shifting in terms of how we were going to fly the Shuttle. I didn't do any of the flight-planning directly. All of the official flight-planning was done by Johnson. Some of it was also done at Rockwell—they had a simulator as well. But I, along with Milt Thompson and some others, advocated that Dryden come up with a Shuttle simulation. There were several reasons for that. One of them, from my point of view anyway, was that a real-time simulator with a pilot in it is one of the best tools for trying to understand the interaction of complex phenomena.

I'd had that experience on all of the lifting bodies, the LLRV, the X-15, the 3/8th-scale F-15, and many other simulations that Dryden had done, so I knew what a powerful tool it was. It seemed to me that we could make better suggestions and perhaps study areas that others had not looked at by having our own simulator. Milt was also very interested in the simulator, because although he was no longer an active research pilot for Dryden, he still had all his piloting skills.

He and I spent many hours in our Shuttle simulator going through what the program was going to try to do, and making our own "what if" studies and misprediction studies for the Shuttle.

Johnson became concerned that if our simulator wasn't using exactly the same information they were using, we would come up with comments or conclusions that they wouldn't understand and couldn't duplicate. I thought at the time that was a little high-handed of them to not want us to have a simulation unless they could define it. Later I could see that this was the most important step in having our own simulator to help the Shuttle program. Eventually, it was agreed that Dryden could have a simulation, but it had to be certified by Johnson as being exactly the same simulation as they had. We had different computers and different software, but our answers had to agree with their answers—which they should. If the simulations were done right, they should agree.

We had our group—first Larry Schilling, then Lee Duke and John Bresina—at Dryden come up with a Shuttle simulation. We ran dozens and dozens of check cases to make sure that our responses to a given input were exactly the same as Johnson's. Every time Johnson did a revision of their simulator, we had to do a revision of ours, then again go through the certification. It was a lot of work for us, especially for the two or three simulation engineers who spent a great deal of time getting it right, running the check cases, and confirming that the output was the same as Johnson was getting. In addition to myself, Bruce Powers and Mary Shafer did a lot of the very tedious check case work. The real value of that simulation was that if Milt and I found a problem, for instance, we could talk to the stability-and-control, the controls, or the simulation people at Johnson or Rockwell, tell them what we'd done, and have them take a look at the same thing. We were communicating very clearly, even though we were using completely different simulators 1,500 miles apart. However, with the same implementation and database, we would get the same response to the same input.

We learned in our simulator what others knew before: that for the Shuttle the bank angle became the primary way of controlling the energy management all the way from about 250,000 feet down to Mach 2. The bank angle basically kept the Shuttle from skipping back out of the atmosphere. The Shuttle was able to use a significant amount of lift and drag to control its energy by putting the lift vector more horizontally than vertically with bank reversals, and adjusting its flight path by zigzagging back and forth along the nominal trajectory to the landing site. So bank angle was a primary parameter, and Milt and I quickly caught on to what was being done there. Using bank angle to adjust the flight path was very important. It's hard to explain how important that is for a vehicle going from orbital speeds to a landing at a given point on the ground where there were some aerodynamic uncertainty, atmospheric uncertainty, and local winds. We could look at all of those issues with the simulator in real time and start to get a feel for whatever it was that those in the overall Shuttle program were worrying about.

Our Shuttle simulator was only from orbit down to landing. It did not simulate the launch portion of the flight. We did have the full simulation of the control system, including the reaction control system jets, as well as all the predicted forces and moments with the tolerances and variations that were documented in the aero-data book. I would say it was really a small percentage of my time that I spent in the simulator. But I learned things faster in the simulator. If I had a question that I didn't quite understand, I could go into the simulator for an hour or two. I then understood the question and sometimes even had an idea of how we might solve the problem.

That simulation was the first one that Dryden did, at least that I'm aware of, that included the round, rotating, oblate Earth so we could do orbital simulations, do our de-orbit burn at a given point, and follow the flight down to a landing at Edwards. I think we practiced landings only to Edwards even after the flight plan started calling for going into Kennedy. The simulator also allowed us to put discrepancies into the atmosphere in terms of density, pressure, and temperatures, then look at what those dispersions would do to our energy management scheme.

The other thing that I used the Shuttle simulation for was to do practice stability-and-control and performance maneuvers. Milt and I would practice the pushover/pullups in the simulator, and we could capture the data. Then I could go off and analyze the maneuvers, even though I knew what the answers would be because of the deterministic nature of the simulator. I did the same things with the stability-and-control derivatives as we practiced the stability-and-control pulses. I think all of ours always were generated by the pilot. In most cases that was Milt, although on a few occasions I did use some of the other engineers as pilots for the simulation.

I could take those maneuvers, put them on our batch unpiloted, non-real time simulations with the fixed input, and analyze them to get an idea of accuracies and sample rates. I then used these results to justify my requirements for getting adequate stability-and-control pulses and adequately specified measurement sensors so that when we did the analysis of Shuttle flight data, we would be able to appropriately update the aero-data book on the Shuttle.

One of the early things that I championed was specifying the sensors, including the sample rates, resolutions, and sensor locations, for things that were vital for the stability-and-control and performance analysis. I spent a lot of time advocating those sensors in 1975 and 1976. It sounds like a small task, and it would be if it were an F-15, an F-18, or an F-22. But for the Shuttle, there was so much competition for a limited number of sensors. This was because of the power they required, along with the weight of each sensors. Finally, there was a limited capability for recording the data from all of the sensors.

After those specifications were in and they were accepted, I became more involved in the Shuttle in a global sense—in terms of the flight dynamics, the performance, the aerothermodynamics, the stability-and-control of the Shuttle in general as well as the ascent work that I was to do for a couple of years starting in 1980.

X

STS-1

Well before sunrise on April 12, 1981, I found myself in the Dryden control room, waiting for activities at Kennedy Space Center to begin. I knew the time lines and I knew the holds beforehand, but my memory is that we had our strip charts ready to go two hours before the Shuttle Columbia was scheduled to be launched on the STS-1 flight, wanting to make sure we had everything that we thought we would need. There was the possibility that for some reason the Shuttle might need to do an abort-once-around (AOA). If so, Columbia would end up landing at Edwards less than two hours from engine ignition. In that case, our activities in the control room would be much different than the ones officially planned.

Wings into Space

As the STS-1 countdown was just prior to zero, the pumps on the Space Shuttle's main engines started, and we could start to see the vapor. Within a few seconds, we started to see the thermal plume from the hydrogen/oxygen combustion in the chamber. Just as that happened, there was a little gimbling on the nozzles. Then, the solid rocket boosters lit, and, almost instantaneously, the vehicle started to lift off. We were now enjoying the show, not worrying so much that something might go wrong, because there was nothing we could do anyway. We might as well enjoy the launch.

It was, from my point of view, a perfect launch. Television, of course, made it very easy for me and millions of other people watching the launch. We were watching a first flight by John Young and Bob Crippen, the first astronauts to be launched into space by solid rocket boosters, as well as the first to be launched into orbit on a winged reusable vehicle. It was also the first flight for a crew with Space Shuttle Main Engines (SSMEs), and for that particular Shuttle configuration. The Shuttle stack looked ungainly at liftoff, but I'd become used to seeing it, so it looked normal to me by then. As the Shuttle cleared the tower, Houston took over. A few hundred feet in the air, Young and Crippen went into their programmed roll which put them heads down and downrange to the east to go into orbit. That's the optimum combination to get the vehicle out of the atmosphere as soon as possible, but not end up having to make a right-angle turn once it gets out of the atmosphere. The trajectory optimizations all come out with something fairly close to what the Shuttle does with its programmed trajectory.

The controller was talking everybody through as the strange configuration got further and further away, until only the exhaust from the solid rocket boosters (SRBs) was visible. The separation of the SRBs from the external tank was easily visible with a telephoto lens. My recollection is that the first time I saw it, I was surprised to see how much plume was still coming out of the SRBs, and I wondered if they had separated at the right time. But that is the way it's supposed to go. We watched on long range optics, and by the time the SRBs dropped away, we could see the high-intensity light coming from the three main engines on the base of the Shuttle. Then, little by little, that began to flicker in and out. Due to the thickness of the atmosphere, the humidity in Florida, and the curvature of the Earth, the camera eventually couldn't pick it up anymore.

We, of course, had continuous information from the control room at Johnson Space Center. They could hear conversations that were not on the public transmissions. The astronauts were going through their various checks, and we had good information that things were nominal all the way to orbit. They did their lob of the external tank, and then they finished circularizing the orbit they were to stay in for two days. Only about 20 minutes had passed from the time it took off until it was more or less in orbit at about 25,000 feet per second.

When it was clear that the Shuttle was not going to abort to Edwards, I felt a great deal of relief. Most scenarios for aborts are not very good, so all of us at Dryden were very pleased to see it in orbit. Then we started hearing some concerns about things that we hadn't noticed on the launch. Specialists in various areas were talking about things that were off nominal, such as overlofting, debris, and other things not quite as expected. None of them sounded particularly serious to me, except when they started noting some of the closeup telephoto shots they had of the launch. They showed a lot of debris, and a lot of it had hit the Shuttle. Some of the debris, I think, was coming out of the trench below the Shuttle's engines, and included various expendable equipment that was nearby, as well as ice coming off the external tank.

The next exciting thing that happened on the flight was the opening of the payload bay doors. If they didn't open right away, the astronauts were coming back to Edwards at the next opportunity. I don't think the doors opened as smoothly as I thought they would. There was probably uninformed concern at Dryden about all of the activities, because although we knew roughly what should happen, none of it had anything to do with our specialty. Once the payload bay doors were open, they had video coverage of the Shuttle. They could see damaged tiles on the orbital maneuvering system (OMS) pods, as well as what looked like dings on various parts of the vehicle.

The astronauts on STS-1 were launched wearing space suits so they were prepared to use their SR-71 ejection seats if something happened at Mach 3 or below. I think Mach 3 on the Shuttle occurs around 90,000 to 100,000 feet, so that probably would have been okay for using the SR-71 ejection seats. However, I remember some talk at the time that there might be some way, once in orbit, that the crew could go outside and take a look to see if anything was damaged on the bottom of the Shuttle, where the greatest heating and aerodynamic forces were going to be if any tiles there had been damaged.

Otherwise, things seemed to be going fairly well, despite little gripes about things that would be fixed before the next flight. I remember being very concerned about whether the Shuttle had significant tile damage on the bottom. Depending on what the damage was, the vehicle could reenter with a significant sideslip. If there were very much more than one degree of sideslip, the thermal protection nosecap would not be the highest heating point, so we would get a burn-through on the vehicle. There were also concerns that if a large enough group of tiles was missing, the heating would get into the wing's aluminum structure and damage it to the point where the vehicle could no longer fly.

I don't know how Young and Crippen felt about it, but to me it did seem risky. There was talk of using our classified military cameras to take pictures and see if there was significant damage to the tiles. Information about our capability for taking pictures of things in orbit was classified. At any rate, I remember hearing less talk as we approached the time for the Shuttle to do its deorbit burn. I assume that they got enough verification with those classified cameras that there wasn't a significant bunch of tiles missing, although I never heard directly that they were no longer concerned.

Heading Home at Mach 28

On the day of the landing for STS-1, I remember leaving home quite early. I knew they were allowing visitors into Edwards. There also were visitors with passes specifically for Dryden. There were also visitors who were supposed to go to other parts of the lakebed to the south and east. I knew there would be a lot of traffic and a lot of people in the wrong place, making for even more traffic. The base security people were going to let only the right people into any given site for viewing the landing. There were also a lot of us who were supposed to be there for work, as well as a large contingent of support people, who took care of the crew and the Orbiter after landing. We also had a large percentage of Dryden people serving as escorts and guides for the public affairs office. They were all proud to be doing this.

This was a big day. It was going to be a crowd bigger than anything we had seen before at Dryden. I had seen this commotion and heard about it on the Apollo flights, so I knew of many sad stories of people who were in a traffic jam somewhere because they'd waited a little bit too late to start for their destination. I didn't want to be one of those people for two reasons. The most important one was that I wanted to watch the Shuttle come in. The second reason was I actually had official duties in the control room and might be of some use in case of certain contingencies. It's hard to interact and save the Shuttle, but there are always questions that the sooner they're answered by somebody, the better off everyone is.[1]

I had the duty of being at the strip charts for the flight dynamics parameters from the Orbiter's operational instrumentation system. The parameters from that system were telemetered to the ground, and at Mach 14 the telemetry signal would be coming into Vandenberg. It was then merged with our data, and we'd have it from Mach 14 all the way down. Rich Maine was also with me in the control room, along with Milt, Archie Moore (who was Dryden's range manager at the time), Al Carter, Joe Weil, Dick Day, Bruce Powers, and Tim Horton, who were also working on the Shuttle. There were a number of people from outside Dryden who were there as well. We also had our world-class control room technicians who not only assured us of the Shuttle real-time data, but also had always provided Dryden with real-time flight data and communications for all of the research aircraft flights I had worked on over my entire career. They always came through with this vital data, without which we could not proceed.

At any rate, I don't remember the people as much as I remember the event. Rich and I had been through the time line dozens of times, putting forth our own uncertainties regarding at what Mach number the initial bank maneuver would take place, where the various bank reversals would take place, and when, in terms of the clock, we would be able to hear the Shuttle emerge from blackout. We had displayed Greenwich Mean Time in the control room so we had our own little time line, independent of the official person who announced all events from Houston.

Having left home many hours before the Shuttle was due to land, I spent some time working in my office. I noticed a lot of other people who also had duties who decided that an extra half-hour or an hour of sleep wasn't going to do them any good, so they might as well get here before the traffic made it impossible. I don't know actually how many people were required to be at Edwards that day, but there were large Air Force groups involved in rescue and such contingencies as security and closing the base to any air traffic during the reentry. There had to be a couple of thousand people there on official duty from Dryden, the other NASA centers, the various contractors supporting the Shuttle, and the Air Force along with their involved contractors. In terms of Kennedy Space Center, that may not be a big number, but for any operation at Edwards, it was the largest by far of any that I had seen.

After a while Rich and I decided to go to the control room, get into position, and get into the flow of things to be ready for the actual deorbit burn. On the earlier orbits, they'd had conversations through Hawaii, I believe. This time, on final orbit, Guam was going to have the last contact with the crew until we heard from them either through Vandenberg or Dryden communications. We believed, based on the antenna patterns and the altitude, that we would pick up data from the Orbiter's operational instrumentation (OI) and development flight instrumentation (DFI) systems at about Mach 14.[2]

They did the 2-minute-40-second deorbit burn with a delta-V of about 300 feet per second with the OMS engines. It was about 60 minutes from the start of this deorbit burn to wheel stop after touchdown. The vehicle then went to the next event that the entry community talks about-entry interface, a useful point from which to time things. The entry interface would occur about 25 minutes or so after the deorbit burn ended, just after the Shuttle passed Guam. That's roughly a half hour before it was to touch down at Edwards, so it would still be quite a ways out. Entry interface to an airplane guy is simply a mark in the sky, there being nothing at all special about it.

The Shuttle was going close to 25,000 feet per second. As we would discover after we had obtained all of the onboard data, the first time that we can see any significant deceleration on the Shuttle is when it reaches an altitude of about 350,000 feet. Then, we can start to see microdeceleration on our aerodynamic coefficient identification (ACIP) instruments, which are very sensitive. So at about 350,000 feet we start to see about one tenth of a pound per square foot of dynamic pressure. That's what I define as the place where interesting things start to happen. The first place I try to extract aerodynamic data from the Shuttle is about 350,000 feet. It's an interesting coincidence to me that 350,000 feet also was roughly the maximum altitude the X-15 flew. Of course, the X-15 had much lower dynamic pressure than the Shuttle, because dynamic pressure is a function of velocity squared. In thousands of feet per second, roughly 25 squared is 625, and 4.3+ squared of the X-15 at 350,000 feet was roughly 19. So the ratio in dynamic pressure at that condition is more than a factor of 30 lower on the X-15 compared to the Shuttle.

We knew that about 47-48 minutes after the initiation of the deorbit burn, or 12-13 minutes before wheel stop after landing, we were going to start receiving data, so we were watching the clock. Depending on antennas, orientation of the vehicle at the time, and various other things; we knew that the Mach number might be a little bigger or smaller than Mach 14, but it would not be much different if the Shuttle was still on its planned entry trajectory. All of the significant parameters that we were interested in, even at low sample rate and poorer resolution than we desired, were going to be telemetered and we'd be able to see them on the strip charts.

Consequently, we were well prepared for that approximately 47-48-minute point after initiation of the deorbit burn. We had our own little conversations in the control room. We were paying attention, but we knew we weren't going to hear anything very interesting for awhile. The mission announcer made reports on the Shuttle's position and its altitude. We kept hearing those and that was encouraging. He was getting those from the planned profile obviously, as nobody was in contact with the Shuttle at that point.

We heard the entry interface announced. Then with the Shuttle at 350,000 feet, 44 minutes after the start of the deorbit burn, and about 16 minutes before wheels stop, the announcer in Houston said, "We should be leaving blackout and reestablishing contact with the Shuttle at any second now." Rich and I looked at each other fairly shocked, wondering if we were two or three minutes off or had missed a calculation somewhere. We stared at our strip charts, listening for communication, but there was nothing but silence. That was the worst thing that we could have experienced at that point, because if we don't hear anything it could mean there's nothing to hear from. Since this was a first flight on a very ambitious full flight-envelope entry, it was very scary. Whatever my heart rate had been before that, it went up a great deal more, because I started having this feeling that this was not going to be a good day. Then, probably a minute later, the announcer corrected himself and said that we should acquire data in 30 seconds.

That made us feel better, but it gave me a bad couple of minutes. Then, the strip charts made a noise as they started responding to telemetered signals and the needles started scratching back and forth across the paper. At various places in the control room, displays came alive, and it was clear that we were getting data. The Shuttle was going the speed it should be going, it was at the altitude and the angle of attack expected, and it was at the position we expected for Mach 14. The Orbiter was off the coast of California, north of Santa Barbara, and coming in to land at Edwards. The first data was actually received at Dryden, not at Vandenburg, because one of our tracking gurus, Jack Kittrell, had set our antennas at a lower angle than officially stated. His intuition based on other research flights, including X-15s, gave him a hunch on where to look.

Then we had something very real to do. We monitored what are called "bank reversals," [3] (Appendix F) which are energy management maneuvers. [58] We knew where they would be. They vary a little bit, depending on how far the Shuttle is from its landing point compared to its Mach number and altitude at any given time. Those are adjusted for any given orbit. There were no planned maneuvers for stability and control or for performance evaluation on STS-1. Each time the Shuttle entered or left one of those bank reversals, we would get aileron and yaw jet data, and then, at the last one, there's a possibility of getting some rudder as well. For the most part on that mission, we were watching for the bank reversals. The initial bank was up at Mach 24, which, of course, was before we had communications from the vehicle. The next one was just before we received the first telemetry data. We saw the last two bank-reversal maneuvers and got to see that we had the control positions, angle of attack, roll rates, yaw rates, and pitch rates that we were expecting. Things were looking quite nominal to us, which was a very good thing.

Milt and I had made a deal that if everything was going well at Mach 2, since we didn't have any duties that required us to be in the control room, we were going to leave our strip charts and go out on the roof and see if we could spot the Shuttle. We went out, and just about the time we got in position, Milt pointed up and said, "There it is." I saw it shortly thereafter. We watched it come over Edwards, go out toward Boron, then circle around to get on the heading alignment circle (HAC). (Also called the "heading alignment cone.") It was early enough in the morning that we had good sun angle on the white Shuttle. It was a clear day, so we could see the Shuttle when it went over the base. Then as it went into the heading alignment circle, we could see the mostly white vehicle.

We could see the very familiar pushover. They were going to land on the dry lakebed on runway 23, which is about 26,000 feet or 5 miles long and only about a half a mile from Dryden, so it can be seen very well from the Dryden roof. (Appendix E) The Shuttle looked as if it were coming a little bit south of directly at Dryden. Then, it did its pushover to 280 to 300 knots and came down at a glide slope angle of something around negative 20 or 22 degrees. It's a larger vehicle than the X-15 or any of the lifting bodies that we've had, so we got a better feel about how it was doing in terms of altitude and position with respect to the runway. Young came quite a way down runway 23 and then did his pullout. As he leveled it out, probably 100 feet in the air, he lowered the gear, then came down and completed the landing. We had been able to see the Shuttle from roughly Mach 2 on down to touchdown and landing rollout. (Fig.1)

As it turned out, he landed three thousand feet long, but the vehicle did not do a slam down as we'd seen on the X-15. It landed on its back wheels and went down the runway for a fairly long period. Then it did what they call a derotation and put the nose gear down on the runway. I had a nice view of the landing, and I wanted to do something to celebrate, because I'd been working on it for at least nine years and working on the project intensely for five years. The success of STS-1 was quite a relief for us, and we had a feeling of euphoria that lasted for many hours, if not days.

I can only report on the excitement that there was on the roof and along the taxiway on the ramp out behind Dryden. It was a huge celebration. The national anthem was played. It was a very special moment, not only to me but obviously to the half million people who were watching it at Edwards. I think most people there who had decent vision got a good view of it. We, of course, got the double sonic boom as it came overhead, caused by the compression shock at the front

Fig. 1. The Space Shuttle Columbia touching down after orbiting the Earth on STS-1. (EC81 15104)

and the expansion shock at the back. We heard both of those, and it sounded like a double boom with the Shuttle. It was a very special thing to hear, although it was well past overhead when we finally heard it because the boom was probably generated at Mach 1.5 to 2 at 60,000 to 70,000 feet. The shock wave moves down to us at roughly 1,000 feet per second, so there is a delay of a minute or so before we hear it.

Those who couldn't see the Shuttle descending, knew something was coming when they heard that double boom. Of course, they could watch everybody else pointing and hear everybody explaining to each other where it was, what it looked like, and how great it was. It seemed everyone was busy living the moment.

The other thing that I learned from STS-1 was that as the Shuttle became subsonic and before entering the heading alignment circle, I could hear a rumble typical of a large transport or a commercial airliner. I thought, "Oh, oh. Somebody has really messed up, because the airfield and its airspace is closed." There were no other aircraft allowed anywhere near the Shuttle's flight path. My feeling was that somebody had managed to slip through, maybe to get a better view of the Shuttle landing.

That was my reaction for the first ten or fifteen seconds, before I realized it was actually the Shuttle that I was hearing. The low-frequency rumble was a noise I associated with engines, but it's actually the noise that the airframe makes on every airplane. On STS-1, it was the aerodynamic noise of the air flowing around the Shuttle. I remember commenting on the noise to Milt, "That's the Shuttle making that noise." He puzzled about that for a minute, then said, "I think you're right." I've explained this to many people over the years, and they're people who know aerodynamics better than I do. Their initial response is, "That's really interesting." Their second response is, "Of course, you would expect that, wouldn't you?"

Debriefs

After savoring the jubilation of the thousands of people who were near me and my own particular excitement, and not knowing exactly what to do to celebrate, we eventually went back into the control room. We stayed in the control room where we had a good view from all of the video displays of the various stages of "safing" the vehicle. We stayed there until John Young and Bob Crippen walked down the stairs and did a walk around of the vehicle. Later that day, after the Shuttle was brought up to the mate/demate area, we went all round and under the Shuttle, doing our own inspection. We were feeling some of the tiles on the bottom of the Shuttle to see what they felt like after reentry, when someone in charge of the vehicle got quite upset with us and sent us out of the area. We didn't know the rules and had no right to be in the area. The official who sent us packing was doing the right thing, because the state of the vehicle needed to be closely controlled until it had been fully documented, especially after the first flight with the ceramic tiles.

We had strip charts from the control room—which meant we had information to take back to our offices to go over. This included not just Rich and me but also the other people like Al Carter in the loads and the structures area and the people with the heating parameters. There were two reasons to take the data back and study it. There would be several postflight debriefs. The primary one was where Young and Crippen were debriefed. That's the one that most people remember. The only real feedback given during the postflight was if they brought up any issue that was not clear to them or they didn't quite understand what had happened. Anybody who had the information in the postflight crew debrief would be able to say, "Well, we saw this on the strip charts," or "We would expect that," or "We're puzzling over the same thing," or whatever.

At the postflight, I remember each of the astronauts going through the events before the deorbit burn, talking through the deorbit burn maneuver, and commenting as they came down. I don't remember that they talked much about anything really spectacular that was unexpected, seemed more intense, or that might tend to frighten the crew. They went through each of the sequences. They were not actually flying the Shuttle, but they were in a position where if something wasn't going as expected, they would be able to take over command of the Shuttle just by moving their controls. The vehicle came in on an autoland trajectory. They had studied what all of these would be because they were entering on the orbit that was planned when they launched, but as the Shuttle descended, they were more or less following along, watching that the Shuttle was doing what it was supposed to be doing. Of course, once communication was established at Mach 14 or so, they were also in contact with the capcom.

You would think that John Young would be euphoric, but he never showed it. I thought Bob Crippen acted pretty darn pleased, while John, in his normal low-key casual manner, acted as if it was just another day at the office. Even though John didn't show it as the rest of us did, I suspect he was very excited, and certainly well satisfied with coming back to Earth and sitting at the debriefing table talking about what had just happened.

They went through the things that they saw as either out of sequence or unexpected, partly to get it written down. The debriefs are recorded, then later are transcribed. The transcription of a debrief may have action items for people to investigate further. It's also a good text for future crews to read because they're going to be going through much the same situation.

The only thing I remember that John Young was concerned about was landing long. He felt that the vehicle had a lot higher L/D than was predicted and that he'd been flying in the simulators and in the Shuttle Training Aircraft (STA). I don't remember too much about the debriefing other than that comment.

Because they had been in space for two days, prior to that crew debrief, they had been in the medical facilities being checked out, probably lying on couches and reacclimatizing themselves to gravity. The debriefing that I remember was quite some time after the landing, and I suspect there had been other debriefs, prior to that one, that occurred in the medical facilities, certainly anything regarding physiological effects. In this case, the first flight was more a blur to me than I like to admit. It just seemed such an up day, that everything was wonderful.

Once we settled down, Rich and I took the strip charts with the telemetered stability, control, and performance parameters. This might consist of four to six sets of strip chart data, and there are eight channels per strip chart. But having worked on lots of other research aircraft flights, we had a process where we went through and read off the time on the strip charts, which was in a coded form rather than in the nice clear numbers as it is these days. We had to look for special coded marks and then work back and forth from that until we found something that we could identify as a specific time in terms of hours, minutes, seconds, and milliseconds. Since it came out of Dryden's control room, it was the same set of time codes that we had used in the past for research flights.

We spread the strip charts across a large table, lining them up in one place to study a maneuver. For instance, if we were looking at a bank reversal, we would line up the strip so we could look at all of the parameters of interest to the flight dynamics and stability-and-control community. We could do an initial assessment, just looking at the strip charts. Along with that, of course, we were also selecting portions of the flight for more intense analysis. Since there had been no planned performance or stability-and-control maneuvers on STS-1, our primary task was to read up the bank reversals above Mach 2.

Any time we got motion on the controls, we'd identify it as a maneuver. For that particular flight, the start of the bank reversal showed up as aileron deflection with many reaction control system (RCS) jet firings until it got to the right bank angle, then the opposite aileron and opposite yaw jet firings for the end of the bank reversal. We would consider those as two maneuvers for each change in direction of the bank reversals. We could then look at them and start to ponder whether the Shuttle flew more or less the way it had been predicted to fly. The real analysis had to await the arrival of the data in a digital format, because with stability-and-control derivative extraction, we really need to have everything very well time-synced.

Rich had written a program to combine all the data from all the systems so that this would be a straightforward task, but it was going to be awhile before we actually got the digital data. The first look that we got at the flight data was from Mach 14 on down. The data we were looking at on the strip charts was of lower quality than we would be getting later from the aerodynamic coefficient identification (ACIP) recorder. The ACIP sensors and tape recorders were attached to the wing spar, so it was quite a rigid attachment, but it was also quite close to the structure that would have the highest heat. It never had high heat, because the system functioned as planned. The ACIP package flew under the payload bay, attached to the wing spar, on many flights of Columbia until well after the Challenger incident. It also flew on Challenger from its first to its final flight. Even though the vehicle was going to be operational after four flights, we still had to continue to refine the database for the simulation and control system designs and better define the center-of-gravity placards for future payloads.

The developmental flight instrumentation (DFI) and recorder, which weighed over 10,000 pounds, were going to fly for just the first four flights, because that system was considered necessary for certifying the vehicle as operational on its fifth flight. The DFI had a large number of temperature thermocouples, surface pressure measurements, and other information more detailed than the operational vehicle would want, things that were identified and had won their way into the master measurement list. The data was time-tagged as to when the sample was made. We could reconstruct in a specific time what all of the measurements were on the DFI, so all of this data could be merged later.

The DFI data was in addition to the operational instrumentation (OI) system and the general-purpose computer (GPC) system. The OI and DFI data were telemetered to the ground during those early flights, so we had some data from Mach 14 on down. The ACIP data did not end up with its own transmitter and its own downlink. We had to wait to see high-quality data until that data made it all the way through the system. As a result, the data we were looking at was of relatively good quality, but it was not good enough for the types of analysis that we would need to do, although it was certainly adequate for studying to see if there were any particular anomalies from Mach 14 on down.

I don't remember anything that looked particularly strange to me. We could see that as the Shuttle went from Mach 2 down to Mach .7 or .8 that it experienced an increase in buffet, which was expected. Because of the angle of attack the Shuttle was flying and the lack of fineness to the vehicle profile, it was expected that we would see low buffeting in the transonic region. Later the project was to identify a more specific part of the buffet termed a quarter hertz oscillation. I don't remember spotting that when I was looking at the original strip charts. Others at Rockwell or Johnson may have, but I only started to see the phenomenon as being of more interest after we got the complete merged data set several days later.

In all of the programs that I'd worked on at Dryden to that point, the data was always acquired in one of two ways. One of those was from onboard recorders on the vehicle. The other source was from the telemetered data received by the antennas at Dryden and copied onto magnetic tape. If we had the opportunity, we always recorded it both places, because if for some reason there was a malfunction on one of the two ways, we would have the other one as a backup.

Most people who haven't worked on data very much assume that we would get our best data from the onboard recorder. While that's true in some cases, all other things being equal, I normally prefer the recorded telemetry data, because we were going to be playing it back with the same recorders on which it had been recorded. Any idiosyncrasy in the recording mechanism would reappear as we played it back, so I always preferred the downlinked data. In this case, of course, all of the parameters were recorded onboard, and ultimately that would be the data set the whole Shuttle community would be using.

There are data ports on the Shuttle where the technicians can command the recorder. They actually play portions of the recording on the tape backwards into a device that was made specifically for that. Playing it backwards avoids damaging the tape by rewinding it unnecessarily and then playing it back. The data was brought to the Dryden range and we copied it for other users, primarily Johnson Space Center, which had the responsibility of coming out with the official data set for all Shuttle participants. Because Dryden was a flight facility and we were used to doing this, we got our own copy, called dubs, of the data from the onboard recorders to merge with the data that we got on the downlink.

So we had two sets of data to choose from, those that had been downloaded from the Shuttle recorders and given to us as dubs, and those that had been recorded directly from the telemetry. The way we processed data was that Dryden would make copies compatible with the program Rich had written for doing the data-merging. That took quite awhile because we had to wait for the vehicle to be in a position where they could download the tapes. It took several days to process the tapes at Dryden. Then, Rich ran his program that merged the data. As a result, we at Dryden actually had the first completely merged set of data from STS-1. The official computer-compatible tapes that were the responsibility of Johnson were found to have some bugs in the program, so Rich's very good but relatively simple method of accurately merging all of the data became the official data tape for the users that I worked with. Our unofficial way of getting the data became the official system. It was blessed by the project people at Johnson since they knew their tapes were not adequate for all users. We gave them our merged data sets, and they distributed that data.

The first big disappointment on STS-1 was the discovery that the onboard DFI recorder had no information on the reentry. I don't remember whether it didn't turn itself off, and ran out of tape before reentry, or it snarled the tape. At any rate, there was no onboard entry data, so it was fortunate that we had gotten the telemetered data from Mach 14 down. The data that Dryden had received and recorded during the entry was going to be the only DFI data that the program would be able to use to get an understanding of any unexpected events or try to refine prediction.

The ACIP recorder worked on STS-1, and we used the onboard ACIP data since it was not telemetered. As a result, we got that complete merged data set from Mach 28 or so on down. I think the set that I had was from about 600,000 feet on down to touchdown, so it covered well before the so-called entry interface at 400,000 feet. I was interested in that other data, because I had some other things I wanted to study. The firing of the pitch, roll and yaw reaction control system (RCS) jets prior to getting into any sensible atmosphere was of great interest to me. I wanted to learn how the RCS jets would perform and what their characteristics would be in terms of providing pitch and roll without any aerodynamic effects. That would give me a much better feel for the actual forces and moments generated by the various jets in terms of rise times. Since each RCS jet is a little hydrazine rocket, it does not instantaneously put out its full force.

Similarly, the pressure in the rocket chamber trails off when the burn is over.

I also was hoping to be able to extract the moments of inertia of the vehicle when it was more or less still in orbit, when there would be no aerodynamic effects to account for. That worked to a certain extent, but I don't think the numbers I got were actually as good as the ones that were done by the official system, which took every screw, rivet, and chunk of metal in every component and catalogued their exact location. We then computed the three-dimensional center of gravity—x, y and z—and the total weight of the vehicle and all of the moments of inertia.

The centers of gravity and the total weight could be double-checked by weighing the vehicle, but they had no other way, on the ground, of checking the moments of inertia. I wanted to check those by knowing what the forces were of the various jets, viewing the response in the x, y and z axes, and then come up with a complete set of moments from inertia. I did verify that the official computed version of the inertias was probably within a few percent of what I was getting when the Orbiter was at extremely low atmospheric density conditions. My number was not always the same. I had a variation of a few percent, so I knew that my number wasn't exactly right, but it did give me a lot of confidence that the inertias that were computed before flight were reliable.

This was important because the only way we could compare the flight data with the predicted data—whether it's wind tunnel or some computational technique—was to have correct moments of inertia and weight for the flight vehicle. The primary numbers are extracted from dynamic maneuvers. To nondimensionalize them, the weights and inertias are required. Then, we can put them in a form that is exactly the same as the numbers that were measured in wind tunnels.

If we have moments of inertia that were off by five or ten percent, all of our derivatives for a given axis would be off by the same amount. Let's say that the moment of inertia about the X axis was off by 5 percent. Then we would expect to see all of the roll coefficients—roll stability and control derivatives—to be off by five percent as well. The only way we have of telling that is if we consistently get all of the roll values lower or higher than predicted. All of those would be too low or too high if we had an inertia problem. So an independent verification of inertia was very important to me, because the way the data was handled in the simulator, in updating the data book, or in redesigning the control system was with dimensionless derivatives. These required as good a measurement as possible of the mass and moments of inertia.

So once we had all of the data, we were able to look at a winged space vehicle as it made its first flight through the entire atmosphere from hypersonic flight to landing. We were the first people to get a look at this data, partly because Dryden was well prepared to do this job and worked very hard to get this data. Rich and I had good data, and I'm sure Al Carter and the others using Dryden data did as well, quite a bit sooner than the people outside of Dryden. The other people knew the DFI recorder had failed, and they were going to wait quite awhile. But in terms of merged data sets, we had a very good set. As a scientist and a researcher, I found it was a very exciting time. For the first time, I was looking at winged vehicle flight data from well before it encountered a significant number of air molecules. For the ACIP instruments, we could see the very tiny initiation of the drag buildup in the longitudinal (AX) accelerometer and the normal (AN) accelerometer.

The first inkling that the atmosphere was having a significant effect on the Shuttle was when I started to see a consistent reading of one or two counts on the AX and AN accelerometers. The general-purpose computer system had a derived dynamic pressure which was later corrected slightly as we got more information. There was a large effort to make the meteorological best estimated trajectory (BET) data set that attempted to define the state of the atmosphere in terms of temperature, density, winds, and wind shears from Guam to Dryden. Some of the data they used was a couple of days old. Virtually none of it was taken in the exact place of the Shuttle's entry corridor.

Nevertheless, these guys knew the atmosphere pretty well. It was one of the few times that I did quite a bit of work with the National Oceanic and Atmospheric Administration (NOAA). The NOAA supplied people who were experts on the atmosphere to work with the various people within NASA tasked with the job of coming up with the best estimated trajectory tapes, including estimates of the winds, temperature, density, static pressure, and speed of sound, along the Shuttle's actual flight path. These were used as an independent truth model. We had our Shuttle-derived data—estimates of dynamic pressure derived from the force predictions and the actual drag data. From those two sources we ultimately could find where the dynamic pressure, which we call Q-bar, came off the stop, so to speak. This is what I alluded to earlier—that at about 350,000 feet, we were getting about a tenth of a pound per square foot (psf). Our ability to measure the dynamic pressure for lower values than that was at 360,000 feet, when we started to see a bias in the toggling of the first data bit. There was probably 0.01 psf or a few hundredths of a psf at 360,000 feet, but it was magical watching everything start to come off the stops at about 350,000 feet. [Appendix F]

Our first bank maneuver was still quite a way away at that point. When the Shuttle reenters from orbit, its speed is actually increasing when there is no drag. It is losing altitude or potential energy and trading it for kinetic energy with a small increase in speed. The Shuttle's speed increases until the small dynamic pressure increases the drag force and it then slows the rest of the flight beginning at 290,000 to 300,000 feet. The atmosphere is still not thick enough to slow it very much, and the first bank maneuver occurs at about 250,000 or 260,000 feet. [Appendix F] So there are 100,000 feet of descent between a tenth of a psf and when the Shuttle gets down to 12 psf, which is the condition for the first energy management bank maneuver. The Shuttle banks so that its lift vector pulls the vehicle to the side, rather than having it skip back out of the atmosphere. If the Shuttle did not bank, and continued a 40-degree angle-of-attack reentry, the vehicle would eventually get enough dynamic pressure (which results in higher lift) so that it would skip back out of the atmosphere. That's not what we wanted it to do.

First of all, this would be terrible for the energy management, but it also would add a great total heat load to the flight. This skipping maneuver was actually proposed by Sänger in the 1940s as a way of skipping along the top of the atmosphere. It's a doable thing, and this is what the DynaSoar was going to do. But the Shuttle did not want to skip. It wanted to get to Edwards and land. The energy management scheme rotates the vehicle in bank and holds that for a length of time. If the Shuttle banks to the right, then the lift vector points more to the right. The vehicle, instead of heading in a straight direction, tends to turn very slowly off to the right. Once it has done enough of that, it does what we call a bank reversal, and the vehicle rolls from on the right side to over on the left side. [Appendix F] Now the flight path is curving back to the left and eventually will get the vehicle back more or less headed where it was before the first bank maneuver. The energy management is basically modulating the lift vector to the right and to the left and in a normal direction to the flight path so that the vehicle enters the atmosphere in a smooth fashion and with enough dynamic pressure to continue to decelerate, but without the heating becoming too great. Eventually, after the last bank reversal around Mach 2, it will be ready to pick up the heading alignment circle and on around to do the landing. [Appendix F]

Once the dubbed data tapes were ready, we were given copies that we could then put on the digital computer and merge with all of the other data streams. The data was also run out on strip charts for us, with all the parameters that we requested. In that process, I was startled to see what the first bank maneuver at Mach 24 looked like. It was absolutely nothing like the maneuver that had been planned, practiced, and studied. It was supposed to be a smooth bank as the Shuttle slowly rolled to one side. [Appendix F] What we saw was one of the largest maneuvers that we've ever had on the Shuttle. It had an angle of sideslip of plus or minus three degrees and fairly significant roll rates and yaw rates as well. [58]

The thing that startled me when I first saw this was that I knew the whole thermal protection system of the Shuttle had been designed to fly within only plus or minus one degree sideslip or beta. That's where the stagnation point, where the flow appears to stop and the hottest point, would remain on the carbon-carbon nose on the Shuttle, which was capable of taking something around 3,000 degrees F. But if that stagnation point were to move off of the nosecap, it would be on some of the high-temperature tiles, which would fail at a lower temperature than the nosecap. I knew that if the stagnation point went off the nose cap, and certainly if it stayed off the nosecap, it would burn right through the ceramic tiles. Because this was an oscillation, it didn't do that, but it did momentarily put that stagnation point on either side of the carbon-carbon nosecap. This was actually about a 40-second oscillation. If I had been along for the ride, I probably would have passed out when that happened, knowing that the next thing that would happen was it would get really hot in there.

Rich Maine and I were certainly the first people to see the full significance of that first Shuttle maneuver. It was a very unplanned, unexpected maneuver. In fact, of all of the Shuttle analysis I have done, it was by far the most startling maneuver that I've observed. At that time, we had to quickly understand the data to prepare for the next flight. It was clear that some of the predicted stability-and-control derivatives had substantial errors. This dynamic maneuver would not have occurred if the stability-and-control derivatives had remained within the variances and tolerances from the aero-data book that were specified for the control-system designers, so something was way off from what we had expected. If it had been much further off, we could have lost the Shuttle on that first flight. In fact, with a less robust, less sluggish control system, they probably would have lost the vehicle because the misprediction was quite large.

As we went through these initial strip charts, it was hard to concentrate much on any of the other things that we were seeing. In addition to the wild oscillation at Mach 24, I had another somewhat serious concern. To trim the vehicle at the planned 40-degree angle of attack, 16 degrees of the available total of 21 degrees of body-flap deflection was used instead of the seven degrees predicted. [Appendix F] This could also bode ill for the Shuttle and the body-flap material.

For someone like me who had spent his career in flight dynamics, this looked like a serious problem, but not as startling as the Mach 24 oscillation. Other than the wild oscillation and the large body-flap deflection, everything looked more or less as planned. I won't say they looked exactly as planned and predicted because there were some differences in the stability-and-control derivatives throughout the flight envelope. But to me, as I looked at the strip chart, the problems had to be with the aerodynamics on the Shuttle.

[1] I use the term "Mach" or "Mach number" in two different ways throughout this discussion on Shuttle flights. During the flight, all of the systems and the data on the Shuttle use what I call GPC Mach number, which is generated by the onboard general purpose computer. The GPC Mach number is merely the velocity in feet per second divided by 1,000. In most cases, this is fairly close to the true Mach number, assuming that the speed of sound is about 1,000 feet per second. In most cases when I mention Mach numbers associated with flight events, I will be using the GPC Mach number. This is consistent with all of the data occurring in real time, and it is the only Mach number that is available until some time well after the flight data has been obtained. Once the flight data is merged with the best estimated trajectory data, the true Mach number can be calculated. This calculation is based on the true speed of sound, which is primarily dependent on the constituents of the air and the absolute temperature. I use the true Mach number in the description only when the difference between the GPC and the true Mach numbers is of some significance.

For example, when the Shuttle is in orbit, the GPC Mach number is about 25, but the true Mach number is about 12 or 13. The true Mach number usually peaks between 350,000 and 250,000 feet, at a value of Mach 27 to 28. In discussing such events as the Mach number at entry interface, I will use the true Mach number because it will be roughly the same on all flights. However, when talking about a maneuver identified from the flight plan, the real-time data, or from the onboard instrumentation, I will use the GPC Mach number. In most cases, there is not a large difference between GPC and true Mach numbers below 200,000 feet. However, we always use the true Mach number to update the aero-data book because, for comparing data from flight to flight, or from a flight to a prediction, the true Mach number is a very important parameter, while the GPC Mach number is only an approximation to it.

[2] Appendix F shows the time history of the key flight parameters from 23 1/2 minutes after the start of the deorbit burn to wheel stop. The discussion of STS-1 in this chapter and the next is augmented by Appendix F.

[3] The Shuttle program frequently refers to these as roll reversals. Since the term "roll reversal" has a completely different meaning in the aircraft community (described in chapters 2 and 3), I will always call these maneuvers "bank reversals."

XI

Analyzing the Data

The Shuttle obviously had done something quite different than predicted. It wasn't the control system that was the problem, so it was the aerodynamic data. Even though it wasn't intended for stability-and-control analysis, the bank maneuver and the bank reversals on the Shuttle were quite good for getting the major derivatives and getting a fair idea of some of the smaller stability-and-control derivatives.

First Looks

The strip charts recordings of all the Shuttle data, including the onboard parameters, were our first look at anything above Mach 14. Rich and I read up all of the bank reversals, the start and stop times, giving ourselves a little room on either end in case we wanted to enhance the analysis once we started. Any place that the vehicle moved or oscillated on the strip charts, we would read it up as a maneuver. We could read up in the neighborhood of 50 events on the flight, including the bank reversals and the small roll, pitch, and yaw jet firings prior to the sensible atmosphere from a flight-dynamics point of view. [See Appendix F for the identification of the following events on the STS-1 time histories.]

Among those maneuvers were two additional ones that were of interest beyond their original intent. Just prior to the vehicle making its first bank maneuver, a big oscillation in the body flap and elevator occurred. [Appendix F] Not much happened on the vehicle. It turns out that that maneuver has occurred on every Shuttle mission on which I've worked. It's due to the body flap being in the wrong position when the hydraulic system is activated. So the flap drives itself to a proper position, and in getting there it creates quite an interesting oscillation of the vehicle. It was of little consequence, although we did spend some time trying to analyze it initially. Later it would prove useful in defining the body-flap characteristics, but the sample rate on the body flap was so low and the actual aerodynamic forces and moments were so low that initially it wasn't a very interesting maneuver. I was a little perplexed when I first saw it. But now that I have seen it on all the Shuttle flights, it's just a way of defining where the interesting stuff is going to occur. The oscillation occurs at about 330,000 feet, and somewhere right after that, the Shuttle starts performing atmospheric flight maneuvers.

The other maneuver that was interesting was a small longitudinal pitching maneuver that occurred at Mach 15. [58] I didn't see any large correlation, and the pilot was not flying the Shuttle at that point. We didn't know where that had come from, so we read it up as a longitudinal maneuver. After many attempts at analysis, I was still completely dumbfounded by it. Eventually we decided that it occurred because of a big gouge in the ceramic tile on the front of the nose's landing-gear door or because of a discontinuity between the vehicle's lower-surface contour and the gear door that had formed a boundary-layer transition wedge that went from nearly the front of the Shuttle to the back. That transition wedge was large enough that it caused a nose-down pitching moment—small but, nonetheless, detectable—and a change in axial acceleration (AX) of about 10 percent. [Appendix F]

That was the first boundary-layer transition event to occur on the Shuttle. It's very evident from the flight dynamics, since there was not a lot of temperature data available at Mach 15. An increase in temperature at a given location was the planned way of determining boundary-layer transition. Because of the failure of the developmental flight instrumentation (DFI) recorder, we cannot absolutely declare that's what it was, but it was certainly the smoking gun indicating that it was the initial boundary-layer transition event. The elevator change that I had noted as being an unexpected control input was actually a reaction of the control system to the pitch-down that occurred when that boundary layer transition wedge formed. It was quite powerful and strong, an interesting and unexpected encounter on the very first flight.

The transition from laminar to turbulent for the boundary layer on the wind-ward side is characterized as an increase in the shear stress at the surface of the Shuttle, causing an increase in convective heating rate with a corresponding increase in the surface temperature. This change is very rapid and, as we learned on the X-15, can be best seen locally as an increase in the temperature sensor (thermo-couple) readings where the boundary layer becomes turbulent. Where the acceler-ometer instrumentation is very accurate and at a high resolution, as it is on the Shuttle aerodynamic coefficient identification (ACIP) package, the transition from laminar to turbulent flow can be seen as vehicle deceleration in the AX instrument. This is probably because the increase in shear stress also can be seen as a sudden increase in skin friction and, therefore, as a deceleration of the Shuttle.

As a technologist, the most exciting time in my work is when I am going to look at what we call "virgin data," data beyond any other ever obtained before. The preceding examples of the bank maneuver and the motion due to transition are two examples of virgin data. It's the first chance to look at something that no other human being has ever seen. There were many such events on the Shuttle. I'd had them on X-15s and lifting bodies as well. However, this one was special to me because of my desire to go from runway to orbit. This was the first look at how practical reentering with a winged reentry vehicle was, and what we could sort out and what we could tease from the data that might be relevant to that overall goal.

I have no idea how many hours a day I worked, but it probably was about sixteen. It wasn't tiring at all because looking at virgin data is the ultimate excite-ment for a technologist. I know that sounds like a nerdy sort of thing, but it is a nerdy sort of thing for any technologist having the opportunity to work on a new type of data—in this case, in a whole virgin flight regime—it's a very special time.

It's one that I remember vividly, and I feel I was very privileged to be able to participate in it.

These time histories that we were looking at from STS-1 were the culmination of the seven years that we'd spent working on that particular Shuttle configuration. The 25,000 to 27,000 hours of wind-tunnel testing done on just the Orbiter itself to make this flight possible, and the efforts of the technologists bringing their experience and expertise to the program to help come up with tolerances, variations, and the predicted values for the aerodynamic coefficients, for all Mach numbers, altitude regimes, and flow regimes culminated in this data. [46] The first bank maneuver looked awful, and the others looked somewhat different than predicted, but they were close enough that we knew the Shuttle was going to be a viable vehicle from the aerodynamic point of view. Obviously, from the aerothermodynamic point of view, it also was going to be okay, but there was, as yet, little aerothermodynamic data above Mach 14 because of the failure of the DFI recorder. From Mach 14 down, other people interested in the heating also were undergoing the same thrill I was of looking at data that nobody had seen before.

The Bank Maneuver Oscillation and Other Issues

As soon as we had successfully merged all of the STS-1 data streams into a single file, the first thing I looked at was the first bank maneuver with the scary-looking oscillation. I'm not one who likes to tantalize myself with reflection when I have the opportunity to go see the real cause. That was the first maneuver that I concentrated on. I had been developing parameter identification techniques since the mid-1960s, one of which was the modified maximum likelihood estimator (MMLE) that Rich Maine and I had recently developed. It was a fancier and more user-friendly program than previous parameter identification programs, but it used the same theory and techniques that I had been using since 1966. Analysis of this maneuver to see what caused it to be so different from what had been predicted is precisely what the parameter identification techniques had been developed to do. The same program also was being used by the Air Force, Rockwell, and Johnson Space Center. It was the primary tool that all of us were using on the Shuttle.

In 1981, computer analysis was quite different from what anybody in the new millennium would think was much fun. It involved using punch cards for the program. The punch cards would be read into the computer. Since our programs at that time were written in FORTRAN, the computer compiled that into what we called a binary or object deck—a compiled form of the code. Then, we loaded the punch cards with the compiled code in with the data. The way we interfaced with the computer was with punch cards. In 1981, the actual flight data was on a magnetic tape read by the same computer, as I recall, was a CDC Cyber digital computer. The raw data itself was processed in what I call the range, which consisted of the people and equipment at Dryden taking care of the communication, tracking any vehicles, and also receiving the telemetered data, and processing it. In this case, they also had received the dubbed tapes, which they converted into a digital tape read into our Cyber digital computer.

We got our data by telling the computer the start and stop time. It started each time slice with the time in hours, minutes, seconds and milliseconds. Then it had the value of each of the 100 or so parameters we were primarily interested in that were reconstructed to have occurred at the same instant within a millisecond, so the data had what I would call a time frame. If things were sampled at different sample rates, the program that Rich Maine wrote would interpolate all of them to be at the same sample rate. In the case of the data we were looking at, the highest sample rate that we had was from the ACIP package which ran at 174 samples per second. [59] Other parameters were at fifty, twenty-five, twelve and a half, five, and one sample per second. For no reason that I knew, one of the most important parameters—angle of attack—was only available at one sample per second, so we had to interpolate that parameter and, by some fashion, make it appear as if it had been sampled at 174 samples per second.

There was quite a range of sample rates in this merging, but by the time we looked at the data, using our own engineering judgment, we had reconstructed all of it as if it had occurred at 174 samples per second. The computers we had in those days were very slow compared to the ones we have now. We first needed to take a closer look at the data, which we did at the highest sample rate. Then, we thinned the data down to 50 samples per second. We were confident that we'd get all the information we needed if we did all of our analysis at 50 samples per second. We could do the analysis of the data file itself either by reading it in on punch cards or by reading it off a magnetic tape. Since we were going to be using somewhere in the neighborhood of 300 to 1,000 data points and over 100 parameters, depending on the maneuver that we were looking at, it was clear that we wanted to do all of our analysis from magnetic tape.

In 1981, the Cyber computer was very fast, but it didn't have a very big memory—only 128,000 sixty-bit words, roughly one meg of memory in today's computers. The program itself was loaded in binary form, and we could also store it on a removable disk. We also could write our data files to those removable disks so we didn't have to keep reading the tape, so we normally kept the object form of the program and the digital data on the removable disks. Then, when we wanted to run the program, we would load the disks along with a set of control cards into the computer. It would execute the instructions in the object code. Since the Cyber had such small memory and slow speed by today's standards, the key to writing good programs in 1981 was to write the most efficient code possible to keep the program small so that the small size of the memory didn't become a problem.

To run the MMLE program, we had to do something that was called overlaying. A small part of the main program always had to reside in memory. Then, we would partition the program into pieces so that it would not need to swap the partitioned part of the programs any more often than necessary. (It would take a great deal of time if the computer ended up reading in each program segment as it went through the code.) However, even doing that, analyzing a maneuver would take the computer about 20 minutes to take one cut at it, compared to less than five seconds on today's personal computers. We then might want to change something and run it again. We might look at an important maneuver dozens of times, because in addition to getting the stability-and-control derivatives, we also might want to find out how sensitive it was to changes so that we could spot the culprits causing the vehicle in certain instances to fly so peculiarly.

The task was made a little more complicated in that the bank maneuver was not ideal for extracting stability-and-control derivatives. However, it was more than adequate for finding a large change, such as the one the Shuttle was obviously exhibiting from mispredictions in its aerodynamics. The results of the analysis were that two terms were quite different than predicted. The actual flight dihedral effect proved to be larger than was predicted for the Mach 24 to 26 regime. The dihedral effect was off quite a bit, but it was not the primary reason the vehicle flew so strangely.

The real culprit was the rolling moment due to the yaw jets. [52] The value was about the right magnitude but opposite in sign than predicted. The tolerances and variations of the aero-data book really hadn't covered that as a realistic possibility. The yaw jets are little hydrazine rockets that fire for anywhere from 80 milliseconds on up. The jet in the orbital maneuvering system (OMS) pod is firing into the air flow around the vehicle, more or less along the axis of the wing. The yaw jets are on the OMS pod near the tail, and they fire out to the right or left. The left yaw jet fires to the left, and the right jet fires to the right. They also point down a little bit to give a little roll, and part of that was in anticipation of interference effects.

The yaw jet effects are broken down into three pieces. The primary piece is the reaction to the thrust of the rocket as if it were in a vacuum—what would be the thrust of the rocket. The moment here, of course, is the thrust of the rocket times the distance between where the rocket is and the center of gravity where the vehicle will rotate. The second component is called the jet impingement, the effect of the spent propellant impacting with whatever momentum it has onto the surface of the vehicle. The third piece is called the yaw jet interference. It is basically how much the firing of this jet into the flow going over the wing and aft of the OMS pod changes the forces and moments on the vehicle.

The yaw jet interference turned out to be the culprit in this case. The interference effect had been mispredicted. The major part of the misprediction was attributed to the wind tunnel being unable to duplicate the low static pressures and densities that occurred in flight. Another part of the uncertainty of the prediction was due to not being able to model in the wind tunnel the real gas effects at near Mach 24-that is, the heating and dissociation of the atoms into various ions, and just what the flow chemistry is at that point. That couldn't be modeled because at Mach 10—the highest Mach-number tunnel available-there is not significant dissociation. Consequently, the real gas effects either weren't included or were not properly accounted for in the predictions for the Shuttle. We learned from this very first maneuver to improve our understanding of what these interferences are as a reaction control system (RCS) jet is fired into the air flow. After months of analysis and speculation, it was decided that the main culprit was that the low static pressure and density had not been duplicated in the wind tunnel. Although the real gas effects were deemed minor for yaw jet interference, they would become responsible for other anomalies.

For this particular shape and jet, we then had some hard data indicating what that interference actually was, so the computational codes used to predict the characteristics of a flight vehicle were updated with this as a benchmark to be tested against. As a result, the computational fluid dynamics (CFD) codes should be able to predict the effect of firing the yaw jet into Mach 24 flow much better than they had before.

Another major concern was that the body flap was at the wrong deflection angle by a large amount from about Mach 25 down to around Mach 15. [60] Only then did it start to get closer to what had been was predicted. [Appendix F] The way the Shuttle's control system works in the longitudinal mode is that for every Mach number there's a given elevon position (called the elevon schedule). The body flap is moved to trim the vehicle at the scheduled angle of attack. At the Mach 25 condition, the elevon was at the proper deflection and the Shuttle was at a 40-degree angle of attack, but the body flap was at 16 degrees down as opposed to the seven degrees down that it was predicted to be. The total deflection of the body flap that was possible was 21 degrees, so it was within five degrees of the maximum deflection.

Whatever was causing that error, if it had been a little bit larger the Shuttle would not have been able to trim at the 40-degree angle of attack, and that first entry might not have gone nearly as well. That was a very troubling result because something significant was wrong in the pitching moment characteristics that determine the trim. There were four potential contributors in the coefficient of pitching moment (CM): CM alpha, the pitching moment due to angle of attack; CM delta elevon, the pitching moment due to the elevon deflection; CM delta body flap, the pitching moment due to the body flap deflection; or CM zero, the pitching moment bias. If CM zero is wrong, it means the center of pressure location has been mispredicted. We were hoping that wasn't true, because the center of pressure is a very important component in any design. All design elements—the center of gravity, where the wings go, what the control deflections are—are based on the combination of predicted location of the center of pressures throughout the flight envelope. It's a compromise. But if it was a CM zero problem, the center of pressure was ten inches behind where it should have been. This may not sound like a lot, but its consequences represent a huge misprediction, even on a vehicle the size of the Shuttle.

Whatever the source was, it was something that needed to be resolved. We were sure that on future flights we were going to see this again, because there was nothing that we really could do to change it. The only variable that we had would be to change the angle of attack or change the elevon schedule, which was the chosen elevon position versus Mach number for each flight condition. Looking at the flight data on STS-1, we didn't have very good information on getting CM alpha, CM delta elevon, or CM delta body flap.

The problem with the body flap being deflected more than planned is that it gets hotter than predicted because it's down more into the flow field, the body flap already being a very hot part of the vehicle. The other problem is that deflecting the body flap more lowers the hypersonic L/D, and that cuts down the cross-range capability. This was a serious problem that we needed to fix. It probably wasn't going to put the Shuttle in jeopardy, but the materials on the body flap might have to be changed.

Our panel proposed obtaining programmed test inputs (PTI) for the next mission, STS-2, in the longitudinal mode in the Mach 24 to Mach 15 region so that we could figure out if this was a CM delta body flap, a CM delta elevon, a CM alpha or a CM zero problem. One of the difficulties encountered is that the Shuttle program viewed the body flap as merely a trim device that wasn't expected to be very important or move very fast. When we were making our instrumentation request at the beginning of the program, we had asked for the same resolution and sample rate on the body flap that we had on all the other controls. That had been denied, and we were given a very coarse resolution on body-flap position, and, even worse, we were stuck with a one-per-second sample rate.

The first thing we did as a panel was to advocate for increasing the sample rate on the body flap. We knew that would be a tough fight because that would be a significant change to the way all of the data was handled. However, we needed that higher sample rate to get very accurate body-flap positions. The Shuttle body-flap implementation had another problem, the slow movement of the flap at only ten degrees per second. Trying to do a pulse of two or three degrees at ten degrees per second, isn't really a pulse but a ramp. A ramp input is not a good way to get control derivatives. We knew it was going to be a struggle to improve our understanding of the body-flap derivative, but that was the only way we were going to resolve whether this really was a center-of-pressure location problem or a problem with one of the three stability-and-control derivatives.

For STS-2, we requested two body-flap doublets which were really two body-flap pulses that were sawtooth-shaped, because it moved so slowly. One doublet was up around Mach 24, and the other one was down in the region of Mach 15. Even with a low resolution signal, this would start to give us a better feel for what the correct CM body-flap derivative would be. The other thing that we requested was to make the reentry with the elevon deflected more down in the flow. This would then presumably reduce the required body-flap deflection so it wouldn't be so far down into the flow, and we could check where the trim point was and get a cross check on the ratio of CM delta elevon to CM delta body flap. There were many others who wanted these maneuvers for aerothermodynamic and aerostructural analyses as well. Those two body-flap maneuvers were approved and put into the STS-2 flight plan.

The other body-flap information we could use was the trimming that happened when the Shuttle started to get very small aerodynamic loads and the control system is activated. [Appendix F] This was the particular longitudinal oscillation that, as I mentioned earlier, occurred at 330,000 feet. [59] We could use that as a maneuver. It was quite a long maneuver. I think it's about 40 seconds long. It starts at a dynamic pressure of around two or three pounds per square foot. We had broken that maneuver up on STS-1 and had done quite a bit of analysis to get a rough idea of whether we thought the CM body flap was grossly mispredicted. Our result was that we didn't see that it was grossly mispredicted, nor did we see that the CM delta elevon was grossly mispredicted. However, these were preliminary results to help those in the program have confidence in where they could move the elevons to a more downward position so we could raise the body flap.

From a stability-and-control point of view, the other thing noted on STS-1 was a slightly divergent lateral-directional oscillation at a quarter of a hertz or a quarter of a cycle per second. It occurred about Mach 1.7 or 1.8, then disappeared around Mach 1.2 or so. It wasn't a big problem in itself. But it was unexpected and we didn't know what might make it grow larger or under what situations it might be undamped. It was obvious that the control system was working against this quarter-hertz oscillation. If we were to have a degraded control system for some reason, the quarter hertz could have diverged at a much faster rate and caused significant problems, so a great deal of effort was put into understanding that issue. We didn't reach any firm conclusion, partly because the phenomenon was obviously driven by the aerodynamics themselves, not by the controls.

The source of the oscillation wasn't clear. Since the Shuttle was in the transonic region and at a significant angle of attack, where we could expect buffet, a good guess was that shock waves on the surface of the body were moving around. As the vehicle moved, the shocks moved and resulted in an oscillation. I don't think we ever completely documented the cause. However, after getting more maneuvers on subsequent flights, and seeing the quarter-hertz on all of the flights, we'd noticed that the roll due to aileron was lower than predicted in the transonic region. We suspected, since this was primarily a rolling motion, that the control-system gains might need to be increased. After we got better data on later flights, the control system was changed, those gains were increased, and the problem disappeared. So the increase in gain for the roll rate feedback quite likely was the solution.

There were other theories in terms of the speed brakes. On the Shuttle in the transonic region, they are deflected at quite a wide angle. There was some suspicion that there could be oscillatory motion from the transonic flow—the mixture of transonic and supersonic flow coming off the speed brake. If that was the source, we could fix it by increasing the damper characteristics of the control system. Eventually, the concern about the quarter hertz oscillation went away.

Other areas of concern were the temperatures. We didn't have much temperature data from STS-1 above Mach 14, so we didn't have good boundary-layer state information. Therefore, most of the work in the thermal and the aerothermodynamics area really occurred at Mach 14 and below. However, even at those Mach numbers, some areas were found to be hotter than predicted, and others were found to be cooler than predicted. [60] The one area of particular concern to me was the heating on the OMS pods, because if it was a problem, getting the lower angle-of-attack entries that we need to give us the larger cross range that we needed to be able to land back at the launch site would be more difficult. If we launched from Cape Canaveral, we needed to have a cross-range capability of about 1,300 miles to be able to return to the Cape. That's basically how far the Earth rotates at Cape Canaveral in the somewhat less than two hours from launch on an abort-once-around (AOA) until the Shuttle can land.

The Air Force had a hard cross-range requirement for their Vandenberg launches, so it could launch into a near-polar orbit for payloads and military astronauts. The Air Force needed to be able to demonstrate a cross range in the neighborhood of 1,300 miles. The only way we were going to increase that cross range was to reenter at a lower angle of attack, which increases the L/D. So the likelihood of increased heating on the OMS pod was a little unsettling. If temperatures were a little hotter than predicted, the thermal protection system in that region might have to be redefined for the Shuttle to be able to enter at the lower angle of attack.

The final area that, at least in my mind, was of significance from the first flight was landing long. I didn't think it really was a big deal, because I had been impressed that after launching from Cape Canaveral and orbiting the Earth 36 times, the Shuttle could land within 3,000 feet on the right runway going the right direction. That seemed a fairly good validation to me. I think that perhaps the reason it got more emphasis was that John Young, an outstanding pilot and a perfectionist, was disappointed that the vehicle had landed long. He felt that it was due to the vehicle having higher L/D than predicted. We studied all of the data and continued to study it on other flights, and we determined that the expected degradation of the L/D due to surface roughness of the tiles had been overstated. We'd reduced the subsonic L/D given by the aero-data book from what we'd seen on the ALT program to compensate for the increased roughness of the actual tiles. It turned out that the roughness of the real tiles was not significantly greater than that of the fake ALT tiles.

The speed brakes also may have caused John to land long. John had retracted the speed brakes during the preflare as planned. Since the speed brakes provided more drag than had been predicted, the vehicle accelerated, which contributed to the Shuttle landing long when he retracted the speed brakes. For some reason, the speed brakes were better drag devices on the Orbiter coming back from space than they'd been on the ALT program. I don't think anybody understood completely why that was true, but it's been demonstrated on quite a few flights that we do get more benefit in terms of a drag modulation device from the speed brakes on the orbiting Shuttle than we did on the ALT. There probably are some minor differences. But the ALT program was four years old by that time, and it was difficult to go back and see if we could track down what was truly different between the ALT and the STS flights. The fake tiles and the real tiles were made of different material, and the assumption that the L/D would be reduced was not correct. Consequently, the increased roughness may have affected the flow aft of the vehicle, making the speed brakes actually more effective. Nevertheless, I don't think anybody has a firm explanation of why that's true. It's certainly unexpected, and it's a little disappointing that we can't explain it. But since we know what it is on the orbiting Shuttle, it can be included in the aerodynamic model of the aero-data book on the simulators and on the Shuttle Training Aircraft (STA) so that other pilots would not be affected in the same way as John Young was on STS-1.

The prediction for the boundary-layer transition, one of the primary sources of heating on the windward side on the bottom of the Shuttle, was also an issue. The later the boundary layer transitions from laminar to turbulent flow, the lower the total heating will be. By "later" I also mean at lower Mach number, because Mach number decreases monotonically from Mach 28 down to landing speed. The DFI data above Mach 14 was missing, but I think the consensus, as we moved toward STS-2, was that the boundary layer transitioned later than predicted. But once it did it, it transitioned from the rear of the vehicle to the forward part much more rapidly than had been predicted. I think the explanation for that would be that the roughness of the bottom of the vehicle dominated the state of the boundary layer, which was one of the possibilities when the aerodynamicists had been doing their prediction. They didn't know how to weigh the roughness of the diagonal tile lay-up on the bottom of the Shuttle. Looking at it from a distance, or even from up close, it looked relatively smooth. I think the biggest discontinuities were about a tenth of an inch, but evidently those were enough to change the way the vehicle boundary-layer transition took place.

The six issues that came from STS-1 that would be dealt with for STS-2 or other flights later in the program were: (1) the state of the boundary layer, how it transitioned, how that affected the heating; (2) the quarter-hertz oscillation, and whether it was going to be a nuisance or something significant; (3) the higher-than-predicted heating on the OMS pods; (4) the apparently higher-than-expected subsonic L/D of the vehicle; (5) the nine-degree anomaly in the body-flap position; and (6) the misprediction of the rolling moment due to yaw jet. This misprediction was the only one of the six that absolutely had to be dealt with for the next flight, STS-2.

If that initial bank maneuver hadn't been just barely damped by the control system, it could have ended the Shuttle program right at that point. I remember that long after STS-1 John Young was being asked about how he had felt while sitting in the cockpit, doing the first maneuver three and a half times faster than anybody had before in a winged aircraft and seeing the vehicle do something totally unexpected. In fact, his pilot display had a limit of plus or minus 2.5 degrees on the angle of sideslip, and this one had gone plus or minus over 3 degrees, so it had actually pegged and sat on the peg awhile, going back and forth as he watched it. John's reaction was that it certainly got his attention. However, I never got a sense of what he really felt or really thought then. He'd been through quite a few incidents in his career, including going to the Moon twice, and I'm sure all of them were scary, but it seemed to me that this one would have been among the scariest. However, John is very low key, and so his answer of "it certainly got our attention" was probably about as big an admission as we would get from him that he did notice it and took it very seriously.

The way that we dealt with that for STS-2 and up through STS-4 was to change the flight-determined stability-and-control database with what we had learned on STS-1. These changes were put in all of the Shuttle simulators, the primary change being the rolling moment due to yaw jet. There was no possibility of changing the flight-control system for at least a year to a year and a half, so the only way to deal with it was to put it in the simulator and develop a manual way of performing the same maneuver so it would not end up in the scary oscillation we'd seen on STS-1.

Getting Ready for STS-2

The next flight, STS-2, was going to be with Joe Engle and Dick Truly. They and other astronauts worked in the simulator to come up with a technique for getting the Shuttle to do the first bank maneuver without the oscillation. A technique was developed, and it was input manually on STS-2, -3, and -4. [60] By STS-5, the flight-control system had been changed to incorporate the updated simulation and flight-control database provided by those of us working the on extraction of stability-and-control derivatives. On STS-5, the system once again automatically did that first bank maneuver with the control system without the astronaut involved.

We continued to meet weekly as a panel on telecons, discussing various aerodynamic results and some of the aerothermodynamic results. We then came to more or less of a consensus as to what our recommendation was for any modifications that we might propose for getting a database that might work a little bit better for STS-2. We didn't actually revise the aero data book database, but we did pass on what we'd learned. The people working on the Shuttle in all other areas did the same. So the issues of the long landing, the heating anomalies, the mispredicted center of pressure, and the rolling moment due to the yaw jets were worked officially up through the system to be incorporated into the flight plan for STS-2. Those of us working in the performance and the stability-and-control area, and to a lesser extent in the aerothermodynamic area, were going to get programmed test inputs (PTIs) on STS-2. We'd done some of the PTIs on ALT at Dryden, on the 3/8th-scale F-15, [51] and on the HiMAT vehicle.

Those working on the Shuttle program had to decide when those maneuvers would be performed during the reentry. They had to be spread through out the reentry, making sure they did not interfere with the bank reversals or any of the energy management issues. We made our request as to where we thought the maneuvers should be made to get the best data. My input was in the neighborhood of 20 or so flight conditions to get both lateral-directional and longitudinal inputs, and that was from velocities near those for entry interface all the way down to subsonic maneuvers-of which we already had a few for the ALT but not enough to do a really good database.

The panel recommended quite a few maneuvers throughout the flight. In practicing in the simulator for STS-2, we had the 12 PTIs that would only occur if the vehicle was within what they called the nominal region of the flight. In other words, if they were about to initiate a bank maneuver they would cancel the PTI. Likewise, if the vehicle was outside some constraints in terms of Mach number and angle of attack they would bypass the PTI. It had to be okayed by the Shuttle's commander, so if the PTI was included, it meant the crew was comfortable with doing the maneuver. Then, the control system got to decide whether it was comfortable doing the maneuver. Those were the ground rules. I certainly didn't disagree with them because those are safety-of-flight issues, and I never wanted a data request adding risk to a flight.

During that period, I think Joe Engle was practicing more than the PTIs already automated in the system. He was doing what on the ALT program we called ASIs, or aero-stick inputs. I don't know officially how that happened, but I know when we actually flew STS-2 that Joe gave us a lot of aero-stick inputs, along with the PTIs which greatly enhanced our ability to improve the Shuttle's aerodynamic database. The ASIs were not as good, because the control system fought against some of the maneuvers that we tried to do. What the people at Johnson did for PTIs was that they would run them through the simulator and come up with a scheme of how the control inputs would go. They would settle in on the amplitude, then they would have switch points in time where the program started the pulse, the time it stopped the pulse, and then the time it started the second pulse, and the time it stopped the second pulse.

So each PTI was a series of five or six singlets that were one half of a doublet. These were single pulses of varying widths or time durations at various points along the maneuver to enhance the information on the aileron, rudder, and RCS jet derivatives. When the Shuttle commanded an aileron input, the control system fired a jet almost immediately to try to counteract that. This tended to mask a lot of our ability to estimate the effectiveness of the aileron control. The flight dynamics and simulation people at Johnson came up with a set of PTIs for a series of four or five different Mach numbers, both longitudinal and lateral-directional. Then, they sent the simulator results to us, so we could look at them and see if we thought they were good maneuvers for us to do a parameter identification analysis. Later in the program, they bypassed that step because most of what Johnson had given us was considered fairly good by all analysts.

The second Shuttle flight, STS-2, was planned to be five days or so in duration. It would have the two pilots with ejection seats, but in addition to the reentry from the planned mission, they also had to practice the return-to-launch-site (RTLS) maneuver, the landings across the Atlantic, the abort-to-orbit and the abort-once-around; in none of those cases would we get any PTIs. PTIs would only be done if things were going well at the time of reentry. So they had a huge workload, doing all of these maneuvers in addition to the usual crew-training requirements. They also had to practice the ascent and the on-orbit activities. Part of the reason the flight was going to be five days was that they had more testing to do on-orbit than they'd done previously in terms of maneuvering and changing orbits.

After several tries, STS-2 finally managed to fly in early November 1981. The first thing that happened was that they lost one of their fuel cells, the source of their electrical power. The ground rules called for shortening the flight to two days if that happened. Some other malfunctions occurred during the flight, which was to be expected with any new vehicle. On the day that it was scheduled to return to Edwards, we got into work well before we needed to because we expected tens of thousands of spectators for this landing, too. To get on the base, spectators needed passes to get to the Air Force Flight Test Center, Dryden, or the east shoreline of the lakebed, which is where most of the public was.

STS-2 planned to come in on runway 23, so we went through all of the same things we had before in terms of being in the control room ahead of time. To me, the big issue always was getting the payload bay doors closed. The crew had to close those doors after they'd gone from darkness into sunshine, and then back again, on each orbit for several days, so it was nice to know that the doors were closed and secured again. The crew then did their reentry burn, rotated the vehicle, and set it up for a reentry at a 40-degree angle of attack. According to all of the things that we heard on the Shuttle network that was in our control room, it sounded as if things were going as planned.

We knew once again that we would not be getting any real-time data until Mach 14 or so. A lot of us didn't believe that there would be a total communications blackout on the Shuttle because it didn't have the ablatives. It only had the ion sheath formed when the Shuttle passed through the air. The Air Force sent out an aircraft capable of receiving the Shuttle's signals, and it received the downlink at Mach 18. At any rate, the Shuttle community always referred to the vehicle coming out of blackout, when it's actually coming out of loss of communication because it's not within the range of any ground-based antenna.

We listened to the mission announcer, as we had before, and he didn't scare us this time by announcing radio contact with the crew any earlier than we thought that he should. We got our data from Mach 14 on down, and we could see all of the pulses that we were getting. All of them were PTIs or bank reversals. We could see we were going to have a gob of data to work on. That data was recorded in the same fashion as it had been for STS-1. However, the crew had some difficulties. They were dehydrated, largely attributed to low water production because of the failed fuel cell, among other things, but they still managed to obtain all of the ASIs and PTIs. They were not always in the right mode in the Shuttle because the crew was being overworked and getting overly tired, but they were doing their best to get the most data that they could. We watched the strip charts as the Shuttle went through our maneuvers, and at Mach 2 we went out onto the roof to watch the Shuttle come in. It was a partly cloudy day, so we didn't get to see as much of the landing as we had for STS-1.

The Shuttle can't fly through rain, but it can fly through clouds as long as there are no raindrops. The impact of rain on the ceramic tiles will damage them because of the difference in the speed of the rain and the Shuttle. It's like hitting it with little hard objects. It was the first unpowered landing I'd seen at Edwards that actually came down through the clouds. Up until that time, we always had criteria for flying any of our experimental airplanes, such as the X-15s and the lifting bodies, that included all visual flight rules (VFR), which meant we couldn't have clouds. If we had clouds, we didn't fly. This was the first flight that I'd seen that emerged from the clouds as it came down.

The landing looked good to me. (Fig. 1) Listening to the issues afterwards, I learned that there evidently had been some problems with wind shears and the mode that the vehicle was in. I think the landing was short by close to half a mile. I knew Joe Engle could put that the Shuttle down anywhere he wanted to, so I assumed he'd been busy and hadn't had a high priority on landing at any particular point on the runway. As it turns out, he was having problems with dehydration, as was Dick Truly, so their goal was just to make sure they got the vehicle back on the ground safely. Despite the dehydration and fatigue, they had given us a tremendous amount of data.

After coming down from the euphoria of watching the Shuttle landing at Edwards-hearing the national anthem, the big crowd, and all of that—we went inside, watched the Shuttle activities through crew egress, then got our strip charts, and went down to the office, and started going through them in the same fashion we had on STS-1. This time, we had 15 maneuvers to look at, plus all the bank maneuvers and bank reversals and some other subsonic input as they came around

Fig. 1. The Space Shuttle Columbia over Edwards AFB prior to landing on STS-2. (EC81 17537)

the heading alignment circle and in for landing. We marked up our strip charts and prepared for receiving the magnetic tape of the same data and later receiving copies compatible with our digital computer so we could merge the data.

Everything seemed a little faster on the second flight, so we weren't doing as much inspection. We knew more of what we wanted to look for by the second flight than we had on the first flight, but it still took us several days to get the dubs through our range facilities and into a form that we could use. This time the recorder for the development flight instrumentation (DFI) had worked, so the aerothermodynamic guys were ecstatic. I was doing a little aerothermodynamic work on the side, trying to see if I could get some of the heating rates from the surface down into the vehicle. There were some sensors that I could use, and I wanted to see if I could get dynamic information of the transients on heating going into the structure.

The good news was that the DFI had worked perfectly and we had good data. The bad news was that the recorder for the aerodynamic coefficient identification (ACIP) had failed, so we didn't have the high-quality data that we had fought for so hard. Here we had more maneuvers than we were to get on any other flight by at least a factor of two, and we were only going to have low-quality data to analyze. That was disappointing. Although we could learn a lot, we just couldn't learn as much. We wouldn't get the highest quality estimates and we wouldn't have the confidence in them that we would have had with a higher sample rate, a higher accuracy, and the higher resolution system that the ACIP provided.

It turned out that we had 12 PTIs on that flight. They were all lateral-directional, however. We also had about ten ASIs. Those were of great interest because all of them occurred above Mach 15, and many of them were longitudinal maneuvers. We hadn't had very many longitudinal maneuvers from STS-1 because the STS-1 maneuvers had been primarily banking maneuvers, and we don't get much information on the longitudinal characteristics from banking maneuvers. In actuality, the below Mach 14 data that we saw initially had only PTIs—there were no ASIs. All of the ASIs that Engle did were above Mach 20. We could see the end of the last body-flap maneuver from STS-2 as we acquired the data in the Dryden control room.

The ASIs that Joe gave us actually put him at an even higher altitude and higher Mach number for doing a controlled aircraft flight with manual inputs than any pilot had done before or since. (Appendix G) He gave us two ASIs in pitch, one from 327,900 to 319,200 feet at Mach 27.4, inertial velocities from 24,600 to 24,610 feet per second (fps) at a dynamic pressure from .3 to .5 pounds per square foot (psf) [event 1]. The second ASI was from 282,100 to 279,600 feet and Mach 26.8, inertial velocities from 24,605 to 24,598 fps at a dynamic pressure from 3.1 to 4.2 psf [event 2] while the pitch jets were still active with the elevons active. He also gave us two lateral-directional ASIs before the roll jets quit firing above 10 psf. (This was changed to 20 psf on later flights.) The first lateral-directional ASI was from 273,000 to 269,800 feet at Mach 26.3, inertial velocities from 24,586 to 24,570 fps and a dynamic pressure from 5.1 to 6.0 psf [event 3]. The second one was at about 264,000 feet at Mach 26.3, inertial velocity of 24,540 fps and a dynamic pressure of about 7 psf [event 4]. After one more longitudinal ASI at about 256,000 feet [event 5], he performed the bank maneuver from 253,900 to 247,000 feet at Mach 26.2, inertial velocities of 24,443 to 24,000 fps and a dynamic pressure from 13 to 19 psf [event 6], rolling the vehicle over to a right bank angle of 80 degrees. These six maneuvers (events 1-6 in Appendix G) were all higher and faster than any pilot has maneuvered a winged aerodynamic vehicle either before or in the more than 20 years since. It is my opinion that any of those maneuvers gives Joe Engle the speed and altitude record for flying a winged vehicle. (Fig. 2)

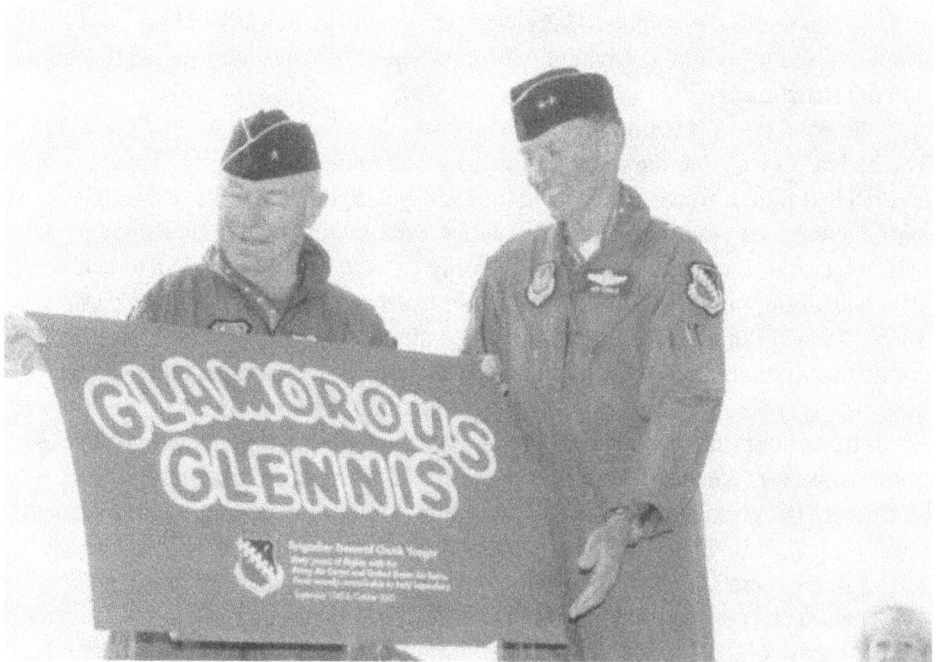

Fig. 2. Chuck Yeager (left) accepting a plaque from Joe Engle after Yeager's last supersonic flight as a pilot on October 26, 2002. Yeager first flew supersonically on October 14, 1947, and Engle had set the speed and altitude record for the manual control of a winged vehicle on November 14, 1981 on STS-2. Both aircraft landed on Rogers Dry Lake. (EC02 244-38)

The pitch jets cut off at 20 psf, but for that flight the only way we could get pitch jet information and find out if we had the same interference effects that we had seen on the yaw jet was to request some maneuvers there. Those were excellent maneuvers. Most of what we learned in terms of interference from the pitch jets was determined from those maneuvers. There was a strong impingement effect also on the down firing jets in the pitch. This was of interest to us because it was not an interference effect. It was an impingement effect that we could look at as well.

So, on STS-2, we actually had three body-flap maneuvers, one each at Mach 21, Mach 17, and Mach 13, and a POPU with the elevon only in the Mach 20 region, which was about a 40-second maneuver. Then we had a total of eight ASIs (five longitudinal and three lateral-directional) and the 12 PTIs. We had 24 total maneuvers, plus the bank maneuver and the four bank reversals which gave us actually ten more maneuvers, so we had 34 intentional maneuvers to look at, and then, partly because of the vehicle being in the wrong modes, we got other incidental deflections as well. Later in the program, we quit using the data from the bank reversals for updating the aero data book because they weren't as good as the PTIs. However, half of the maneuvers after STS-2 were still bank reversals, so we continued to analyze them and spent a good deal of time in perfecting the analysis of the bank reversals.

Now that we also had the heating data from the deorbit burn to touchdown, we found that the boundary-layer transition on STS-2 occurred very quickly. It started at the back of the vehicle, and within a few seconds it was most of the way to the front of the vehicle. However, it occurred at a lower Mach number than had been predicted, which, as would be expected, ended up with lower heating on the bottom of the vehicle than had been predicted. [60] We also could see that the temperature at the 55 percent station on the wing where we expected to see the shock impingement heating was running a couple of hundred degrees higher than predicted. [60] On the OMS pod, particularly during the POPU, the temperature was also running a couple of hundred degrees higher than predicted, which was not good news. [60] As the crew lowered the angle of attack for the POPU, the heating went up even more. So the higher heating on the OMS pod and the body flap looked like issues that needed to be addressed to be able to get the cross-range and the full envelope capability of the Shuttle well in hand.

The flight crew did a good job on everything, giving us every maneuver we wanted and doing them well. The most PTIs we got on any flight was on STS-2, and we also had eight ASIs on that flight. Engle and Truly did the final bank reversal as well as the initial bank maneuver. The next-to-last bank reversal was done automatically. So Engle did two bank reversals—one that he flew through at Mach 26.2 and the last one. He also did quite a bit of flying, coming into the heading alignment circle, and we could analyze that data and his unplanned maneuvers, a real bonus for us as well.

However, during the debriefs after the flight, we learned that in addition to the dehydration and other issues, the crew had been too rushed in trying to get through with all of our maneuvers. The crew was right on the ragged edge the whole time during reentry. That was probably when Shuttle management made the statement that only four or five PTIs should be made per flight. Not knowing the state of future crews, they took that very conservative approach. All future flights were going to be under greater scrutiny in terms of keeping the number of PTI maneuvers for stability-and-control derivatives to a lower number. We also would not be getting the same amount of data on future flights as we had on STS-2.

Because the quality of the STS-2 data was not as good as it would have been if we had the ACIP data, it took more effort to do the analysis. Most of the measurements that we used in the analysis, such as the gyros and the control positions, the resolution and the sample rates, were one to two orders of magnitude poorer than the same measurements from the ACIP system. [59] So for the small maneuvers, there was no hope of getting any information that was very useful. The larger maneuvers, because of the reduced sample rate and the reduced resolution, were not a whole lot better than the bank maneuver and the bank reversals from STS-1 with the ACIP, when compared to the planned maneuvers for stability and control on STS-2.

STS-2 Results and Conclusions

We were disappointed, but there was nothing we could do. Although we tried many ways to get the information off the ACIP recorder, there just was no information for us. That led to a lot of intense analysis of what were poor to marginal maneuvers from the data quality. The maneuvers themselves were excellent. Joe Engle always attempted to get us as much data as he could and still fly the airplane very well. We'd seen this earlier on the X-15 and on the ALT program. That the vehicle landed short was the result, I think, of being in the autoland mode in the longitudinal axis and Joe not realizing it. I think his approach speed was 20 or 30 knots lower than planned, so that would make him land short. He was also landing into somewhat of a head wind, and the guidance algorithm that the Shuttle was using at that time did not include the wind effect. Consequently, all of these things tended to make him land short. I'm sure he was very disappointed in that, being such a perfectionist and a good stick-and-rudder man. There really had not been any airplane that he hadn't been able to fly. His experience with STS-2 in no way reflected on his abilities—in fact, quite the reverse was true. In a bad situation, he had managed to get us every bit of data that we wanted and land the vehicle within a half mile of the planned landing point.

So with STS-1 and STS-2, we had one landing that was three thousand feet long and then one that was nearly three thousand feet short. We knew that the Shuttle was capable of better landing precision than that. As the program matured and we got better information on the subsonic L/D, the effect of the speed brakes, and accounting for the winds both aloft and near the landing, precision-landing capability became much better with the Shuttle.

The trends that we'd seen on STS-1 in rolling moment due to yaw jet were verified with the maneuvers that we got on STS-2. Of course, the Mach 26 bank maneuver was flown by Engle and looked much different than the one we had before, but we could apply the same analysis techniques and learn a good deal from that information. Furthermore, the body flap did not deflect as much as it had on STS-1, so we had two sets of trim data with known center-of-gravity locations that we could look at for the ratio of CM delta elevon to CM delta body flap. With the body-flap maneuvers, even though they were very poor, we did have elevon, pitch and angle-of-attack motion that we could once again analyze. It was looking like the body flap and the elevon effectiveness in pitch were not going to be too far off from the values in the aero-data book. [59] It was starting to look as if the main problem was a misprediction of the center-of-pressure location. It turned out, in the final analysis, to be a ten-inch location error in the center of pressure, and the primary culprit for the misprediction was real gas effects. [60]

The rolling moment due to yaw jet was primarily due to the inability of the wind tunnel to simulate even Mach 10 conditions at high angle of attack and low static pressure. With the low static pressure, the yaw jet firing into the partially separated region had a much different effect than was seen in the wind tunnel, even with the enhanced computational results. We did find that the down firing pitch jets from Engle's ASIs were very much affected by impingement, and this reduced their effectiveness compared to the up firing jets firing into a similar partially separated region. [59] So the story was starting to make more sense, but we would need more data from more flights to have an ironclad argument. It was pointing in that direction for explaining the two biggest aerodynamic anomalies that occurred on the first two flights: the center-of-pressure location and the RCS jet interference. The tendency to have less dihedral effect at hypersonic speeds also was validated with the information from STS-2, so that started to look like an issue as well, but not one as big as the center-of-pressure location problem and the jet interference with the flow around the Shuttle.

We again met as a panel, and each person went through his analysis, having a lot more data now and a lot more discussions as to how we should send it forward. There was going to be an update called a "FAD"—a flight assessment delta correction. It would be added to the data book values after STS-2, the first time we were doing that. It was a big effort for all of us on the panel to get that information together and into the proper channels so it could go back and forth between us and the other researchers and end up as the final FAD-2. [52] In addition to that, of course, each individual, as well as the consensus from the panel, was advocating maneuvers for STS-3, which were being worked through the system exactly the same way as for STS-2. We requested that the sample rate on the body flap be increased to twelve-and-a-half samples per second for STS-3. The higher sample rate was approved, so we would be able to enhance our body-flap effectiveness estimates.

However, we were going to be limited in the number of PTIs that we could have. We knew we'd be getting much less data to analyze, but we would have the better data from the ACIP recorder on future flights to help with the quality of our analysis. We were proceeding much as we had on any other envelope expansion that I had done at Dryden during my career. With the Shuttle, though, it was a more complex flight regime, and the effort involved a much more complex series of meetings, presentations, and reformatting of the data and analysis and refining of our arguments than I had seen before.

XII

Becoming Operational: STS-3 Through STS-5

There were several months between the end of STS-2 and the STS-3 launch. For the flight test panel, which included the people participating in the parameter identification, the primary task was to come up with a flight assessment delta (FAD) to update the aerodynamic design data book (ADDB), usually referred to simply as the aero-data book. Rather than re-formatting the tens of thousands of points that were in the book, we were going to update it with the flight data to reflect what we had learned from the first two Shuttle missions.

Writing the Book

All of the aerodynamic forces and moments were functions of Mach number, angle of attack, elevon position and body-flap position, and the atmospheric density characteristics at the altitudes that the Shuttle flew. That information, even though we had data from two flights, was not in quite the same configuration or flight conditions as were explicit in the aero-data book, so Rockwell was in charge of taking all of the stability-and-control and performance data on STS-1 and STS-2 and comparing them with the aero-data book values for the corresponding Mach number, angle of attack, body flap, and elevon deflections and altitude. Rockwell prepared the FAD summary information—the difference in the forces and moments between the predicted and the flight data—for all of the data maneuvers we had from STS-1 and STS-2. They were primarily from STS-2, of course, because we had a lot more intentional maneuvers on that flight.

There were four sets of analysts, including Joe Baumbach and Jeff Stone at Rockwell; Doug Cooke, Rick Barton and Dave Knipe at the Johnson Space Center; Paul Kirsten and Dave Richardson at AFFTC; Rich Maine and me at Dryden; and Bill Scalion and Hal Compton at Langley. We got together and came up with a consensus on each data point for each maneuver. Each stability-and-control derivative for each maneuver was compared in this fashion. Then, we came up with a fairing for each coefficient for every maneuver that we called the FAD-2, which stood for the flight assessment deltas through STS-2. This assessment also reduced the "variations," the official term the Shuttle program used for the uncertainty in the stability-and-control derivatives.

This information was then presented to the aerodynamics panel and the controls panel, which included Charles Unger of Rockwell, Joe Gamble at Johnson, along with other people from Rockwell, Johnson, Langley, and Ames. They were then given an opportunity to question the findings and the fairings and our confidence in them. Of course, the aerodynamics panel, which I was also on, questioned the quality of the flight maneuvers and the quality of the wind-tunnel data on which the aero-data book was based. It was a very complex process, but it was a very good process. I think the FAD-2 was a definite improvement for the aero-data book.

But we needed more flight data so that people could start honing their tools, particularly the control-systems people and the flight-planning people, so we would not have the kind of dangerous event that we saw with STS-1's initial bank maneuver. Looking at that data, they would be able to see data voids where we had a higher priority for more stability-and-control pulses and more performance maneuvers. These maneuvers would be used to relax some of the limitations on the Shuttle's flight conditions and on the allowable center-of-gravity locations. The allowable center-of-gravity location was critical for certifying the future payloads the Shuttle was committed to carry. That was the process that we worked on, going from STS-2 to STS-3.

STS-3

When STS-3 was launched in March of 1982 with Jack Lousma and Gordon Fullerton as the crew, we knew that they were not going to land at Edwards. Our lakebed was wet, and the program was still constrained from landing on a concrete runway, so the primary landing site was switched from Edwards to the Northrup landing strip at White Sands, New Mexico. This was quite disappointing to me, because I enjoyed watching the Shuttle land, and I enjoyed all of the activity that came with that—being able to look at the Shuttle, see what kinds of damage it had on the tiles, see what the scorch marks looked like, look at the flexible reusable surface insulation (FRSI) and those kinds of things. My primary assignment for the Shuttle was to determine the stability-and-control and performance characteristics for each flight. We'd made our pitch for maneuvers and had managed to get eight PTIs, all lateral-directional maneuvers, and one longitudinal ASI on STS-3.

Since in the event of an abort, the Shuttle was going into White Sands on the first orbit or any subsequent orbit, we did not come to Dryden for the once-around. I did go to Dryden to watch all of the prelaunch and launch festivities on NASA's television system. There were also people at Dryden who were in charge of maintaining communication with the Shuttle and receiving data from the Shuttle when it was within the range of the antennas at Dryden.

We watched the launch. Then, about a week later, we came in to watch the landing at White Sands. We got to see the data at a slightly different Mach number because the Shuttle was on a trajectory to land at Northrup, so we didn't actually get as much real-time data on that flight as we had on STS-1 and STS-2. The data from anywhere on the communication network was relayed into our control room so we could watch it. We knew that White Sands had quite a different kind of lakebed than the one we had at Edwards. The one at White

Sands is made of gypsum, which is very smooth and powdery, unlike the silt and dust characterizing the surface of the lakebed at Edwards. The winds at the Northrup Strip on the planned landing orbit were too high, and there was blowing dust on their lakebed, so the crew remained in orbit an extra day until the winds and the dust were deemed acceptable. Then they made the deorbit burn and headed toward a landing at Northrup.

My impression watching as the Shuttle got very close to the ground was that it did a couple of pitch oscillations that were very reminiscent of the ALT-5 landing. The Shuttle appeared to land on the second oscillation, going about 220 knots. It is supposed to land at about 190 or 195 knots. The Shuttle also made a very hard landing a half mile short of the intended touchdown point. It had a fairly high sink rate, so it looked to me as if we were going to have a repeat of Fred Haise's landing on ALT-5. However, the Shuttle landed before the PIO was as fully developed as it had been in Fred's case. Then, right after the Shuttle's main gear touched down, it started doing the de-rotation much too early. The Shuttle got down to almost being level, and then all of a sudden the nose popped back up in the air more or less where it should have been. The rear wheels never left the ground, but the nose went back up. Then, a few seconds later, it again started through the derotation. This was still too early, at least compared to STS-1 and STS-2. Then, the nose wheel came back down, and the vehicle rolled out to a stop.

My recollection is that I had been afraid this might happen. The fixes to the control system after ALT-5 hadn't changed the aerodynamics in any way. All we had done was use filtering to hide the PIO tendency of the vehicle from the pilot, making it easier for him to do the landing. However, if he ever gets outside the effective area of the PIO-suppression filter, he experiences the true aerodynamic characteristics of the vehicle. All of a sudden, the pilot ends up with a completely different airplane. That makes it very difficult to keep from doing the PIO, and this was always a concern. The fix to the control system after ALT-5 was a good thing to do, but it still meant that if the pilot's gain got really high, he could still get into a PIO. The only way to get out of that was to lower his gain, and the best way to do that is to release the stick. So my first impression of the STS-3 landing was that the Shuttle was in the beginning of a PIO, although the flight ended in a fortunate landing.

If we'd seen the same thing as we'd seen on ALT-5, I don't think the landing would have ended up as it did. I think that on ALT-5, the Shuttle was going about 170 knots, the STS-3 landing was 50 knots faster. Since White Sands is 1,500 or 1,700 feet higher than Edwards, the actual velocity relative to the ground was even a greater difference than that, but it was over quickly, what I saw and what I reacted to at that time burned into my memory. After going through the strip charts—the ones that we had immediately and the ones available later after the dubbed tapes came to Dryden—it looked like there had been several things going on. They had a wind shear at White Sands, and the guidance, navigation, and control GNC system in the Shuttle at that time didn't handle wind shears very well.

The Shuttle was in autoland mode through most of this approach maneuver, and it was not right on target. Jack Lousma thought that the automated system was performing the maneuver at too low an initial velocity and that it was too flat a maneuver, so he took over control from the autoland system just a few seconds before landing, which is another problem. That's the worst time for the pilot to intervene, because pilots, as I've said many times before, sample with the stick. They don't know they do it, but they put in little tiny inputs, to feel the vehicle is responding. If the pilot is in control of the vehicle coming around the heading alignment circle, or cone, he has a very good feel for the flying qualities of the Shuttle. As he comes around after doing the preflare maneuver, the flare, and feeling for the ground with the wheels, he's fairly familiar with the system and quickly compensates for any differences that he may have felt after his training on the fixed-base simulators or the Shuttle Training Aircraft (STA).

In Lousma's case, however, I don't think he had enough time to learn the real flying qualities of the Shuttle, and that's when he made, in my opinion, overly large inputs and got started with this oscillation, that I'm not sure could be called an official PIO. But to me it was the beginning of a pilot-induced oscillation because his gain was very high and this was the first time he was in control of the vehicle close to the ground. I don't believe he had enough time to learn the Shuttle's flying characteristics, and since his gain was very high, he was not learning what he needed to know. He was seeing the Shuttle more as Fred Haise had seen it on ALT-5.

The quick derotation, the rotation back up, and then back down onto the ground was due to his gain being too high and his lack of familiarity with the sensitivity of the vehicle and the quarter-second lag between stick inputs and the response of the vehicle. Most of the pilots I've watched practice for the Shuttle move the stick with pulses rather than with smooth control input that you normally see with pilots of high-performance fighters. Looking at that, it seemed to me that this was still a result of Lousma's oscillation at landing, because he did land hard, and he was going far too fast. He knew all of those things, so his gain was still high. He didn't keep the nose up long enough, and then when he decided to bring it up again, he brought it up very fast. Then he decided that was perhaps too much and pushed the nose back on the ground earlier than called for in the flight plan.

The rest of the data, from deorbit burn down to just prior to landing when Lousma took over, was a very good flight. We got six of the eight planned PTIs and one longitudinal ASI. We had good thermal data, and we had our ACIP data. The PTIs on that flight were the very best maneuvers we had to date, because we hadn't had any PTIs or good parameter identification maneuvers on STS-1 by design. On STS-2 we hadn't had good quality data, even though the maneuvers had been very good. On STS-3, we had good maneuvers for parameter identification, and we had a very good data system. We also had twelve-and-a-half samples per second on the body flap.

We now had the problem with the first bank maneuver firmly in hand. This problem primarily was caused by the interference effect being mispredicted in the wind tunnel for the rolling moment due to yaw jet, because we were not able to simulate the very low density and static pressures and the vehicle's partially separated flow in the wind tunnel. Now it also looked like the other big anomaly was caused by a misprediction of the location of the center of pressure (CP). The CP location being different than predicted was going to have an impact ultimately on the center-of-gravity placards placed on the Shuttle for various payloads. On STS-3, we again saw the transonic quarter-hertz oscillation. To some extent, it looked a little bit worse on STS-3 than it had on STS-1 and STS-2. The whole flight was also flown at a very small apparent sideslip angle due to a lateral offset in the center of gravity, some miscalibration on some of the control surfaces, or perhaps more tile damage on one side than on the other. On STS-3, we not only had quite a bit of low amplitude oscillation, but also it was large enough that the control system fired the yaw jets during the oscillation, making it look a little more interesting than it had before.

For the most part, it had very much the same characteristics that we'd seen, and we suspected that it was due to the elevon being less effective in roll and different in yaw than predicted. We had a transonic PTI maneuver that would be flown on STS-5 and several other flights in the future, but it would be that long before we finally could resolve what was causing the quarter-hertz oscillation. It was not a serious problem, but it was one that we wanted to understand in case there were circumstances where it might become a serious problem.

From an aerothermodynamic point of view, we were also seeing higher temperatures on the OMS pods, on the leeside of the vehicle, and at the bow shock intersection on the leading edge of the wing. In contrast, there were lower temperatures generally on the bottom of the vehicle, primarily due to the boundary layer transitioning at a lower Mach number and moving forward more quickly than predicted. So the aerothermodynamic issues remained something that we wanted to understand, because they might call for a different configuration of tiles or for thermal blankets on succeeding flights or in the construction of additional Shuttles.

There were a couple of other things of interest regarding the STS-3 landing. It still holds the record for the fastest touchdown by the Shuttle. It also holds the record for the longest rollout, at two and a half miles, and came very close to going off the end of the runway at the Northrup Strip. It was fortunate that Lousma landed short of the intended touchdown point because, otherwise, he might have gone off the end of the runway. His long rollout was due to two effects. The Shuttle's touchdown velocity was too high, and he derotated too early which reduced the drag and therefore reduced the deceleration.

The other interesting thing that happened on that flight turned out to be a consequence of the differences between the two lakebeds. When the Shuttle lands at Edwards on the lakebed runways, it puts up a huge rooster tail of dust. Some of that dust gets up into the cracks and crevices of the vehicle. When the Shuttle landed at Edwards on the first two flights, it was dirty. Nevertheless, I didn't hear very many complaints about getting it cleaned up, probably because the NASA Dryden people and the Air Force people at AFFTC had a lot of experience with the lakebed landings and they knew what kind of dust it was, where it would get into, and how to clean the vehicle up. At Northrup in New Mexico, the gypsum dust from the lakebed is a very fine powder. The landing at White Sands produced the same rooster tail and the same tendency of the dust to settle on and inside the vehicle. The Shuttle has some compartments that open after it gets into the lower atmosphere to cool the vehicle. Among other things, it vents to the outside so the pressure equalizes. Of course, landing with those vent doors open means that dust is going to get in through them and in through the wheel wells and any other cracks and crevices not fully sealed. It got all over inside the electronics and the hardware inside the Shuttle. They really had to tear the vehicle apart and go through and clean the very fine dust out.

After that landing at White Sands, the program managers made a decision that only under totally unavoidable circumstances would they ever want to land the Shuttle at that lakebed again. Not that it wasn't a good lakebed, but it just made a huge cleanup mess for them. It took many months and a huge number of work hours to get the Shuttle cleaned up and recertified for STS-4. It was not planned to be a huge problem in the future, for STS-4 was the final flight in the flight-test program by definition. The STS-4 flight was going to land on the concrete runway at Edwards, and after demonstrating that, future flights would be scheduled to land at Kennedy, weather permitting, with Edwards as the secondary landing site. However, White Sands remains an emergency third alternative for landing, so if the weather at both Kennedy and Edwards is not adequate, the Shuttle would still land at White Sands if it needed to land somewhere in a hurry.

We refined our analysis of STS-3, our panels meeting to discuss any anomalies and differences that we saw in the data. There was no FAD-3. However, if we saw something that we thought was unusual, we passed it on to the other panels so they could take a harder look at what might be a concern. We, of course, were now looking at the proposed PTIs and at least one or two POPUs for STS-4, so we were busy between STS-3 and STS-4. Even after three flights, it was still a very exciting environment for me. I was working on a vehicle that went from orbit to landing, and the next flight was going to land on a concrete runway. In consideration of my desire to see flights go from runway to orbit, an important thing to demonstrate was precision landing on a concrete runway.

STS-4: Hail to the Chief

The STS-4 flight was going to be a very special event for the Shuttle program and for those of us lucky enough to attend the landing. It was launched at the end of June 1982, and was a seven-day mission. It was going to land at Edwards on the Fourth of July. Because we knew this far enough ahead of time, we built a viewing stand (Fig. 1) for President Reagan to watch the landing at Edwards. It would be a very big fourth of July celebration for us if things went as planned. The tower is about a mile from the concrete runway, and they built the viewing stand near it for President Reagan and some of the people traveling with him. The one-mile distance gave a very good view of the Shuttle at landing.

Those of us working on the Shuttle assembled for the launch of STS-4, in

Fig. 1. Nearly complete viewing stand built next to the flyby tower on Rogers Dry Lake for President Reagan to watch the STS-4 landing. (ECN 20027)

case there was a once-around abort. The crew was T.K. Mattingly and Hank Hartsfield. Things went smoothly on the launch, and there was no hint that it would be coming into Edwards on the first orbit. Dryden is an excellent place to watch all of the happenings associated with a Shuttle launch because we have such good video and audio coverage of the flight. We are never in the dark as to what's going on.

For the Fourth of July landing we knew we were going to have a huge crowd of spectators. I'd heard estimates that ranged from 500,000 to 800,000 for the number of people attending that landing. (Fig. 2) The expected crowd and the general confusion of who got in what gates with what kinds of passes convinced us to go in very early for the landing. Usually when I'd go in early for a landing, I would do some work in my office prior to going up to the control room to get ready for the landing. On this particular day, the Secret Service had asked us to vacate some of the offices on the first floor. The President was going to be down the hall from us, where they had a room for him and his contingent prior to going out to viewing the landing of STS-4. I'm not exactly sure what they did with our

Fig. 2. Some of the recreational vehicles parked in the official east lakebed viewing area preparing for the STS-4 landing. (EC82 18986)

offices, but the whole hallway was closed to the public to maintain security for the President. He came in the building, and I know that was the pathway by our office to take him from the main part of the building down into the area where he would relax prior to going out to the viewing platform.

Since we didn't have our offices to go to, I went up to the control room even earlier and took some reading material and things to work on. It was going to be a big day, because, in addition to the normal landing and postflight activities, President Reagan was going to welcome the astronauts back and give a speech. This was the Fourth of July, so we anticipated some big announcement.

The STS-4 flight was launched into an orbit with an inclination of 28.5 degrees, the lowest energy orbit from Cape Canaveral. The previous three flights had been launched at an inclination of 38 or 40 degrees. The Shuttle, as it did its normal burn, would be tracking north to get to Edwards. Since Edwards is nearly 35 degrees north, it would have to come about six and a half degrees north to land at Edwards, so it was going to need to have a substantial cross range, but nowhere near what it needed to do for a once-around back to the launch site. This was going to be the first time that it had to go that far in cross range, and it was also going to land on the concrete runway, a first for the Shuttle Orbiter.

During the reentry, the first lateral-directional PTI occurred around Mach 23.

We had four more lateral-directional PTIs and a longitudinal ASI. Then the crew did two pushover/pull-up (POPU) maneuvers—one about Mach 13, another about Mach 8-to get information about the vehicle's lift and drag and the thermal environment on the Shuttle during an off-nominal flight trajectory. Normally, the Shuttle stays at a 40-degree angle of attack until it gets down to around Mach 10, then it gradually decreases. (Appendix F) To get the required cross range for a once-around landing, the Shuttle needed to fly at a higher hypersonic L/D, which required a lower angle of attack. Unfortunately, this results in higher heating on the OMS pods, the side of the fuselage, and the leeside of the vehicle. These are the portions of the vehicle having less thermal protection than the lower surface.

The POPU maneuver at Mach 13 gave us the opportunity to see data from 35 to 45 degrees angle of attack. Normally it would only be obtained at 40 degrees, so this was important information. The POPU at Mach 8 showed us the lift-and-drag data and the performance and heating characteristics of the vehicle in the Mach 8 region off the nominal angle of attack. We got good data on those maneuvers.

Watching in the control room, we started to receive data in the Mach 13 to 14 region. We acquired data at the beginning of the first of the two POPUs, so we got to see most of that maneuver, a couple more PTIs, and then a POPU around Mach 8. I wanted very much to see the landing with the festivities, so, at Mach 2, I went out on the roof. Milt had been there with me for the previous two landings, and he was there again for this one. That this was the first Shuttle landing on the concrete runway may not have seemed like a big deal to the unindoctrinated, but STS-4 had a lot of the same aspects that had given Fred Haise problems on ALT-5. Landing on the concrete runway, the pilot's options are not as wide open as they are landing on lakebed runway 23, which is five miles long. The concrete runway, runway 22, is a little less than three miles (15,000 feet), but it has a 1,000-foot overrun.

Landing on the concrete runway 22, however, there is a threshold the Shuttle would need to clear, a fairly steep area at the beginning of the runway. There's also a step at the end of runway 22, so it is better if the Shuttle stops before reaching that. The Shuttle is going to roll out in the neighborhood of two miles on a three-mile-long runway with about a 1,000 feet of overrun, or about 16,000 feet. The pilot doesn't want to eat up too much of the runway, so he has to put it down somewhere between 1,000 and 5,000 feet from the beginning of the runway. However, 1,000 feet is getting too close to the beginning of the runway, and anything beyond 5,000 feet is going to leave less than 10,000 feet of concrete runway. Consequently, landing on the concrete runway is a harder task than the previous lakebed landings. Of course, Mattingly also knew that President Reagan and a half million or more Americans were going to watching the landing. If anything was going to raise his gain, this landing would do it.

As it turned out, Mattingly took control of the vehicle on the heading alignment circle as he entered it and flew manually all the way in through the landing. So he didn't run into the problem that Lousma had on the previous flight, trying to get in the loop too late. Mattingly did a nice approach and ended up landing 2,000 feet short of his intended touchdown point, which I'm sure the President appreciated because it gave him a little better view of the landing, since he was a little closer to the actual touchdown and abeam of the Shuttle as its gear was lowered.

It was a very nice landing and a very good flight. Landing short seemed to be a characteristic of the Shuttle that wasn't well understood by any of us. We believed that with the known lift-and-drag characteristics and the vehicle's flying qualities that the pilot ultimately should be able to land it very close to where he wanted. The Shuttle's speed brakes are quite good for controlling its landing speed, so they help in energy management and in achieving a touchdown point. The pilot dove at the runway at something a little over a negative 20-degree flight path angle, but this gives the pilot a good aim point, as we discovered many years earlier with other vehicles. Consequently, if the pilot was high on energy, he would add speed brake; if he was low on energy, he would reduce the speed brake. It works very much like a throttle, except there is a point where there's no more "throttle" available in retracting the speed brakes. Once they're fully closed, the pilot can only lower the angle of attack as a way of increasing his speed.

I remember my anticipation of this landing, having seen Fred's problem with landing on a concrete runway. I was hoping not to see it again in the STS-4 landing, but it did add to the drama for me personally while watching the landing. As it turned out, we were watching from the rooftop at Dryden about two-and-a-half miles from the runway, so we had a really good view. The landing looked good from where we were. I didn't know his true aim point so I couldn't tell within a 1,000 feet or so how well he'd done. It was only in the analysis of the data and in the postflight debrief that that became apparent. After the Shuttle stopped, they drove the President down to meet the crew at the Orbiter. He did an informal welcome to them at that point, and then a couple hours later he did a formal one with all the trappings that go with official functions where any President is involved.

When Columbia touched down on the runway at the end of STS-4, it was the last of three Shuttles to arrive at Edwards. Four days earlier the Challenger, the third Shuttle that was built, had arrived at Dryden and been mated with the Shuttle carrier aircraft. The Enterprise was still there from the ALT program. For someone who had devoted 20 years of his career by then to the aerospace industry, it was a very exciting Fourth of July—all three of the Shuttles, a very successful landing by T.K. Mattingly, and the President on hand. It was one of the most up days that I've ever seen at Dryden, everybody in a very festive mood. (Fig. 3) Even though we were all working on the fourth of July, I think all of us felt that it was a privilege and something that we wanted to do.

After watching for a half hour or so, I returned to the control room. There, on video, I watched the President greet the astronauts. Then I gathered my complement of strip charts and headed to a conference room, since my office was still

Fig. 3. Aerial view of the Dryden Flight Research Center showing some of the crowd gathered for President Reagan's welcoming of the astronauts after the STS-4 landing. (EC82 21128)

off limits because they didn't want a nerd engineer interfering with the official duties of the President. In the conference room, Rich and I looked over the data that we had from Mach 13 or so on down and identified the maneuvers. In our own minds and conversation, we could compare what we had to what we had already seen on the first three flights. We were starting to get a feel for what the data was telling us. This was a useful effort, but since we knew we were going to get better data, better time-tagging, and higher resolution information when the real data arrived, we just did a preliminary analysis from the strip charts.

We found out a few hours later that the recorder for the developmental flight instruments (DFI) had once again failed. The only thermal data available was from Mach 13 on down. Fortunately, that covered most of the POPU at Mach 13 and all of the POPU at Mach 8, so we did have the lift-and-drag data that we needed to increase the cross range to the required 1,300 statute miles needed to do a once-around from either Kennedy or Vandenberg. Losing the DFI data was quite significant, because there were over 4,500 parameters on its recorders, and these were the primary thermal data for defining the aerothermodynamic environment and the thermal environment of the Shuttle. In addition to the surface-based temperature sensors, there were also sensors throughout the internal part of the vehicle. There were also a lot of special sensors that were only going to be on Columbia and only for the four test flights.

Consequently, losing the data on the DFI was a frustrating experience for everyone involved in the aerothermodynamic area. I was trying to look at some of those maneuvers, particularly the POPUs, to see if I could get a parameter identification technique to model the transients of the heat from the high temperature flow into the vehicle. Originally, when it looked like the boundary layer was going to transition slowly, I thought that might be another thing that I could model. However, on most flights, the transition was so quick from laminar to turbulent, so that idea didn't work out very well.

After we had looked at the data for an hour or so, it was time to go outside and try to get into position to see the president welcoming the astronauts and hear the president's speech, which was going to occur right next to the ALT Shuttle Enterprise. Outside, I noticed there was an absolutely huge crowd of milling people. I knew there was a lot of security—not only base security and Dryden security, but also the Secret Service at this point. You couldn't always identify the Secret Service people, other than they were always talking into their wrists, looking the wrong direction, and usually wearing sunglasses. I had thought I was going to be quite a ways away from the president, but Marie Fullerton, Gordon Fullerton's wife, spotted me and helped me talk my way up to near the front row, near the stage where the astronauts and the president were going to meet. Marie was obviously an astronaut's wife, an astronaut who had flown on the previous Shuttle flight, and so nobody really gave any resistance to us working our way up to the front of the stage. That's how I got to watch the whole thing from probably about 50 feet away. (Fig. 4)

I remember it as a big day for me. I'm not big on politicians, but the president of the United States is the President of the United States and everything that goes with that. As a patriotic American, I found the event to be a very patriotic thing happening on the Fourth of July with a President who espoused very patriotic feelings at every occasion. We had a nice welcoming ceremony, and then the president gave a very nice speech, but he did not make any commitments for further space activity—neither the space station nor the fifth Shuttle that we thought we needed at that time. The ceremony didn't last a long time—probably a half hour or 45 minutes.

Then, the President was whisked away, and the astronauts were whisked back out to the medical facility for more testing and probing. The rest of us milled around and wound down after what had been one of the most exciting days that the aerospace community had seen, at least in my career. I had been very excited by X-15 flights, by the first Shuttle flight, by the first lifting-body flights—both the M2-F1 and the M2-F2. But those were a more personal kind of excitement. This ceremony with the President, more than a half million other observers, all three existing Shuttles, and all of the enthusiasm of a Fourth of July was very special. The Challenger atop the 747 Shuttle carrier aircraft (SCA) taxied by, went down to the main runway and took off, heading for the Kennedy Space Center. Then, the SCA pilot came back around with it and did a slight wingover maneuver as it passed over the Columbia and the President formally greeting the

Fig. 4. President and Mrs. Reagan are shown at the podium as they officially welcome T.K. Mattingly and Hank Hartsfield after STS-4. Only a small portion of the crowd is shown, gathered next to the Enterprise for the ceremony. (ECN 18995)

astronauts. Everybody got a good look.

That was just the icing on the cake for the whole event. It was a big event and something that I wish more people could have participated in. Things like that made you feel good about your work and about your country. I wish those would go on more often. They were very good for me, and I think they're probably very good for Americans in general.

Although the DFI recorder data had been lost, the ACIP recorder worked on this flight, and we had very good data. After getting the whole data set, we had excellent data on the initial bank maneuver and the subsequent four bank reversals. This was the last time that the commander was to fly the first bank maneuver. On STS-5, the control-system modification would fly, once again an automated maneuver, as it had been on STS-1.

This flight also showed the quarter-hertz oscillation. We had a PTI scheduled on STS-5 and several subsequent flights to try to identify the cause of the quarter-hertz oscillation to get it fixed. Between STS-4 and STS-5, we came up with a FAD-4—a further modification of the original Shuttle data book to put increments on all the forces and moments as a function of body flap and elevon deflection, Mach number, angle of attack, and altitude region. That took more work, because in addition to the collaboration among various analysts, we worked more closely with the flight controls, the simulation, and the aerodynamics panels so they could see our process, question our results, and make us show why we were putting in particular changes. For the most part, we hoped it eased the job for the people working on the control systems. In addition to changing the best guess value of each stability-and-control derivative, we also reduced the variations, that is the uncertainties.

The reduced uncertainty also should have made the control problem easier. Occasionally, however, we would come up with one that moved the mean value away from the predicted value and actually made it a little more difficult on the control system, so finalizing FAD-4 was a collaboration involving a lot of interchange, one-on-one conversations, conversations within a panel, and joint panel discussions with representatives from the other panels. Then we prepared for STS-5, which was considered the first operational flight because probably 90 percent or so of the objectives of the program had been met by the first four flights.

The DFI recorder in the payload bay was to remain there for one more flight, because most of the DFI data had been lost on STS-1 and STS-4. This was a change in plans, but it could be accommodated. We had by then expanded the allowable flight center-of-gravity positions by several inches fore and aft as well as half an inch or so left and right of the centerline. This was important for the operational phase of the vehicle, although it was more of a bureaucratic definition for starting to fly useful payloads in the Shuttle. Those of us trying to understand the Shuttle's aerodynamic, aerothermodynamic, performance, and structural characteristics continued to work for varying lengths of time. The stability-and-control analysis was going to continue for 15 more years over more than 80 flights.

STS-5: The First Operational Flight

STS-5 was commanded by Vance Brand and Robert Overmyer, and it was the first Shuttle flight to include two additional crew members, Bill Lenoir and Joe Allen, who were, in my opinion, the first two space passengers. The ejection seats were deactivated on the Shuttle, so they were all in the same boat. It could be argued that some of the previous capsule crews had been passengers, including Soviet cosmonaut Valentina Tereshkova, who was not a pilot. On the Voskhod 1 flight, there had been three people on board, and two of them were really passengers, if not all three of them.

The STS-5 mission had originally been planned to land on a lakebed runway so we could do some further testing and so the Shuttle could also make a crosswind landing. Every time we had what looked like an opportunity on earlier flights, management would claim the cross wind was too high and divert to more of a head-on landing into the wind. However, for STS-5 the lakebed was wet, so the Shuttle was scheduled to land on concrete runway 22. (Fig. 5) We came in for the launch because Dryden was still the once-around-abort site. We came in ahead of time and watched the launch, and then we went back to our regular duties, there being nothing for us to do for STS-5 at that point.

The landing was going to be in the early morning. I remember going in on that November day. It was really cold and quite cloudy. I had the feeling that the Shuttle probably wasn't going to land that day at Edwards because it just wasn't very good weather. It wasn't typical of the good weather that we get 75 percent of the time in the winter and almost 100 percent of the time in the summer. However, the crew continued to press on, made the deorbit burn, began to come down, and did their maneuvers. There were eight PTIs scheduled, two longitudinal and six lateral-directional. The crew aborted two lateral-directional PTIs after

Fig. 5. A scorched Space Shuttle Columbia and its reflection in the water from recent rains. The vehicle is being towed after the end of the STS-5 mission. (EC82 21081)

the first bank maneuver, for it had consumed more than the budgeted amount fuel for the reaction control system (RCS), putting the remaining hydrazine below the desired amount. Prior to STS-5, we'd had aero-stick inputs (ASIs) that were both lateral-directional and longitudinal, but all the PTIs had been lateral-directional maneuvers prior to STS-5. Also on STS-5, we finally got our Mach 18 POPU with the DFI recorder still in place. Fortunately, the DFI recorder worked on this flight, so we got good data on it as well.

Once again, we didn't acquire telemetered data until Mach 14 or so and watched the various PTIs and bank reversals. On this particular flight, I waited a little longer to go out on the roof because we had a PTI scheduled for about Mach

1.6. I wanted to see that we got that maneuver, since it would help us start to unravel the quarter-hertz mystery. After we got the maneuver, I left for the roof to watch the Shuttle land. I had my coat on this time, because it was landing in November. It was cloudy, bad weather for watching a Shuttle landing. However, it turned out the weather wasn't that bad. There was a cloud cover over Edwards, and going out on the roof late saved me from being in the cold as long as I would have been otherwise. I would not have seen anything earlier anyway. I didn't see the Shuttle until it got onto the approach maneuver. It came below the clouds, and at dawn I watched it come in and land.

It was an interesting landing. There had been some clouds in the area for Engle's flight, but it was much cloudier for STS-5. In fact, if I hadn't been earlier watching the data and the ground track from inside the control room, I would have had no way of knowing it was coming into Edwards other than the signature double boom. After the Shuttle came down through the clouds, it made what looked like a perfect landing on the main concrete runway. (Fig. 6)

I believe it was later determined that STS-5 missed by a 1,000 feet. So over five flights, the landings at the intended touchdown point had been made plus or minus 3,000 feet. Although STS-5 was the best landing yet, we knew we could do better. It just took more practice. I think most of the astronauts to this point had attributed it to the difficulty from within the Shuttle in judging height. Even though the crew is there on the front of the vehicle, they are quite high. The pilot is about 35 feet above the runway when the main wheels touch down in a normal landing. However, the main runway doesn't give much of a feeling of height within five to ten feet from the Shuttle. The pilot is at a higher angle of attack than in a standard airplane, up around eight to ten degrees. (See Fig. 1, chapter 10) Trying to judge the height from that is something a pilot just has to learn from talking to previous crews and practicing in the STA.

Once the landing was over, we returned to the control room, collected our strip charts, and went off and looked at the maneuvers, hoping that all of our instruments had worked. As it turned out, the ACIP yaw gyro had failed. We had fairly good data, but we didn't have our high-quality yaw data, so we had to make do with one of the poorer quality instruments on that flight. We were starting to weight more heavily the results from the PTIs over the energy management maneuvers, because they were superior maneuvers in terms of crispness of the input. The system used to define the PTIs was designed to defeat some of the things the control system was trying to do that kept the Shuttle from getting crisp maneuvers. These were also maneuvers specifically designed for enhancing the information we needed on specific stability-and-control derivatives. We continued to analyze the bank reversals and the incidental maneuvers throughout the flight program. It was still something to fall back on, in case we were trying to come up with a tie-breaker in terms of additional data.

After STS-5, we did not do a flight assessment delta (FAD), but we did continue to meet as a panel, compare our analyses, and come up with an official set of estimates, although we did not go through the whole verification system as

Fig. 6. The large size of the Space Shuttle Columbia is evident in comparison to the support crew in this view showing the vehicle being towed to the Shuttle area after STS-5. (ECN 22055)

we had with the earlier FAD. Nevertheless, we shared the information we got with the other panels. We got the Mach 18 POPU which, along with the earlier ones at Mach 20, Mach 13, and Mach 8, meant we at least had a single POPU at four separate Mach numbers to give us lift-and-drag and aerothermodynamic data that were off-nominal. As a researcher, of course, I would have liked to have had two or three POPUs at each of those Mach numbers to make sure that we had everything covered, because each of the POPUs had been obtained on different flights and had somewhat different body-flap and elevon settings. There always was a correction we needed to put into it to get really high-quality lift-and-drag results to account for the body-flap and elevator effects. Nevertheless, we had the four high-quality POPUs that helped resolve many of the aerothermodynamic issues.

From the data, it looked like the vortex that was predicted in the wind tunnels to occur at a lower point on the vehicle was actually impinging up on the orbital maneuvering system (OMS) pod, making both the OMS pods and the side of the Orbiter body hotter than predicted. (See Fig. 6) This was of concern, because to do the cross-range maneuvers required for a once-around-abort to Kennedy or to Vandenberg, we needed to have adequate thermal protection on the OMS pods for coming in at a lower angle of attack. Cross range is increased by flying at a lower angle of attack, starting at above Mach 20 down to Mach 6 or 8.

The Challenger was going to fly its first flight on the next launch, and it had the new advanced flexible reusable surface insulation (AFRSI) material on the OMS pods, making the vehicle lighter and somewhat easier to maintain. The AFRSI was replacing some of the low-temperature surface insulation (LRSI) tiles. The Challenger was also going to be launched with a lighter external tank and lighter solid rocket boosters. Consequently, the performance, primarily for launch and payload capability, was going to be higher with Challenger than with Columbia. Columbia's design was more conservative, and therefore, it was heavier and more difficult to maintain. From what we'd learned on Columbia, we could change the requirements for Challenger's thermal protection system. There was less acreage covered by tiles and more acreage covered by AFRSI blankets for Challenger.

We knew the required cross range of 1,300 statute miles would not be realized without having more thermal protection on the OMS pods. Other than that, everything else looked good at the lower angles of attack. The stability and the aerodynamics were behaving fine. During the POPU maneuvers, we were also able to observe some other interesting characteristics of the boundary layer. As the angle of attack varied, we could see a change in the boundary layer from laminar to turbulent and then back to laminar again at the higher Mach numbers. Unfortunately, we didn't get enough POPUs to certify the effect that angle of attack was having on the transitioning of the boundary layer.

All of these were surprises, things that we would expect on future vehicles, but the more we could quantify these factors, the more we would learn. At that point, we were trying to understand why the boundary layer did not transition at high Mach numbers as predicted and why it did so quickly once it started.

Ultimately, it was concluded that this occurred primarily because the Shuttle was dominated by surface roughness on the windward side. Once the boundary layer started to transition, it transitioned all the way forward quite quickly. There could also be an early transition due to a gear door or a tile being way out of line or damaged, so the POPUs were providing good performance information for lift and drag and the thermal information that we would need for a lower angle-of-attack entry to maximize the cross range for an operational vehicle.

XIII

STS-6 to the Loss of Challenger

After having looked at the flight data from the first five flights, we proceeded towards further refinements in our understanding of the Shuttle's aerodynamic and aerothermodynamic conditions for subsequent flights. We had already resolved some of the earlier concerns. We were now confident that the location for the center of pressure (CP) had been mispredicted by ten inches, caused primarily by not modeling real gas effects. We had compensated for the CP misprediction by changing the body-flap/elevon schedule. We had also further refined the vehicle's center-of-gravity (CG) limits for various payloads. We now understood the problem caused on the first bank maneuver on STS-1 and had added corrected rolling moment due to yaw jet and dihedral effects into the control-system design. On STS-5, the initial bank had been done automatically, and it looked like the bank maneuver done manually on STS-2, -3, and -4. [52]

Those two things, now fairly well understood, had been two of our biggest stability-and-control concerns on the first flight. The other concern that we had was the quarter-hertz oscillation. After having looked at the analysis of the 1.6 Mach number PTI that we'd gotten on STS-5, we believed that it was due to the lower effectiveness of the aileron. We accounted for that in the database as a preliminary result and would get more PTIs on subsequent flights. It looked like increasing the stability augmentation gain on the roll axis would solve that problem. The Shuttle would still have the little quarter-hertz oscillation, but it would be of lower amplitude and no longer noticed by anybody.

Expanding the Limits

We were making progress. One of the chief objectives of the work we were doing was to increase the allowable CG limits, which would take many flights before we were finished. Using a process defined to justify additional data requirements called development test objectives (DTO), we had to write up a justification to the Shuttle program management for the maneuvers that we were requesting. The DTOs from the technical panels went through approvals at the various management levels. Once approved, various PTIs were included in the flight plan and practiced in the simulator by the astronauts.

The remaining DTOs that we worked on would involve other aspects of the vehicle. [52] These were primarily the CG placards and understanding the flying qualities of the vehicle subsonically. When the Shuttle was near the ground for landing, the rudder effectiveness appeared to be somewhat of a problem. We also proposed maneuvers designed to help define what was called the wraparound control system, which was to reduce the firing of the yaw jets so that the Shuttle could carry less hydrazine, perhaps as much as several hundred pounds less, which would increase the available payload weight.

Many of the maneuvers we were asked to analyze were things that occurred during the entry that weren't understood. We encountered things such as density gradients, which were large and unexpected changes of the atmospheric density over a short distance at high altitudes. We looked at the dynamic behavior of the vehicle when it encountered them to see if there were any impending problems.

We also looked at asymmetric boundary-layer formation. The boundary layer would form on one side faster than it would on the other side, causing a yawing moment. It tended to make the Shuttle want to fly at a steady sideslip. The control system used the aileron and the yaw jets to trim the yawing moment out, but that ended up using more hydrazine, and we were interested in seeing if we could quantify how big these moments were from the asymmetric-boundary layer formation. We'd seen this on about one out of five or six flights. It was usually small, but occasionally it was fairly large. Opinions varied as to the cause—irregularities in the steps between the tiles, damage to the tiles, material between the tiles oozing out and tripping the boundary layer, or the gear doors. [61] My personal favorite was that it was caused by the gear doors producing a small step.

After STS-5, we had a lot fewer temperature sensors on the surface. We would tend to have them more on one side than on the other, so if the early asymmetric-boundary layer transition occurred, we didn't always have a good picture of what it looked like, other than the resulting forces and moments on the vehicle. I developed a code to calculate that moment. With the few thermo-couples that I had, I found that in most cases if either the front gear door or the main gear doors were not quite tight up against the body, that would cause the boundary layer to trip. A gear door, of course, because it is a door and opens on hinges, tends to have long edges parallel to the local flow. It turned out that on the Shuttle they're not of major concern, at least when the vehicle is healthy. The moments are small enough that the yaw jet firings they cause are within the fuel budget allocated for reentry.

STS-6, with Paul Weitz and Karol Bobko, was the first flight of the Challenger. It was the second Shuttle vehicle, the one that President Ronald Reagan had acknowledged as it dipped its wing to him as it flew off for Florida. From an observer's point of view, the most interesting new effect that occurred prior to the STS-6 landing was the PTI at about Mach 1.5, which could be seen by the contrails initiated by the firing of the yaw jets. These could be seen by a series of contrail dashes that were formed on either side of the vehicle as first the left and then the right yaw jets fired. These intermittent contrails made it possible to see the Shuttle as it flew overhead by making a large contrail that indicated its position. To me, this was very reminiscent of the X-15 flights when the pilot would vent unused propellant, creating a similar effect for the chase pilots to locate the X-15.

This flight was more or less like the others, except it did have the improved thermal protection system and the advanced flexible reusable surface insulation (AFRSI) blankets on the OMS pods and other areas of the vehicle. The AFRSI had replaced most of the low-temperature reusable surface insulation (LRSI) tiles. The main new thing that came from the STS-6 flight was the vortex that we had been studying, which made both the side of the Shuttle and the OMS pods hotter than predicted. The vortex was impinging on the OMS pods at the angles of attack at which the Shuttle was reentering. We found on STS-6 that the AFRSI was not strong enough to put on the OMS pods, and maybe not for some of the other upper surfaces for the lower angle-of-attack entry required for the 1,300-mile cross range. With the long range optics, during the heading alignment circle, we could see where the AFRSI material had been scrubbed off the fronts of the OMS pods.

That was a problem that needed to be fixed, and it involved more wind-tunnel testing and a redefinition of the tiling on the OMS pods for future flights. The near-term solution was to put the LRSI tiles back on the OMS pods. In support of understanding the limitations of the FRSI and AFRSI, Dryden used its F-104 flight-test fixture to evaluate the material under inflight load-testing conditions 40 percent higher than those evaluated during the simulated Shuttle flight. [62]

On STS-7, the Challenger with Robert Crippen and Rick Hauck was supposed to make the first landing at Kennedy. However, because of weather conditions and vehicle problems, it landed at Edwards. It was the first five-person crew to go into space, and it was also the first U.S. spaceflight to include a woman astronaut, Sally Ride. The Challenger landed on runway 15, (Appendix E) which is six miles long, and a little less than a mile from Dryden. From the ramp behind Dryden, we could see the approach, and we were fairly close to the touchdown and very close to where the roll out ended. Runway 15 is on the dry lakebed, and my recollection is that the end of the runway was a little soft and the Shuttle dug in a little bit at the end of its rollout.

There was much anticipation about the landing of STS-8, because it was going to be the first night landing. It also had a night launch. It would be the second orbital flight for Dick Truly and his fourth flight in the Shuttle, including his two ALT flights. The interesting thing about the mission, other than collecting the usual maneuvers that we needed, was the night landing. This was to be the first low L/D max unpowered landing in the dark for any vehicle that I am aware of. I didn't know what to expect. It was going to land on runway 22, as the concrete runway was designated for landings to the southwest. It is a little under three miles from the Dryden ramp. We were restricted as to where we could be for that flight because of nighttime security and worrying about the people who might not be hoping for the Shuttle to do well. During day time landings, we could see the Shuttle quite well as it came into coming into runway 22. It looks quite large, even though it's two or three miles away. At night we didn't expect to see it very well because of the darkness of the bottom of the vehicle and the low amount of light. We normally don't see other airplanes at night, except the standard aircraft lights, unless they actually cross the Moon.

I remember Milt wondered what the night landing was going to look like. We watched the maneuvers from the control room, then went outside with the others. It was quite surreal because Edwards Air Force Base was essentially dark, and there were few observers other than essential personnel. There were buildings with lights, but the lakebed itself was dark except for the exceedingly bright xenon lights placed at the end of the runway. They were very powerful and showed all the way to the other end of the runway, so the runway was well lit, but not the sky. The lights on the runway would pick up the Shuttle only when it came down into the light.

The Shuttle gets very hot during flight, but by the time it gets down to Mach 2, it's cool enough that there's no chance of seeing the vehicle other than by using infrared cameras. We went outside on the ramp (no one was allowed on the roof for this flight) and heard the double boom, so we knew we were in the right place. Then we heard the rumble of the Shuttle but couldn't see it. We were all staring at the runway, listening to the various calls from the Shuttle and from the mission announcer, so we knew it was getting close to the ground. Then, an ethereal Shuttle suddenly appeared at the edge of the light, quite close to touchdown. (Fig 1) It was a very gray color. It wasn't black and white as I was used to seeing, the xenon lights making the bottom of the vehicle look more gray than black.

It appeared out of nothingness and flew in its normal manner right down onto the runway. We could see the Shuttle going down the runway on its back wheels

Fig. 1. The Space Shuttle illuminated by xenon spotlights during the night landing of STS-8. (EC83 26527)

as it always did, then the slow derotation of the vehicle as the nose wheel came down and the pilots did their braking. I think they had another brake failure on that landing. They seem to have had them on most of the Shuttle flights prior to STS-8. Then the Shuttle slowed down and stopped. It was quite a great success. It was a nice landing. The pilot hit his touchdown point very closely and did it in the dark. Anybody who's flown by instrument flight rules (IFR) knows that landing in the dark is just a matter of trusting the force. Truly and Daniel Brandenstein had flown it with a great deal of confidence. As they came into the heading alignment circle, they could see the lighted runway very distinctly, and so the landing was quite easy.

I've heard the same comment from other pilots on other programs that we've done at night for various reasons, sometimes for flying classified missions to avoid being seen. They find that night-time flying is actually a good deal easier, particularly at Edwards, because the only lights that you see are the ones you expect to see. You don't have the clutter of lights from cities, freeways, and things like that. The first night-time landing of the Shuttle was fun to anticipate and fun to watch, and fortunately it was also quite routine.

STS-9 was the second and last flight of John Young in the Shuttle. He and Brewster Shaw were taking the Columbia up on a Spacelab mission for its last flight before they refurbished it at Palmdale. STS-9 was also the first six-person crew to go up in a vehicle. This time the Shuttle was in a 57-degree inclination orbit, making it visible occasionally as it went over Edwards and Lancaster at night in late November and early December of 1983. We sometimes could catch sight of it on one of its orbits. It would go over points between 57 degrees north and 57 degrees south latitude twice on each orbit, enhancing the chances of seeing it, although it was not as bright as some of the later Shuttle flights that went over Southern California a half-hour prior to sunrise. The Shuttle was inverted as it passed over, so the white TPS on the top faced down and there was good reflection from the tiles.

On occasions when Shuttles have gone over in the clear desert air just after sunset or just before sunrise, it is by far the brightest object in the evening sky except for the Moon, much brighter than Venus ever is. It's very easy to see, so bright I've really wondered if I'm looking at the Shuttle, as it appears to be at a lower altitude. At any rate, during STS-9, they did a very successful Spacelab mission. They extended the flight a day, and then the problems began. One of the general purpose computers (GPC) quit, delaying the landing. They also had some failures in the inertial measuring unit (IMU) and perhaps even some auxiliary power unit (APU) issues.

The vehicle waved off on landing at Edwards on the morning attempt and chose the afternoon pass; I think it landed about three in the afternoon. It landed on runway 17, which is in the middle of the dry lakebed and is the furthest out from Dryden, but we still got a very good view because it was doing its landing right abeam of Dryden. We went outside and watched it land. After we returned inside, we were watching the video. On the television screens we could see small flames coming out by the base of the vertical fin, which was something we'd never seen before and a little unsettling. We were wondering how big a problem this might be and what the consequences would be. Then the mission announcer came on and said something like, "Those flames are not a fire," which was not terribly comforting because those flames did look like a fire to the rest of us. It turned out they had an APU fluid leak and they were just the flames from that. They did put it out, and there was minimal damage. It didn't really do any major damage to the Shuttle. The Shuttle was due for a long lay-up, anyway, to take care of some problems that had not been dealt with quite the way they wanted initially when they built it. The overhaul was also to upgrade the tiles and the avionics, and refurbish the vehicle.

After STS-9, the designations for the various Shuttle flights started to get very confusing, because up to that time the flight number and the STS number were the same. However, they had scheduled various payloads that were not being flown in order because they were behind schedule, so to make it more obscure as to whether the payload or the flight was late, management went to a new designation system. The STS-9 mission became "41-A." The "4" meant it was scheduled in fiscal year 1984, the "1" meant it was launched from the Cape, and the "A" meant it was the first payload scheduled in that fiscal year. Those of us doing the support for the vehicle, to make our databases more understandable to us and make sure we weren't missing pieces, continued to call each subsequent flight by an STS flight number corresponding to the mission number.

Consequently, what was STS-10 to us was officially known as 41-B, and it was Vance Brand's second flight, along with Hoot Gibson as the pilot. It was the first flight to actually succeed in landing at Kennedy, and that changed our mode of operation a little bit. We came in and monitored the landing with what was available on the network. Partly because we were interested in it and partly because it came over a different part of the world as it landed, it was going to be an interesting landing. As it turned out, they did have a very nice landing at Kennedy, and we got our usual collection of maneuvers. The main thing I remember about the flight, other than it being the first landing at Kennedy, was that the roll rate instrument failed on our ACIP package, so we once again had a somewhat degraded data set to look at. But we were getting more and more maneuvers and starting to see clusters of information falling nearer and nearer to each other, so we were removing some of the uncertainty of the database and were getting a better feeling for what the aerodynamic and aerothermodynamic characteristics truly were. We were still in the process of expanding the center-of-gravity placard.

STS-11, officially known as STS-13 or 41-C, was Crippen's third flight and the second one that he commanded. His pilot was Dick Scobee. Dick had been a good friend to both me and my wife Mary Shafer, a Dryden handling qualities engineer, as well as to Dryden and everybody that worked with him. (Fig. 2) He was just a super guy, so Mary and I decided we would go to Kennedy for the launch. We paid our own way to the launch and got viewing passes to be able to watch it. We also got to go up to the Shuttle a few days before they fueled and did a close inspection of the vehicle prior to launch.

Fig. 2. Dick Scobee landing the X-24B on October 21, 1975 during his first lifting body flight. (ECN 4916)

It was going to be a very special launch for us, because we were so pleased for Dick Scobee to finally be able to go into orbit and not that long after being selected as an astronaut. The people selecting crews obviously thought very highly of Scobee's capabilities as well. The VIP area was further back for the Shuttle launches than it had been for Apollo 11. A Shuttle launch is not as loud and not as bright as the Saturn V launches were, but for anyone who hadn't seen a Saturn launch, as Mary hadn't, it was still quite a show. I think she was just as enthused after watching that launch as I had been after watching Apollo 11. I was almost as enthused. It wasn't people going to the Moon, but it was somebody I knew fairly well going up in a vehicle that by then I had worked on for 12 or 13 years. I very much wanted to have it be a good launch, a good mission, and a good landing.

It was a very up time for both me and Mary. We enjoyed our trip to Kennedy, but waved off on staying for the landing which we also could have done, since the Shuttle flights were getting long enough that you had to spend a couple of weeks there if you wanted to see both. We wanted to see him land, but we knew there was the possibility that he might land at Edwards anyway, so we returned to Edwards and monitored the progress of the flight.

STS-11 also had a couple of other firsts. It was the first direct orbital insertion launch that the Shuttle had done. This has usually been done since that time, due largely to the improved understanding of the vehicle, having more confidence in it, and having more performance by having lightened the Shuttle, the SRBs, and the external tank. All of those things were working toward making it a more viable vehicle.

The other thing that I remember very vividly was the attempt at rescuing Solar Max, (Fig 3) which had been launched earlier and had languished in orbit for some time because of malfunctions in several components. The idea was to see if the Shuttle could go up and capture a satellite, repair it, and send it on its way. The manned maneuvering unit (MMU) system was used for the first time on that flight. The two mission specialists were Pinky Nelson and Ox van Hoften. Ox is bigger than most of the astronauts, and he's the kind of guy you would want handling a satellite that on earth weighed thousands of pounds.

Fig. 3. Space walk performed to rescue the Solar Max satellite on STS-11 (41-C). (EC 28503)

They groped with how to get hold of Solar Max so they could repair it. Some of the initial things that they tried didn't work out as well as planned. They innovated, and they eventually captured it, repaired it, and it went on and did its mission. A lot of their space walks were shown on cable television, so we watched a good deal even when we weren't at work.

As the time approached for STS-11 to deorbit for Kennedy, the weather there was worsening in a hurry, and I think they knew it would be unacceptable by the time the Shuttle arrived. Since the weather was not expected to improve at Kennedy, they were redirected to do the deorbit burn to come on into Edwards two hours after the decision. I had been watching the weather at Kennedy, suspecting that it might be one of those times when the weather would block them going there and they would end up coming into Edwards.

Since Mary and I knew Scobee, we wanted to be there for the landing. It was quite early in the morning, probably four o'clock or so, and I was listening in. They had better coverage in those days than they do now. They said that they had decided to come into Edwards, so Mary and I were more or less ready. We were going to go into work early, so we drove out and got into position. I didn't monitor the flight in the control room because I wanted to watch it land. Since it had been scheduled to go into Kennedy, there wasn't really much in the way of official duties that we would have had. But we wanted to watch the Shuttle land, and it was still dark as we got into position. We took our Nikon and tripod in hopes of getting a decent picture. As the Shuttle came around, we could hear the double boom. Even in the early dawn we could see the Shuttle because it was high enough up that the Sun was shining on it, so we watched it come down. The Sun was just starting to brighten the horizon as the Shuttle came out of the heading alignment circle on its approach to land.

Mary, being the camera person in the family, had her 200-millimeter lens. She was tracking the Shuttle as best she could with her tripod, clicking away. It was a nice view because the vehicle was landing on runway 17, which brings its whole flight path perpendicular to the direction we were looking. As the Shuttle was on its approach, it was diving and going south, coming from our left hand-side, so we got a very good view of it. As it was pulling out of its preflare maneuver, it went between us and Leuhman Ridge to the east. Mary got one picture of it that was very special. (Fig. 4) It was a dawn sunrise picture taken towards the

Fig. 4. A dawn landing approaching Leuhman Ridge as STS-11 prepares to touchdown at Edwards AFB. (EC84 29856)

east with the Shuttle in silhouette with the bright sky behind it. In fact, the picture was so good that the photo lab took her negative and it's in Dryden's archive. The picture appears in many places around Dryden, including the library.

There were very few people who actually got to see that landing because most people weren't prepared to come in to Dryden that early. By then, people had seen enough landings that they didn't make as big an effort as we had made to get there. I probably would have made the effort anyway, because I was still deeply committed to the Shuttle program, but with Scobee on the flight, we really did want to see it. (Fig. 5) We knew his family and friends at Cape Canaveral would be disappointed about not being able to watch the landing. Only a few people would be coming in early, either because they had jobs to do with the Shuttle or they were tourists like us. We gave copies of Mary's picture to Dick Scobee, along with others of his launch. I'm sure there were only a few people who saw both the launch and the landing.

Dick Scobee was very taken with the pictures, particularly of the landing. He signed a copy and gave it to us. We printed some up for him to give to his friends and for his own mementos. I think we gave some to Crippen as well. Scobee had been such a good friend, and he was so appreciative of the pictures. On many occasions over the next year or so when we had contact with him, he'd bring up

Fig. 5. The crew of STS-11 (41-C) after the landing at Edwards. From left to right in the center foreground are Ox van Hoften, Dick Scobee, Terry Hart, Pinky Nelson and Bob Crippen. (ECN 29752)

how thoughtful it was for us to give him copies of pictures of his landing—that there were very few photos of any kind of the landing.

STS-12 (officially STS-16 or 41-D) was the first flight of the Discovery, while STS-13 (officially STS-17 or 41-G) was Crippen's third flight in command. Crippen was also the first to fly two Shuttle flights only two missions apart. The main thing I remember about STS-13 is the ACIP. I think the power supply failed, because there was no data at all for the launch, on-orbit operations, or the landing. So we didn't get any high-quality data, but we did record maneuvers of lesser quality which we continued to analyze. The only flights we didn't get maneuvers on were a couple of times when there was a situation on board that made the crew feel that they shouldn't risk the maneuvers, or they didn't have enough RCS fuel to get through them, or it was a DoD mission. On DoD missions, if we did the maneuvers, inferences might be made about the payload and the orbit characteristics. There were very few of those in the first 50 flights where we didn't get data.

The next flight that I have a recollection of was the one I call STS-17, which is officially STS-24 or 51-B. It landed at Edwards on runway 17. The official record showed that there was a good cross wind, but that wasn't based on data. It was based on prediction or something that had happened earlier in the day. I was outside watching the landing, and it was absolutely calm. I remember being involved in many discussions, trying to dispute the claim that it had landed in a 12-or 15-knot cross wind. It is very difficult with a vehicle like the Shuttle to certify what kind of winds it's landed in, because cross winds are measured from the angle of sideslip, and we had no real sideslip measurements other than from the inertial measurement unit and the two air data probes mounted on opposite sides of the forward fuselage. Looking at the data, I could see no evidence of a cross wind at all. I didn't want them certifying to a higher cross wind than the Shuttle had landed at, because it later might cause us to commit to a landing when we shouldn't. This could end up in a disaster if the vehicle was not controllable and the pilot couldn't keep it on the runway. At Kennedy, there is a swamp off the sides of the runway.

One of the first objectives of the Shuttle program was to demonstrate its capability to do a cross-wind landing. Up through STS-17, I claim that we never had seen a cross wind of more than three or four knots. That continued to be an issue throughout the program, and even to this day, we have not demonstrated cross-wind landings above 13 to 15 knots. In claiming that STS-17 landed in a cross wind, they were using wind sensors three or four miles away from the landing site, trying to correlate that data with what was going on where it was landing. It is a hard problem to satisfactorily resolve when we fly as infrequently as the Shuttle does—and rarely in any significant cross winds.

STS-19 (51-F) was the first flight with Gordon Fullerton as the Shuttle commander. Of course, he had landed with Fred Haise on three ALT flights, and he had landed with Jack Lousma on STS-3. I claim that Gordon came the closest to actually flying a Shuttle into orbit, because he had engine and indicator problems during the ascent. The ground controllers at Houston decided that the temperature indicator was incorrect and that Gordon command the Shuttle to go into orbit.

The next flight that was significant to me was Joe Engle's second flight, which was STS-20, officially 51-I. It was good to see Joe get a second flight. It had been three or four years since his first Shuttle flight, and I was beginning to wonder if he was going to get another one. I was hoping his difficulties with dehydration and providing more maneuvers than might have been considered prudent wasn't being held against him. However, if there were any issues, they evidently went away because Joe did fly the STS-20 mission.

Those studying the air loads on the Shuttle wings during many of the ascents noted that they were significantly higher and at different conditions than predicted. I used the same parameter identification techniques on the maneuvers that occur as the Shuttle is oriented for the planned ascent trajectory. Al Carter of Dryden studied these mispredictions, and, along with others, proposed trying to understand these unexpected loads by instrumenting the Columbia's wings with pressure sensors. With these sensors, we could integrate the pressures to get an independent estimate of loads and to understand how we might mitigate the problem. We did not get good agreement between the two methods on STS-24 (or 61-C). We felt that this might be due to the pressure port installation on the Shuttle. A flight program was proposed and conducted with Dryden's F-104 to simulate the gap filler irregularities and the condition of the wing pressure ports themselves. The flight tests showed that only part of the error was due to the pressure port installation. [63]

We also were able to get additional information during the entry portion of the Shuttle flight to study some of the aerodynamic anomalies we had noticed earlier. The ascent loads were eventually understood after further Shuttle flights, and the ascent trajectory was modified to minimize these ascent loads.

Many of the early Shuttle landings were affected by predictions of rain or water droplets in the clouds, due to the damage that water droplets would cause on the Shuttle tiles. To better understand what the real damage would be on the ceramic tiles, Dryden did flight tests using Shuttle tiles and other TPS materials including AFRSI, reinforced carbon-carbon (RCC), and advanced tiles installed on the F-104. These tests found that significant damage did occur, confirming the limitations on flying the Shuttle in rain or clouds containing water droplets. [64]

The Loss of Challenger

Flight STS-25 (officially 51-L) was the launch on which the Challenger was destroyed. It's such a sad memory that I don't want to go into it. The launch failed for reasons that are well documented and, in many ways, unforgivable. To have a good friend and a top-notch-guy that everybody liked such as Dick Scobee and the others die in such a fashion was almost beyond comprehension to me. I still can't stand to see any of the video of that flight. Coincidentally, this was the first launch that I don't remember watching on TV. I'm not superstitious, but I was at Dryden in a classified room giving a classified briefing jointly with others on a program that I was advocating at that time. We had guards outside the door. Suddenly, people were trying to interrupt our meeting, and we were told that the Challenger had exploded. We left the classified area, while management went to do what it needed to do. I went to a conference room where I knew the launch was being carried in real time and saw the pieces of the Challenger raining down into the ocean.

After the tragedy with the Challenger, flights were suspended for more than two years while the problems were sorted out and the procedures and processes were completely redefined, trying to make the Shuttle safer and to avoid the type of incident that had happened. Not violating flight rules was something I had been taught on the X-15 program. It was something that we just never did. We never changed a mission rule on the fly. We aborted the mission and came back and discussed it. Violating a couple of mission rules was the primary cause for the Challenger accident.

We had continued to analyze flights as they became available for stability-and-control and performance and, to a lesser extent, for aerothermodynamic characteristics right up to the time of the Challenger accident. We had combined the flight-test panel and the aerodynamics panel into one panel, so I had continued to participate on a weekly basis from probably 1973 until 1986 at the time of the Challenger accident. We'd come out with flight assessment deltas (FADs). We'd done a FAD-2, a FAD-4, a FAD-6, a FAD-9, a FAD-14, and we were coming up on doing what was going to be FAD-26, which was after the twenty-sixth flight when we would have completed the maneuvers that we needed for what was going to be the final FAD. [52]

Any data we obtained after that we would use to do a full update of the data book, so since the Challenger was STS-25, we had not obtained a complete set of data yet for doing the FAD 26. We had thought about trying just to go ahead and do a FAD 24, but we were missing some key maneuvers. It gave us time to get a reassessment from the aerodynamics community to help understand the general conclusions we had from our data.

To help improve the abort scenarios if a situation similar to STS-51L occurred, the aerodynamics and flight test panel jointly studied the dynamics of separating the external tank or SRBs earlier, while it was still under significant air load. This included wind-tunnel tests and extensive computational fluid dynamics (CFD) studies led by Pieter Buning of Ames. We found that the separation could occur under some aerodynamic load, and this information was passed up the line to Shuttle program management for consideration.

During the period that the Shuttle was grounded, all of the groups made use of the time as an opportunity to advocate additional information that we should try to get from the Shuttle to better understand unexpected results that we had already had or to obtain data to help update our codes. The research work we were doing was covered under the orbiter experiments (OEX) program, which was advocated primarily by Langley with contributions from Ames, Dryden, and Johnson as to what kinds of measurements and analysis maneuvers we could obtain from the Shuttle while it was going through its operational development. 1

The ACIP package was one of the things that had been advocated in OEX, and it also was advocated from an operational point of view, so it made the list. The high resolution accelerometer package (HiRAP) [67, 68], led by Bob Blanchard of Langley, was on the ACIP package, and it made finer micro G vehicle deceleration measurements (accurate to 5 micro-g's). This was to assess characteristics in the free-molecular, the transitional region, and the continuum region of the atmosphere. It looked for decelerations of the vehicle as it descended from 1,000,000 feet down to roughly the beginning of the sensible atmosphere around 250,000 feet or so.

One of the OEX experiments that I was very interested in was the shuttle entry air data system (SEADS), which was led by Paul Siemers of Langley. (Fig. 6) I felt it could improve our ability to get stability and control and performance data. Preparing to define and fly the SEADS experiment, Terry Larson, Dryden's leading expert on flight-derived air data, along with Paul Siemers of Langley, conducted a flight-test program using Dryden's KC-135 Stratotanker with a SEADS-type nosecap installed. [69]

Fig. 6. The SEADS ports are placed in a cruciform configuration on the Space Shuttle Columbia nose as shown in this photo taken after STS-24. (EC86 33311-28)

The Shuttle SEADS experiment had been scheduled at one time to fly on the Columbia on the first flight, but it had been delayed because of the difficulty in drilling the ports into the highest temperature portion of the carbon carbon nosecap. When the Columbia went down for refurbishment after STS-9, one of the things added to it was the SEADS [70, 71] nosecap, that was to fly five flights and give us air data information so that we could determine density and better information on dynamic pressure. We had flown one flight, STS-24 (officially STS-32 or 61-C), which was the first flight of the Columbia after it returned to flight status just prior to the 51-L flight of the Challenger. After the five flights, we were going to remove the nosecap and go back to the standard nose configuration. The idea was that the Columbia was always going to be heavier than the newer Shuttles. Every Shuttle was improved somewhat in terms of weight, avionics, or something as it was manufactured. Of course, there were still three Shuttles left—the Columbia, the Discovery, and the Atlantis. Because the Columbia was the least desirable from an operational point of view, being heavier and a little bit more primitive in some of its systems, we continued to get ACIP data on the Columbia up until at least STS-88.

The Shuttle upper atmosphere mass spectrometer (SUMS) experiment [72, 73, 74] was related to the SEADS experiment, in that it shared one of the ports. The SUMS experiment, also advocated by Bob Blanchard, used a mass spectrometer from the Viking program to sample the atmosphere at the very high altitudes. It was to fly on the same five flights as SEADS, but it had failed on its only flight (61-C) prior to the Challenger accident. It still had four opportunities to fly to get data that would be exceedingly useful for future designs that involved the free-molecular and the transitional flow areas of the upper atmosphere. It also would be related to the measurements that were made with the HiRAP accelerometer. We would learn about the atmosphere from the HiRAP and the SUMS experiment. We'd learn something about density in the lower atmosphere with the SEADS experiment, but we would also get new measurements of the angle of attack, angle of sideslip, Mach number, and dynamic pressure with the SEADS experiment.

The orbital acceleration research experiment (OARE) package [75, 76] was flown later to obtain data at the nano-g level. It was two orders of magnitude more sensitive than the HiRAP, so it was thought that we would be able to obtain even better data in the "atmosphere" on orbit down from 1,200,000 to 400,000 feet. It was a complement to the HiRAP experiment and to the ACIP, the SEADS, and the SUMS programs. The OARE package was capable of 10 nano-g accuracy. With the density from the SUMS experiment, it can accurately determine for the first time the lift-and-drag coefficients of the Shuttle in orbit. The package is self-contained and self-calibrates frequently while in orbit. Its information is vital for determining the exceedingly small accelerations during experiments on the Shuttle, such as crystal growth. The instrument is so sensitive that it is greatly affected by crew activities while in orbit.

The catalytic surface effects (CSE) experiment, led by Dave Stewart of

Fig. 7. The SILTS pod on top of the tail of Space Shuttle Columbia. (EC86 33311-33)

Ames, was another OEX experiment that tried to account for the catalytic efficiency of the Shuttle tiles. [77, 78] The high relative velocities during reentry dissociate atmospheric gases into various molecular and atomic species and free electrons, which then tend to recombine on fully catalytic materials with a resulting increase in surface temperature. The Shuttle ceramic tiles were predicted to have very low catalytic efficiency and, consequently, lower heating rates and temperatures. The CSE showed that the temperatures of the tiles that were intentionally coated with high catalytic efficiency material were substantially higher than the uncoated high-temperature reusable surface insulation (HRSI). Tiles with known catalytic efficiency were placed with thermocouples on the DFI system so that the catalytic efficiency could be measured, and they were carried on the five DFI flights (STS-1 to STS-5).

Another OEX experiment was the tile gap heating (TGH) experiment [79] to measure the effects of the environment in the air gap between two adjacent tiles. This experiment was intended to provide a better model for heating on future vehicles with Shuttle-type tiles on them. Furthermore, if there were problems in some of the gap areas on the tiles, it would help us better understand it for the Shuttle itself. The TGH experiment successfully got data only on STS-2.

The final OEX experiment that was flown was the Shuttle infrared leeside temperature sensing experiment (SILTS) (Fig. 7) led by Dave Throckmorton of

Langley. [80, 81] It was a technique for scanning the leeside of the Shuttle to determine the aerothermodynamic effects during reentry. The experiment was in a fairly large pod at the very top of the Shuttle's tail that which extends a little bit forward of the tail. SILTS was flown on 61-C, but I don't think it was successful on that flight. Then, it flew on the same four flights of the Columbia as the other experiments.

Consequently, at the time of the Challenger accident, the Columbia was in the midst of doing the SEADS, the SILTS, and the SUMS experiments. If you look at photos of the Shuttle, you can tell the five flights on which they were flown, because you can see the ports in the nosecap of the Shuttle and the SILTS pod on the top of the tail. After those five flights were completed during the flight series after the Challenger accident, the Columbia was retrofitted with the standard nosecap, and the SEADS and related experiments no longer flew. (The pod on the tail remained.) However, we had gotten really good research data for honing techniques in the future, and also for better understanding some of the Shuttle data we were getting.

After the Challenger

While the Shuttle program was down from January 1986 to September 1988, various advocacies continued to come forward for improving instrumentation or improving the research capability of the Columbia in particular as well as the other Shuttles generally. For the most part, these advocacies were not very successful. The program was very conservative after the first flight. Then after the Challenger accident, it was even more conservative. Anything we tried to add was just one more thing that could go wrong, from the program management's viewpoint. They were looking at the program from an operational point of view, while those of us at Langley, Ames, and Dryden were looking at it from a research point of view. The Shuttle is an excellent vehicle for getting high-quality research data.

We continued at a very low level to propose from time to time that we should take the Columbia, put a better thermal protection system on the OMS pods, and fly it for lower angles of attack to demonstrate the greater cross range, or find out if there was some show-stopper that would keep the Shuttle from doing that. We didn't know the details of the objections to our proposals. However, we knew that the program was very conservative. The program managers decided to abandon Shuttle launches out of Vandenberg and to abandon trying to retile those portions of the Shuttle that would be subjected to higher heating at longer cross ranges. They elected to go for more alternative landing sites, throughout the world, in case the Shuttle had a problem.

I've always felt that was a major mistake. From my point of view, the only thing the Shuttle needed to have sufficient cross range for landing back at its launch site was just more tiling to protect the OMS pods and the sides of the Shuttle from the vortex impingement heating. That, to me, seemed much simpler than building all the other alternative sites. I think the argument for the other sites was that there are other things that can go wrong than just having the OMS pods get too hot. There would be other kinds of emergencies that would call for coming in without using the required maximum cross range of 1,300 miles. The program managers decided they needed the other landing sites, so they abandoned any attempt to expand the Shuttle envelope to have it come in at lower angles of attack. It would have more landing opportunities at both Edwards and Kennedy on any day if they would have increased the cross range, so to me the decision was much too conservative.

During the time that the Shuttle program was restructuring, two useful programs were sold to help us understand both the Shuttle itself and how we might model some of the phenomena we still only partially understood. One of them was the modular auxiliary data system (MADS) experiment, its primary advocate being Dave Throckmorton. In the case of the Columbia, it meant hooking up again some of the thermocouples on the vehicle so we could once again start to get better aerothermodynamic information to look at boundary-layer transition or asymmetric boundary layer-transitions to try to understand a way to model them in terms of tile roughness, damage to the tiles, or loss of tiles. Since the resumption of Shuttle missions, all of the Shuttle flights have had the MADS package, which is about 20 thermocouples on the windward side of the Shuttle. The MADS package is very limited, but it does help a lot in answering questions about what's going on in the boundary layer.

The other program that was sold is the aerothermal instrument package (AIP) program, also advocated by Dave Throckmorton. [66] It was intended to make a substantial number of measurements on the leeside of the vehicle, where it's hotter than predicted and not very well behaved. At the higher angles of attack, it's in more or less of a localized wake (a region of multidirectional flow), which is very difficult to model. The information that we later got from the AIP package gave us a chance for validating and enhancing some of the CFD codes to improve prediction of the leeside conditions of any future vehicle design.

The AIP program was flown on the Columbia for the same four post-Challenger flights that SUMS, SILTS, and SEADS flew to correlate the SILTS data with the in situ measurements of the thermocouples on the leeside of the vehicle. Those 125 thermocouples were reconnected from the old DFI system. The thermocouples were still there, but the wiring harness had been removed. So it was reattached and recorded for four more flights to correlate with the four remaining SILTS flights to better define the flow fields on the leeside of the wing, the fuselage, and on the tail at a 40-degrees angle of attack. This hypersonic flight regime had been poorly understood.

The AIP and SILTS experiments were also used to try to characterize the flow effect of the yaw jets firing over the wing in rarefied air. This is something that we could never simulate in the wind tunnel. With the data and information gathered from the AIP and the SILTS experiment, we now have some benchmark data to correlate with the CFD programs. [82] We certainly are in a better posi-

Fig. 8. Space Shuttle Endeavour at roll out. (EC91 204-8)

tion now than we were before to make predictions on the leeside of a vehicle, for the predictions now are based on improved understanding.

Return to Flight

Between January 1986 and September of 1988, no Shuttles flew. Everything in the program was reviewed and restructured. New procedures in the processing of the vehicle and in the inspection and testing were added. All of them were reviewed, many of them were revised, and in many ways I think they did make the vehicle safer, although it is so complex that it's difficult to believe that everything will always be found with any process.

As we approached the return to flight in September of 1988 we had only the three Shuttles. NASA had been given permission and funding to build the Endeavour, (Fig. 8) which was going to be used for several new things. New processes, new construction techniques, and other things were put into the Endeavour as it was being manufactured. So each Shuttle was somewhat better than the last one. This was the way engineers build a little and test a little. We couldn't reconfigure the vehicle completely, as we would want, but we could get the little things to increase reliability and safety as the new vehicles were manufactured. Then, as each Shuttle was sent back for refurbishment, as had happened to the Columbia just prior to the Challenger accident, we upgraded everything that was reasonable to upgrade. Each time the refurbished vehicle came back, it was a little closer to the others.

One of the things on the Endeavour that we would be doing was to test and validate the use of a drag chute to help slow the Shuttle after touchdown. It had been talked about prior to the Challenger accident. Another area that had been talked about prior to Challenger was the landing systems research aircraft. This was a Dryden project done jointly with Johnson to test tires, brakes, and side loads, using a Convair 990 airliner with a computer-driven research landing gear system (separate from the Convair 990 landing gear) that was used to get it going as fast as a Shuttle landing and able to apply weights up to 250,000 pounds on the tire and brakes and some of the research gear assembly. [83] Dryden was heavily involved in this project during the 1990s. I think it added a great deal of knowledge about what the Shuttle could and couldn't do and what kind of improvements the program might look forward to in terms of the tires and the brakes.

The mission designations were rethought after the Challenger accident, and they did away with the 61-C type of classification. They went back to having it named simply as an STS mission. It was getting close to what we had done, but unfortunately they didn't do them all in exactly the same order, even early in the program. In the first five or six flights, they flew STS-29 and STS-30 before STS-28, so we continued to use our designation of STS-26 for the first flight after the Challenger accident, which was going to be a Discovery flight flown by Rick Hauck and Dick Covey. Its official designation was STS-26R, the "R" stood for re-flight to make it clear that it was flown after the Challenger accident and the new STS designation system. I will use the official Shuttle designations for all flights after STS-26R to avoid adding confusion.

The early thing that we were looking for was completing the maneuvers. We needed to do FAD-26. We actually took data well beyond the twenty-sixth flight and continued to mature that FAD [52] and worked on it up into the early 1990s, when we finally put out the official finalized FAD. It was actually called FAD-26, and it was also the last FAD that was done. Then, in the mid-1990s, we did a complete aero-data book update, which took a great deal of everybody's time, because we had to rebuild the entire database, changing break points where appropriate. It's hard to compare the new database with the old database, because the different break points are applied to different Mach numbers, angle of attack, or body-flap positions. The work that the panel was doing still was primarily getting the data, analyzing the data, and coming up with FAD-26 and, later, the new aero-data book.

The flight-test panel lasted from about 1975 through 1999. Because of the length of time of its existence, there were many changes in the membership of the panels. The people doing the parameter identification changed also. Rich Maine took a much less active role after the first ten flights. The Air Force Flight Test Center quit participating after the eleventh flight. Over the years, people were added, including Ron Pelly and Al Weiner of Rockwell, Mick Culp and Mary Allen from Johnson, and Fanny Zuniga and Laurie Marshall for Dryden.

The first few flights after the Challenger accident started with a lot of DoD flights. The problem with the DoD flights was that we were not allowed to have PTI or ASI maneuvers on those flights. It was a little frustrating, for we had waited almost three years to fly again and were going to get data only on some of our flights. Our favorite vehicle to fly was still the Columbia, because it now had the SUMS, the SILTS and the SEADS experiments, along with the AIP experiment for its first four flights, designated by the program as STS-28R, STS-32R, STS-35, and STS-40. We got the information that we needed, and it still had the ACIP package. We were getting maneuvers on the Discovery and the Atlantis when they flew without a DoD payload, but we didn't have the high-resolution data system that we had with the ACIP. It was a mixed bag of data that we were getting, but we had hundreds of maneuvers to analyze. The PTIs and ASIs were done, and some of those were with the ACIP package with a high-resolution analysis, and others were with lower quality systems.

We knew which data sets were which, and it helped us evaluate the quality of the data from each flight. In addition to all of the PTIs and ASIs, I had a complete collection of the analyses of all bank maneuvers and bank reversals. There were roughly ten of those on most flights (two each for one bank maneuver and for each of the three or four bank reversals), and then there were always two or three incidental maneuvers. I was still looking at maneuvers, particularly when the ACIP recorder was along, between 600,000 feet and 330,000 feet or so when the Shuttle could just get up to a dynamic pressure on the order of one-tenth of a pound per square foot or so. We had lots of maneuvers, and we were getting a lot more. They weren't always the ones that had the highest priority for us, and we sometimes had maneuvers cancelled because of problems on the vehicle or because the Shuttle was in the middle of a bank reversal when the time came to do our maneuver. However, for the most part, we got the maneuvers we wanted. By the time we did the final aero-data book, which was finally approved in either 1995 or 1996, we had well over 500 PTI or ASI maneuvers that we had analyzed incorporated into it.

We still continued to do analysis after that final data book. Two things that I can think of in particular that we worked on were subsonic maneuvers, particularly trying to understand the non-linearities in the rudder effectiveness. These sometimes made it a little difficult to control the Shuttle precisely if there was a slight cross wind. We had one incident where, for no apparent reason, the commander put in a doublet while he was still moving down the runway. We got quite a startling maneuver and something as we were watching it land that was very unexpected.

Then, starting with the first flight of the Endeavour, we got the first drag chute deployment. (Fig. 9) That flight, STS-49 by the official designation, occurred in May of 1992. Some problems were experienced on the first use of the drag chute. It collapsed and moved off to one side, which was undesirable. We spent a lot of time looking at the analysis of different lengths of risers and how we might get the chute to stay directly behind the Shuttle. The drag chute isn't of much use in much of a cross wind because the wind will blow the chute off to the side, and it will be much less effective and, in fact, might make it more difficult to control the Shuttle to keep it on the runway.

The second flight of the Endeavour was considered an operational flight of the drag chute. It did work better than it had on the first flight, but it seemed a little early to me to call the system operational. They wanted to get it tested and get all of the Shuttles equipped with the drag chute because it did cut down on the total rollout distance. (Fig 10) If the Shuttle was landing with the payload or in marginal weather, reducing the rollout with the drag chute by 1,000 or 2,000 feet would be very important. We continued to look at flights and tried to assess when we actually had encountered a significant cross wind so we could raise the official placard on the cross-wind limits for the Shuttle to land. I think it started out at six knots. It was moved up to nine, then to 12, and, I think, to 15 knots in emergencies. Little by little, we did get confidence that, with differential braking and the use of the rudder at the higher speeds especially, the vehicle was controllable in a cross wind, certainly to 12 knots. I suspect it's controllable on up to 18 knots, but we really haven't encountered that yet.

Fig. 9. The initial deployment of the drag chute (not fully inflated) after derotation of Endeavour on flight STS-49. (EC92 5165-6)

1 A broad treatment of the research done early in the program is given in reference 65. A summary of later OEX Shuttle results is given in reference 66.

Fig. 10. Columbia and its fully inflated drag chute on flight STS-58. (EC93 42263-2)

XIV

Return to Flight: The Shuttle Program in the 1990s and Beyond

Those of us involved with the Shuttle program used the period between January of 1986 and September of 1988 to further refine our analysis and examine our requirements, both for improving knowledge of the Shuttle and for obtaining additional research data. Meanwhile, the other people on the Shuttle program who were trying to address the problems that caused the Challenger accident were exceedingly busy during that period. As the summer of 1988 approached, it became clear that most of the accident-related work had been done. The new requirements had been taken care of. The additional checks and balances and safety considerations had been written down, formalized, and were now part of Shuttle operations.

One of the decisions made during that period was that the first ten or fifteen missions would land at Edwards to allow the program to improve the weather model used to make forecasts at Cape Canaveral. There were a variety of runways available at Edwards. Having had much success in the past, with confidence we proceeded slowly and carefully. Using Edwards as the primary landing site, of course, meant mating the Shuttle to the SCA and flying it back to Kennedy after each landing.

STS-26R

I'm not sure what the process was for selecting the next crew to fly the Discovery for STS-26, but those of us in the program felt that the things that had caused the accident on STS-25 with the Challenger were well understood and well thought out solutions were in place. The program had done a superb job in addressing those issues. If there were any problems, they'd come later. However, any organization that has not flown for two years and nine months has to get back to flying status in a very conservative, very deliberate fashion, which is the way I felt we were moving to the next flight.

When it was obvious after the Challenger accident that it would be years before the Shuttle could fly again, a number of the experienced astronaut pilots left the program. Some of them were getting older. Some of them were wanting to go on to other careers. Gordon Fullerton left the program and came to Dryden in that period to become a research pilot. Joe Engle also left during the downtime. Among the astronauts picked in 1966, by 1986 he'd been in the program for

20 years. He'd had his two ALT flights, then STS-2, and in August of 1985 he had flown STS-20. The other Apollo-era astronaut who left was T.K. Mattingly who had flown STS-4 and STS-15, which was called STS-20. His last flight was about a year before the Challenger accident. They had all been in the astronaut program since the mid-1960s. Vance Brand was the only Apollo-era pilot still listed as an active astronaut at the time of the first flight after the Challenger accident.

The selection of the crew for STS-26 probably was based on experience with past flights and the confidence of the astronaut's office in the people who were going to fly it. Regardless of the reasons, Rick Hauck and Dick Covey were chosen to be the commander and the pilot of STS-26, the return to flight of the Discovery. Rick Hauck had flown with Bob Crippen on an earlier flight and had been the commander of STS-14 (officially STS-19 or 51A). Dick Covey had been the pilot with Joe Engle on his second flight. Off the top of my head, that's the only time I can remember anybody flying pilot twice before getting a flight as commander on the Shuttle. I suspect that even though they were more confident in the Shuttle in many ways than they had probably been prior to the Challenger accident, they did want some experience at the controls. The commander, Rick Hauck, and the pilot, Dick Covey, and the three others in the vehicle all had previous flight experience. The first mission was to launch a tracking and data relay satellite (TDRS). Once in its final orbit, it was to provide communication with future Shuttle missions throughout most of the flight.

One of the other things that happened during the two years and nine months that the Shuttle program didn't fly was that a lot of important missions had backed up, because there really wasn't an alternative launch capability for very large satellites. The Galileo and the Ulysses interplanetary probes were back-logged and so were quite a few Department of Defense (DoD) missions. The Shuttle was still the primary way of getting the larger DoD satellites flown because of its stated 50,000-pound payload capability. Since DoD had been the main supporter of the large payload bay and the high weight capability, the DoD needed the Shuttle to catch up on their launches.

Of the first 12 flights after the Challenger accident, five of them were DoD payloads and two others were Shuttle-support payloads. The first and third flights were to launch TDRS satellites. Then, there were four important NASA missions that needed to be launched. The Galileo had been waiting a long time. Also there was the Ulysses, Magellan, and the Hubble Space Telescope. On the eighth mission after we started flying again, the Shuttle was scheduled to recover the long-duration exposure facility (LDEF), meant to find out what the wear and tear on various kinds of materials and components and electronics were in space. The long duration turned out to be a very long duration, because it had been up over six years when it was picked up.

So there were a lot of people throughout the country who had payloads that had not been able to be orbited. Shuttle management went through whatever processes it had in place to allocate the payloads. Sometimes the order of the payloads was based on whether there were problems on the payloads. If there were problems, it would be skipped. That's how some of the STS numbers got scrambled. I think for safety reasons TDRS was given high priority to be able to ensure good communications and good data acquisition for the Shuttle and other experiments. The TDRS wasn't used just for a Shuttle, but the Shuttle program was the primary user.

At any rate, for us, the whole launch process started all over again with our watching the Discovery slowly move to the launch pad. The media coverage was very good once again, because the country was anxious to see it succeed and for the U.S. to get back into the manned space business. The Cold War was still going on, and the Soviets had the space station Mir in orbit, so there were many reasons that a lot of people, including myself, were quite anxious to see the Shuttle get back on flight status. I had a second reason: wanting to obtain more data and learn more about the phenomenology associated with hypersonic aerodynamics and aerothermodynamics. Nevertheless, it was a very tense time, because I think virtually everybody had seen the replays of the Challenger explosion. It was one of the saddest things I've seen the nation go through.

The country was hoping that all the Shuttle problems were solved. Watching the Shuttle program be completely dissected, reassembled, and recertified for flight had been nerve-racking. An outsider looking in can never be sure whether all the problems have been fixed or whether there is going to be a reoccurrence of the accident. NASA couldn't afford to have any kind of an accident on the next flight, because a lot of NASA's credibility had been lost in the Challenger accident. Some very serious decisions had been made too casually, and that had resulted in the Challenger explosion. It's a very dangerous business with a 4.5-million-pound vehicle that is mostly fuel and oxidizer. The simple fact is that you're going to have problems occasionally, but you hope the problems will be acts of God rather than acts of the wrong people making the wrong decisions and not being aware of how important those decisions are.

I went out to Dryden on the morning of the launch, even though we didn't have the monitoring function on STS-26 that we'd had on earlier Shuttle flights. We already had most of our maneuvers that were important, and the safety-of-flight maneuvers were well in hand, so we didn't feel that it was appropriate to bring up the strip charts. But we did want to watch the launch from Dryden's control room because the communications and telemetry were available from the tracking net. Dryden's tracking and communications are up 24 hours a day when a Shuttle is in orbit. I went out to Dryden to watch the preliminaries. I was very nervous. I thought the problems that had caused the loss of the Challenger had been solved, but there were still problems that could occur for some other reason. Any time that a process that had been firmly in place is redone, there's always the danger that something will be done in the wrong order or left out if the right person wasn't there when the checklists were made.

By 1988, most of the people who had originally worked on the Shuttle, designing the systems and the interfaces, and writing the checklists and the mission rules, were on other programs or had retired. As a result, the Shuttle personnel didn't have the depth and the breadth of experience that the original personnel had in 1981. I had confidence that the Shuttle was going to work, but I was nevertheless very nervous as it lifted off.

I remember the "go at throttle up" call. As the Shuttle nears the maximum dynamic pressure region on launch, the crew throttles back to keep the vehicle from accelerating as fast, and that lowers the maximum dynamic pressure on the vehicle. Once they're through that region, high enough in the atmosphere that the atmospheric density is decreasing fast enough, then they go to what they call "throttle up." They throttle up to 104 percent of the engine's designed thrust. It was just after the throttle up call that the Challenger had exploded. Of course, until they studied all of the frames from all of the cameras, they had not realized that there were indications well before the explosion that were showing how bad things were.

So we watched the video from all of those cameras. We had many of the scenes available in our control room, and we could glance at each one. We all knew where to look on the solid rocket boosters (SRBs) for the gases escaping from the O-ring. As the Discovery went through that throttle up, I probably didn't breathe for the better part of a minute until I saw the solid rocket boosters separate. That happens at about Mach 4 at 130,000 feet. Once the solid rocket boosters are released, a tremendous number of possibilities for explosions have been eliminated. However, hydrogen and oxygen are still in close proximity, and any of the three state-of-the-art rocket motors and turbopumps can have a failure. But those are monitored in a different fashion. Once the solid rockets ignite, there's no way to put them out, no way to shut them down, and no way to get rid of them until burnout.

Seeing the solid rocket boosters separate from the vehicle was a big relief. Because the Dryden control room had access to the full network of the Shuttle views, we got to see it go quite a ways downrange. We could see it until it was just a little dot flickering as more and more atmosphere came between the Shuttle and the tracking camera. When we got past the transatlantic abort and to the main engine cutoff, we all started to relax a little bit, because all of the things that we feared from Challenger accident had now been passed, and the orbit was achieved. It turned out to be relatively uneventful from my point of view. There were lots of little problems on the Shuttle, but nothing serious.

After the crew deployed the TDRS satellite, they did housekeeping and checked the Shuttle out in the zero g environment as they always do. When the time came for reentry, we didn't have the control room set up for monitoring the strip charts, but we did have the voice communication, so we could hear all of the commentary, the deorbit burn, and the interval where there was no communication. Even today there's still a gap in the TDRS satellite coverage that's about five or ten degrees of longitude where there's no communication.

We started getting data through our system around Mach 13 or 14. We went outside, as we had done before, at about Mach 2 and looked for it going overhead at Edwards. It was going to land in the middle of the dry lakebed on runway 17, which at over seven miles long is the longest runway at Edwards. The Shuttle did a fairly uneventful approach. At Dryden, we're more or less perpendicular to runway 17 as the Shuttle goes by, so we get a nice side view of the Shuttle as it is doing the preflare maneuver and then the rotation and flare. (Appendix E) It seemed to float above the ground quite a ways before the vehicle touched down on the lakebed runway. Because the lakebed has dust on it, as the Shuttle flies just above the runway, dust is kicked up prior to it actually touching down. That's true of any vehicle. When the wheels actually touch the ground, the rooster tail of dust appears as the vehicle rolls down the runway.

A landing on a lakebed runway has a shorter runout, because there's more friction between the Shuttle wheels and the dry lakebed. The wheels are compressing the lakebed an eighth to a quarter of an inch. There were still questions about the brakes on the Shuttle, so if the wind had been out of the southwest, the Shuttle would probably have landed on runway 23, the five-mile-long lakebed runway on which STS-1 and several of the early Shuttle flights had landed.

It was good to be back in business. (Fig. 1) Even to this day, when I watch a Shuttle launch, and I almost always watch them, I still watch the Shuttle very

Fig. 1. Vice President George Bush welcoming the STS-26 crew back from the first flight after the Challenger accident. The ceremony took place in front of the main Dryden building by the plinth-mounted X-1E. Some of the Vice President's security is on top of the building. Astronaut Pinky Nelson and then NASA Administrator James Fletcher are at the Vice President's left. (EC88 196-8)

carefully as it goes through maximum dynamic pressure, through the throttle-up call, and another minute or so until the SRBs separate. If I'm away from Dryden, watching it on cable or TV, it's very frustrating, because once the SRBs separate, commercial television returns to some mundane news or a commercial and I don't get to see the rest of the flight. But that's probably a good thing, because it means that things have become routine again.

Missions After the Challenger Accident

The first incident of any interest for the program in general was on STS-27R, which was a DoD mission. The Shuttle had significant tile damage when it came back down. It had the most large gouges in the tiles of any flight before or since, and I believe an entire tile was missing on the lower side of the vehicle. [61] I think that's the only time that has happened. I think they traced the damage to

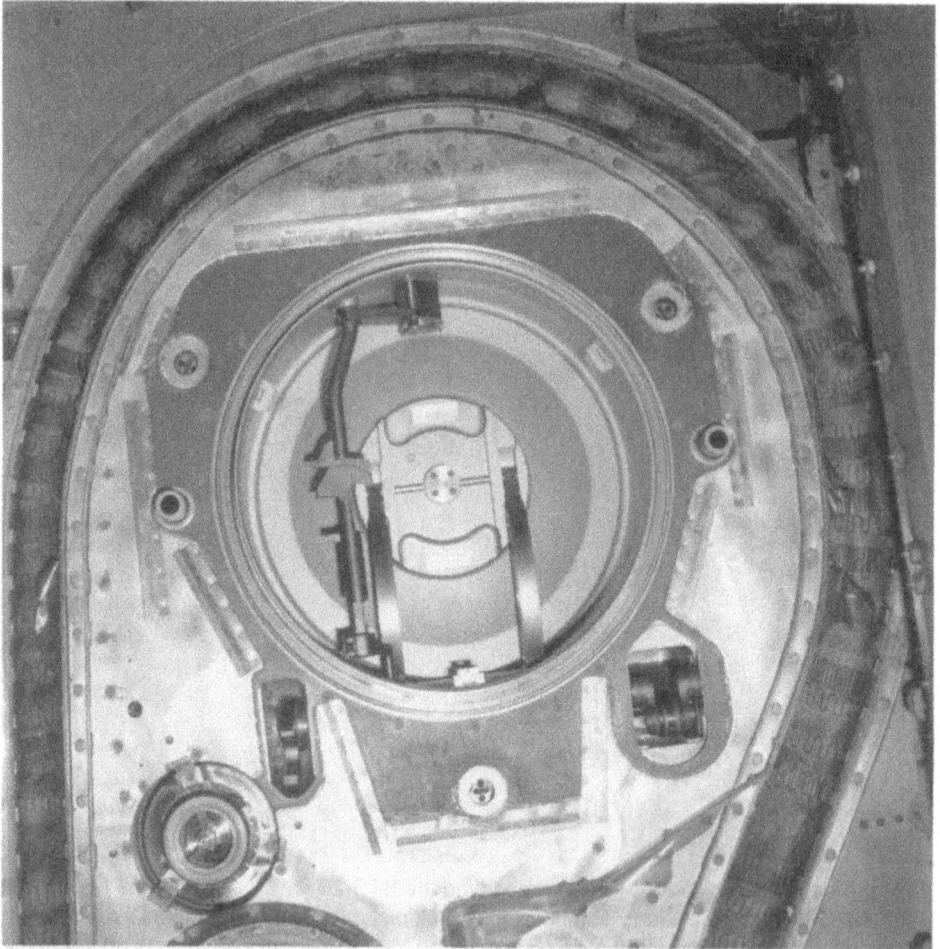

Fig. 2. The 17-inch diameter hydrogen disconnect through which the liquid hydrogen flows from the External Tank to the Space Shuttle Main Engines (SSME). (EC01 130-2)

debris that had come loose as the SRBs had separated. This debris had impacted the bottom of the Shuttle, and I think there were over 600 tiles that had to be replaced, all on the windward side. The next thing I remember was that in 1990, after 10 flights, we started having problems with the 17-inch-diameter hydrogen disconnect, which is the interface between the external tank and the Shuttle. Looking at the Shuttle up close, one can see a big hole in the bottom, which is how the liquid hydrogen goes from the external tank into the space shuttle main engines (SSMEs).

The original problem with the 17-inch disconnect (Fig. 2) had occurred in May 1990 when Vance Brand's final flight was being prepared. It was STS-35, which was going to be the thirty-sixth flight of the Shuttle. While they were going through their preflight checks, they discovered hydrogen in several places inside the vehicle. They eventually cancelled the launch and dragged the Columbia, which is the vehicle Vance was going to fly, back into the vehicle assembly building (VAB) to repair it. They then brought STS-38 out – which Dick Covey was going to fly it on a DoD mission – and it developed a similar problem. They dragged it back into the VAB and brought the Columbia back out and still had problems they couldn't resolve. They swapped vehicles again. The launch window for the Ulysses solar probe was approaching, so it jumped ahead of the STS-35 and STS-38 flights. STS-41 was successfully launched on October 6, 1990, with Ulysses. This was the thirty-sixth Shuttle flight, and a resumption of flight activity after a five-month interruption. It was followed in November 1990 by STS-38, Dick Covey's DoD flight. The main thing I remember, other than it being nervous at launch, was that those of us not around hydrogen very much find it scary when it's leaking into the various parts of the Orbiter itself. However, they have good safety procedures, and they understand hydrogen well at Kennedy Space Center. They were ready to roll, and I think the launch went well. The most interesting thing I recall about STS-38 is that it was scheduled to land at Edwards, but because Edwards had bad weather in November—which can happen—they diverted and landed at Kennedy. That's the first time I can remember that happening, and it may be the only time that it's happened.

A month later in December 1990, Vance Brand made his final flight. It was still officially STS-35, because it had the same ultraviolet astronomy telescope payload as the one they had tried to launch in May. It was the 38th Shuttle flight. It landed at Edwards on the concrete runway. I think all of the flights up to about STS-40 or so were scheduled to come into Edwards, with Kennedy as the primary backup and White Sands as secondary backup.

During the halt in flights due to the hydrogen leaks, we were still trying to get all of our data to finish off for our developmental test objectives (DTOs) that were used to justify our PTI maneuvers so we could give a report to the Shuttle project on the final FAD-26. This was to be followed in 1995 by the complete aero-data book update, so completing the final update was slipping. We knew those were all several years down the line anyway, and a five-month halt just meant they were five more months further down the line. Of course, the hydrogen leaks were a serious problem that needed to be fixed. I think the name of the finished report was "Aero-Data Book Update '95." It was finalized in 1995, but I think we continued to have a few discussions, a few discrepancies or questions asked by others. It was finally implemented and became the official database for all aspects of the Shuttle. The simulation, the design, and all data to be specified in terms of the mean values and the variations were published in that aero-data book update, which was really not quite the culmination of the work that we had been doing since 1981, but it was close. We did some DTOs after that, but the data book was what we had wanted to finalize.

Vance Brand was the last of the Apollo astronauts to fly in the Shuttle. Unfortunately for Vance and probably the rest of us, he didn't get to go to the Moon with the Apollo program. He was to have been the command module pilot for Apollo 18, which was cancelled in 1970. That's part of the reason he ended up on Apollo Soyuz. Joe Engle was also scheduled to be Lunar Module pilot on Apollo 17. Fred Haise was going to be the commander of Apollo 19, before it also was cancelled. So lots of people that had a close association with Edwards and Dryden, and the X-15 and the lifting-body programs, just barely missed the boat on that. Joe Engle actually missed it in a bigger way, because he was scheduled originally to be the Lunar Module pilot for Apollo 17. Then fairly late in the program, management replaced him with Jack Schmidt. I didn't hear any public complaining by Engle on that. I know there was a lot of outcry that we had all of these geologists and scientists who had been through the rigors of training for the Apollo program, and it looked as if we were not going to get any of them to the Moon.

That was probably at least partially a political decision and probably a good decision. It's too bad we didn't have more Apollo missions so more of the scientists could have gone. I've always had a lot of respect for Joe Engle in not making a public uproar over being replaced. It was also interesting to me that Apollo management swapped only Engle and Schmidt. It was actually fairly late in the training of the whole program. Instead of swapping the entire Apollo 18 crew with the Apollo 17 crew, they just replaced Engle with Schmidt, keeping Gene Cernan as Apollo 17 commander and Ron Evans as the command module pilot. Usually the program went to the backup crew for replacements, not to the next crew. That had only actually happened in the case of Jack Swigert on Apollo 13. So, if management had swapped the whole crew, Apollo 18 for Apollo 17, Vance would have gotten to orbit the Moon while Dick Gordon and Jack Schmidt were on the surface of the Moon.

Those of us who had been involved with the Shuttle since the beginning in the 1970s felt that the original astronauts were a special group. We didn't know a lot of the new ones, either by reputation or personally, and there were now a lot more of them. Vance's flight was an end of an era in my own mind, in that we were going to fly only the newer astronauts who had come in after the Shuttle program started.

The first one of the new astronauts to fly after Vance was Steve Nagel. Steve was known to a number of us from having been at Edwards. He was, at least by me, considered a little more of an Edwards guy than some of the others, just because I remember his being around Edwards for a while.

Nagel was the pilot on 51-G (STS-18) in June 1985; later, he was the commander of STS-37. The flight went well, as I recall. The issue was the landing. This flight is an example of the problems that can happen when there are weather issues at both the primary and secondary landing sites. I believe Edwards was still the primary site for that flight and Kennedy was the backup. But Edwards had bad weather moving in on the original day, so they waved off, intending to go into Kennedy. Then Kennedy ended up having bad weather, so they couldn't go in there.

I couldn't believe it when they decided to come into Edwards. I remember commenting to many people that no way was it coming in today. We really don't want to have a landing at Edwards when a front is going through. This can involve heavy rain, which would damage the tiles. Also, the wind comes out of the northwest, and it's very gusty and it's very strong. The Shuttle wouldn't land on the main runway in that situation, because there is a high crosswind. The Shuttle at that time was probably still cleared only for a nine knots crosswind landing. I was upstairs in the control room listening, and I heard them giving rosier pictures of the weather at Edwards than what I'd seen outdoors. There were the right kind of clouds for a front coming in. It was very gusty, and I couldn't believe the decision to pick that particular orbit, at any rate, for coming in. Usually the day after the front comes through it can be cold or gusty, but the wind is usually out of the west or the southwest, which means you can land on the main runway. However, this day it was out of the northwest.

At any rate, they did their deorbit burn, committed to come into Edwards. Because of the wind direction, I think, they were scheduled to land on runway 33, a runway six miles long less than a mile from the back of Dryden. It's a nice place to watch the landing because it comes in from the south. I don't think any other Shuttle flights landed on Edwards runway 33. I was on the roof watching it. I had my jacket on, and it was really gusty. I'm not sure what went wrong with the weather prediction or with the way the information was passed on to Steve Nagel. At any rate, I could see he was way too far south for coming in on 33. I could see him stretching the profile as much as he could to try to reach runway 33. He ended up landing three or four hundred feet short of the runway. The unmarked lakebed is not too bad, so it was not the end of the world, but it didn't look very good to land short of the real threshold on runway 33.

As the Shuttle did its rollout, it came on down runway 33 and stopped fairly close to Dryden. I thought that was a significant flight because I think it made people aware of how important energy management is. If a pilot missed the head winds and the energy management that badly on either the concrete runway at Edwards or the one at Kennedy, he could have a major accident. Landing short of the concrete runway at Edwards, coming in from the direction of runway 22, could badly damage if not shear off the landing gear on the Shuttle, and it could go out of control. It wouldn't necessarily be fatal, but I think it would be the last flight that Shuttle would ever make. We were very fortunate, for they were landing in weather that to me looked unacceptable, except in an absolute emergency with the Shuttle. I think they still had more opportunities for landing. Probably there was a forecast of rain and the lakebed runway might not have been thought to be available the next day. Maybe they were out of options. But my recollection is that they still had another day to wait to come in. The next day would have been acceptable at Edwards and might have been acceptable at Kennedy as well.

At any rate, I know there was a lot of criticism of that landing, both from the program and the astronaut community. Very privately they were quite disappointed that Nagel had landed that short. I also think Nagel was very unhappy that whoever was in charge of clearing him to land into that kind of a head wind had really made him look bad. To me, it was a good lesson. I hope it's been learned and understood that coming into any landing with an unpowered vehicle where a front's in the vicinity is just not something we want to do. We either want to wait or divert to another landing site because of the gustiness, the direction, and the strength of the wind. I think the wind was very strong at altitude, or he would have been able to stretch it and at least make the threshold of the runway.

I haven't seen much written on that particular mission and landing, and that worries me. I have nothing to refer to other than my memory on that particular flight, but it needs to be written down somewhere for the new crews. Any time we are dealing with a vehicle that's going to do a dead-stick landing, we need to be extra careful on the proximity of fronts. Future crews who don't recall the STS-37 flight will not be aware that landing during an approaching front isn't something to be casual about. We really need to take that seriously. There can be little lulls as those fronts come in, but that's all they are. When the front comes, there is going to be very strong wind and a wind shift following the front.

As it turns out, Nagel's landing was the next-to-the-last landing on a lakebed runway to date. That's not because the lakebed runways aren't considered a good thing. I think it just shows the maturity of the Shuttle program, and that it was finally getting into more of an operational mode after the Challenger accident. People were getting more comfortable for the little things that went wrong, and it became a little bit more routine after that—little by little—not in a dangerous sense in any way.

The very last landing on a lakebed runway occurred in December 1991 on STS-44. From the point of view of all of the landings that I've watched, it was my favorite in terms of what I could see and what I could observe. The Shuttle came across the back of Dryden, and I was within about 1,500 feet or so of the landing, about as close as prudent for watching it land.

STS-44 was Fred Gregory's DoD flight, was originally scheduled to go into Kennedy. A few days before the mission was scheduled to end, however, the Shuttle developed a problem with the inertial measurement unit (IMU). I knew the mission rules. The loss of an IMU was considered a serious enough situation that if it could not be brought back on-line, the Shuttle had to land at the next opportunity. It's not a true emergency, it's just the right thing to do. When there is an important component failure, we don't know if it failed by itself or if it's a symptom of other things to come. In this case, we didn't know if the vehicle might lose another IMU. In that case, the Shuttle was close to a single string failure. Since I paid close attention to all of the Shuttle flights, I knew it was scheduled to go into Kennedy. As soon as I heard about the IMU problem, I started looking at Kennedy's weather and at our weather. I knew that if they couldn't fix it, they'd be coming into Kennedy, and the Kennedy weather didn't look that good.

When I heard that they were waving off from Kennedy, Mary and I hopped in the car and drove to Edwards. Dryden was not working on that Sunday morning, December 1, 1991. There were a few people there for the communications, and a few other people who had duties that they were trying to complete. Normally, Sunday at Dryden is fairly quiet. Either not many people felt inclined to go, or they hadn't figured out soon enough that the Shuttle was coming in to Edwards. By then the flights were becoming fairly routine, and it was hard to get information. I had monitored as many sources as I could on radio and TV to figure out if they were waving off at Kennedy. As soon as I heard they were, I knew they had to be coming in to Dryden.

Mary and I went to the back ramp at Dryden and were in place well ahead of time. The other thing I was hoping when I was watching the weather at Edwards was that we were having easterly winds, but we don't have east winds very often at Edwards. I thought as I was driving out that probably meant the Shuttle would come in on runway 04, which is the designation for the main concrete runway when landing to the northeast. Since our wind is normally out of the southwest, aircraft usually land on runway 22, the main concrete runway when landing to the southwest. After I was out at Edwards, however, it became clear that the Shuttle was landing not on runway 04, but on runway 05—a real treat from an observer's point of view at Dryden. Runway 05 is the designation for landing to the east on runway 23. Touchdown on runway 05 is usually within a mile of Dryden, the runway extends to about 1,500 to 2,000 feet from Dryden. Consequently, I knew there would be no obstructions to watching the Shuttle come over supersonically at 50,000 feet, then come around the heading alignment circle to land on runway 05.

It was a very special treat, although we can see the Shuttle land quite well on any of the lakebed runways. The Shuttle is a pretty big vehicle with a large volume, which makes it much easier to see in detail than an airplane of the same length. Knowing that the Shuttle was going to be coming by just south of me one or two thousand feet away, I was really looking forward to the landing. I knew it was going to be close. I knew it would be on its final approach. It would not have much altitude at the time it was closest to Dryden, so I would be able to see the gear extension and the touchdown in great detail. All of that happened, and it was very special. I didn't see more than 15 or 20 people, at least at the Dryden site, watching the landing. So in addition to it being the very best view that we had from any of the viewing sites at Edwards, it was almost a command performance because so few of us got to see it. I didn't like it coming in for an emergency. I didn't like it coming into Edwards because of the disappointment it would be for the crew's families and friends at Kennedy. But as a unique and opportune viewing of the Shuttle, it was a very special landing for me. Although I didn't know it at the time, that was the last landing on a lakebed runway that the Shuttle has made for more than a decade.

Looking back on STS-44 after ten years, the feeling I have was that it seemed to fill my visual field as it went by. I could see the individual tiles and little motions of the Shuttle and its control surfaces as on no other flight. It was 1,500 feet away, a vehicle 120 feet long, which tells you that it didn't actually fill my visual field. But the way the human brain operates, you concentrate on it so much you see nothing but the vehicle. My recollection was of this really large vehicle spanning a good portion of my visual field and of being able to see the motion of the speed brakes and the elevons. I could see the response of the vehicle as it went by and arrested its sink rate. In addition, the aerodynamic noise that I had first heard on STS-1 was much louder and richer, unmistakably coming from the Shuttle. The crew made a very nice touchdown on runway 05, and coasted out with the rooster tail. It was going away from us instead of towards us as it normally would do on runway 23. (Gregory was appointed NASA Deputy Administrator in 2002.)

Summing Up

After watching forty or so Shuttle landings over a period of over 20 years, it's a little hard to rank them. The favorite of all, of course, was STS-1 because of all of the uncertainty and the anguish of it doing a full envelope flight. At least from a spectator's point of view, it was an excellent landing. All of the problems that occurred on the flight had to be sorted out afterwards, but the landing itself was very special.

Another one, as I had mentioned earlier, was STS-8, which was the first night landing and where we essentially got to see something like the Star Trek transporter—having the Shuttle materialize out of nothingness 100 feet or so above the runway. That was a very special one. And, of course, as I had mentioned earlier, Bob Crippen and Dick Scobee's sunrise landing was a very special one to me, both because of the beauty of the scenery and of the landing in the early sunshine and having a friend, Dick Scobee, on board. Those are my favorite ones as of this writing.

A couple of things have happened in the last 60 or so flights that I think were significant. The 100th flight of the Shuttle was STS-92, which landed at Edwards in October 2000. It was special in its own way. A big celebration had been planned at Kennedy, but due to weather it landed at Edwards. When they got it back to

Florida, they did their celebration. Still, the 100th landing was at the Shuttle's original home, Edwards AFB. Another significant flight that landed at Edwards was STS-111. It returned on June 19, 2002 with the space station crew which had just broken the American record for continuous time in space of 196 days.

The other interesting thing about that was before these flights, it had been four and a half years since the last Shuttle mission had landed at Edwards. (Fig. 3) They'd landed at Kennedy Space Center 23 times in a row, which really showed a lot of maturity and some good luck on weather. Forecasting and anticipating the weather at Kennedy has certainly improved. Having situations where they needed to come down at the next possible opportunity also was occurring less frequently.

Fig. 3. The SCA carrying Columbia flys over Endeavour shortly after STS-68 touched down at Dryden on October 11, 1994. STS-68 was one of the last Edwards landings before 23 consecutive landings at the Kennedy Space Center. (EC94 42789-5)

So these last few landings that have occurred at Edwards have been very special to me. It's been over 20 years since I saw the first one, and to see the Shuttle land at Edwards again is a real thrill for me. It spends a week or so being put into condition and being mated to the SCA to be flown back to Kennedy. I have been able to look out my office window at the preparation of the Shuttle and the mating of the Shuttle (Fig. 4) to the SCA about a half mile away for all of those years. Seeing the mated Shuttle/SCA take off to fly back to Florida brings back a lot of memories from the last 20 years.

Fig. 4. The view out the author's window of Endeavour being mated to the SCA using the mate-demate facility after STS-111. This mission landed at Edwards on June 19, 2002.

The other thing that always seemed significant to me that I was very pleased about was that on STS-95, in late 1998, John Glenn, our first astronaut to orbit the Earth, got to fly again. It was 36 years from his first flight to his second one. Although Vance Brand was the last Apollo astronaut to fly the Shuttle, John Glenn is the only Mercury astronaut to fly in the Shuttle. There was some criticism of NASA for sending him up. But I think it was very good for the Shuttle program, for manned spaceflight in general, and for the country. He is a national hero. He did a good job. He showed that if a person takes care of himself, the aging process doesn't eliminate him from things such as being able to go into space. They got a lot of data that they could compare to his earlier data in terms of physiology. I think that was important. The criticism to me, I think, was unwarranted. Jake Garn and Bill Nelson had been up as members of Congress back before the Challenger accident.

So even though John Glenn was just finishing his final term as a senator when, at age 77, he went on his Shuttle ride on STS-95, I thought it was totally appropriate and was just the right thing to do. He was healthy enough and willing enough to go through all of the training and all of the preliminaries that go with a Shuttle flight. They didn't just take him fat, dumb, and happy out of the Senate and stick him in the Shuttle. He was in good shape, and he was ready to go. I think it gave us an opportunity to get some really good data that we had been keeping track of for about 50 years from the start of his military career through the flight.

Lessons Learned

Of the Shuttles' successes I've gone through, most of the lessons learned are obvious. It is the first winged vehicle to reenter from orbit and the first (mostly) reusable orbital spacecraft. But as a researcher looking at what we have learned in terms of helping us build the next vehicle, there are a lot of operational issues. These include the launch, the mission rules, the checklists, and how to reduce the number of people needed to fly the next vehicle. All of those lessons I think are fairly well documented. Also covered are the issues on reentry from an operational point of view and in terms of energy management.

We haven't flown a new manned hypersonic vehicle since 1981. Prior to that was the X-15, which quit flying 13 years prior to the Shuttle's first launch. As a result, we don't have a very good record in terms of pressing on in this area. The big steps, it appears, were taken by the mid-1980s, not only in spacecraft and hypersonic flight but in high-performance fighters and airliners, including the Concorde. All of those things were close to flight, if not in flight, almost 20 years ago. Consequently, at present, I think there's a big lack of commitment to taking enough steps to validate how we can make spaceflight more routine, so more people can participate in it as technologists, as beneficiaries of technology, and as tourists or adventurers.

In the materials area, we obviously learned quite a bit about thermal protection systems. The thermal protection system now on the Endeavor is a lot less conservative than what was on the Columbia for its first flight. A lot of that was learned through the results of what we obtained on the Shuttle. A lot of this was done by ground-testing and materials development over the last 20 years. There are more durable ceramic thermal protection systems available now than were originally on the Shuttle, and their development, in my opinion, is due largely to the motivation and the flight experience that we've had on the Shuttle.

In the area of aerodynamics we certainly learned about real gas effects, including its effect on the center-of-pressure location. [52] We learned about the rolling moment due to firing the yaw jet into the rarified flows at the very high altitudes [52] with the different conditions there that cannot be simulated in the wind tunnel. We're a little smarter on how to do that because we have much better CFD codes and capability now. We can put the things into the CFD codes that were learned both in the real gas area and in the effect of firing reaction-control rockets into hypersonic flows for vehicles at high angle of attack. The Shuttle flight data becomes the benchmark that all relevant predictive techniques must achieve. Those modifications are now in the codes. We also completely documented the Shuttle and updated all of the databases for the aerodynamics derived from the 500 or so maneuvers performed over the life of the program, so we have a very mature database. [52] Unfortunately, it is relevant only near where the Shuttle is flown.

We could learn a lot more if we would do what is necessary to the OMS pods and fuselage thermal protection system to allow us to do a lower angle-of-attack entry to get the greater cross range. With the lower angle-of-attack entries, we could learn a great deal more about aerodynamics and aerothermodynamic, including the phenomenology of the boundary-layer transition and the prediction—both the original prediction and the prediction we might make today in both wind tunnels and with CFD codes. So there's a ten degree angle-of-attack difference between the standard 40-or 41-degree angle-of-attack entry and the one that is down to 30 or 32 degrees at Mach 10 that gives us the large cross range. Those 10 degrees in angle of attack would also provide an immense amount of aerodynamic and aerothermodynamic research data.

So that is something that's still missing that we could learn from the Shuttle as it's now configured, by only improving the thermal protection system (TPS) on the OMS pods and the sides of the fuselage. We still have to demonstrate the full 1,300 mile-cross range needed for an abort-once-around. Extending the cross range could be done on a flight-by-flight basis, successively lowering the angle of attack by one or two degrees and examining the effects on the aerodynamics, aerothermodynamics, and TPS. We might discover there was yet another smoking gun. But that's what research is all about—to discover the unknowns and the unexpected. In this case, if we could fly the Shuttle five to ten degrees angle of attack lower than it's been reentering in the above Mach 8 to 10 range, we would find out if there are more things to be learned in terms of RCS firings, real gas effects, control effectiveness, the hypersonic boundary-layer transition, or other areas.

There is the final area where I would claim that there's been a big research advance based on what we learned on the Shuttle is the aerothermodynamic area. In this case, I include what we learned on the boundary layers [60, 61], what we learned on the leading-edge effects, shock impingement characteristics with the Shuttle [60], the vortices heating the side and the OMS pods of the Shuttle [60], and the effect of the flow on the leeside of the vehicle. [60, 80, 81]. We now have documented this data, and we can use it to benchmark the CFD codes for future vehicles. [82] Most reentry vehicles that we currently consider will reenter at fairly high angles of attack. There are some materials and some ideas for vehicles flying at lower angles of attack [84], but those are yet, I believe, one more generation away than we are now. So the more we learn about leeside flows at high angle of attack, the less conservative we will need to be with the next vehicle.

The final aerothermodynamic area is the data on the effects of catalytic efficiency obtained from the Shuttle. [77, 78] On early flights, we got catalytic information from the ceramic tiles. Because of changes we saw in the ceramic tiles on the first few Shuttle flights, the contaminants made those tiles more catalytic. We really can get good catalytic data only in the Earth's atmosphere from about Mach 12 and above. It's very difficult to do, for we've never found a way to do full catalytic testing of real air in the ground-test environment. As a result, the flight demonstrations with the so-called catalytic tiles on the Shuttle were very valuable, because to my knowledge that's the only real flight data or ground test data which exists that relates to full catalytic efficiency. Looking to possible future vehicles, we could place test specimens of the new ceramic tiles and metallic tiles on the Shuttle to learn more about catalytic efficiency.

We also got data on catalytic efficiency from looking at the contamination of the tiles due to salt spray, spent fuel, and explosive bolt residue on the Shuttle. [77, 78] When it reenters, we get some catalytic information. To my knowledge, however, there's no current data-gathering on that. Most of that data was gathered on the first few Shuttle flights.

There are two main aspects to my view of the Shuttle operation and research. There's the engineering achievement of making an operational vehicle as complex and as ambitious as the Shuttle in terms of its engines, structure, thermal protection, aerodynamic and aerothermodynamic design, and all of the accompanying phenomena that an engineer wants to understand about flying in new regime. The other part of that problem is a much smaller one, and that is the research aspect in understanding the phenomenology of flying this kind of a vehicle both on ascent and on entry. On entry, the phenomenologies that have been of most interest are aerothermodynamic, aerodynamic, and the properties of the material, tiles, insulation, and internal structure. All of those areas have also benefited the research community by helping us improve our ground-test capa-

Fig. 5. The X-40A on its first flight on March 14, 2001. The X-40A is similar to the X-37 orbital vehicle. (EC01 0070-2)

bilities-that is, wind tunnels or thermal test facilities-and helping us improve our computational techniques for understanding the physics we observe when the test is done in the real environment.

Those are very important areas. They provide a significant quantity of research data that allows us to calibrate our ground facilities and our CFD codes [82] so that they will do a better job of predicting on the next vehicle. Doing a better job of predicting on the next vehicle removes conservatism from the design. All of the conservatism that can be removed ends up in something that's equivalent to payload or operability, so there's a very big benefit to the research data that's been obtained from the Space Shuttle and its applicability to the next vehicle.

The recent vehicle programs that have benefited from this Shuttle data are the unmanned vehicles that we talk about today, such as the X-37 (Fig. 5) and the X-38, which are both reentry vehicles, not complete launch vehicles. So the

Fig. 6. The X-43 immediately after launch from the B-52 on June 2, 2001. (EC01 182-20)

predictive capability that we've improved for the Shuttle mapped directly into improving the design and reducing the conservatism on the X-37 and the X-38, or any other unmanned vehicle like them that we plan to do.

The other vehicle that has been looked at and has gotten some benefit from this is the X-43. Some might say, "Well, that's an engine program, and the Shuttle is quite different. The Shuttle flies at much lower dynamic pressures and doesn't have an air-breathing engine." But those same codes that we have calibrated with the Shuttle data at lower dynamic pressures also have been used in the design of the X-43. (Fig. 6) Consequently, there can be a direct benefit even to an air-breather from what we've learned on the Shuttle from the research point of view.

XV

Going Nowhere Fast: The NASP Program

In the late 1970s and early 1980s, I was also spending a lot of time working in the high-angle-of-attack region for the high-performance F-14, F-15, F-16, F-18, and eventually the X-29 aircraft. [85] The 3/8th-scale F-15 spin research vehicle made its final flight in 1981. [51] From there I went on to advocate jointly with Ames and Langley for a high-angle-of-attack test aircraft. I was advocating the full-scale F-15, but through various compromises that occurred in trying to satisfy multiple objectives, we ended up with the F-18, [86] which turned into the F-18 HARV with the thrust-vectoring vanes.

Exploring Hypersonic Options

From 1980 through about 1983, I was intensely involved in Shuttle work in the hypersonic area, supporting the various functions in terms of getting the first flight off safely, analyzing data, and interacting with the other disciplines, designers, and operations people on the Shuttle. By late 1983, my involvement was getting to be more in terms of program support. These were safety-of-flight issues as well as some continuing interests in obtaining more hypersonic data in the aerodynamic, aerothermodynamic, and stability-and-control areas, and in looking at possible Shuttle modifications that might increase its cross range, make it lighter, making it more robust, and increase its operability and viability.

In 1983, the people whose working experience went back to the time of the lifting bodies and the X-15 continued to see a need for a hypersonic research vehicle. Mach 8 up to Mach 15 was the speed range that we were interested in, as well as a more conventional wing-body shape and, hopefully, an air-breathing engine. It would be a combined aerothermodynamic research vehicle with an air-breather test for the primary research focus, but it would have rocket propulsion to augment the vehicle to get it to the flight conditions that we needed to examine. Several industry proposals had been around for many years. Langley had continued to work in the scramjet area after the X-15 ramjet experiment was terminated. So there were people doing work in tunnels and with analytical codes, trying to define an air-breather in high hypersonic-speed regimes.

Dryden was, as a facility, very interested in the next such vehicle. Since we felt that flight research fell under the purview of Dryden, it was incumbent on us to advocate the vehicles that would get to the regions where there'd only been either ground-testing or computational projections. At the higher Mach numbers, it was not just an air-breather issue. It was heavily a materials issue and an aerothermodynamic issue. How hot would the vehicle get? What kinds of materials did we need to build the vehicle so as to maintain its structural integrity? What might be done with the excess heat that was generated while it was in the lower atmosphere at higher dynamic pressures? All of those issues came together as one, in my mind at least.

Along with trying to develop a viable proposal for a hypersonic vehicle, we were also trying to identify partners within the government and industry. These were people who might be helpful in adding knowledge and credibility to our proposal as well as advocates who needed information from a hypersonic research vehicle. This was the next logical step beyond the X-15, now that we had gone through the capsule technologies of Mercury through Apollo and demonstrated access to space for humans in the Shuttle.

In 1983, Milt Thompson had conversations with us and other managers. We were part of the Ames Research Center at the time. The proposals were for a set of candidate vehicles that we might fly hypersonically. Dryden started to look at what kinds of vehicles we could air-launch as we had done with the X-15. We were thinking in terms of a subsonic air-launch, but to get to Mach 15 with a pilot and a large research payload, the B-52 could no longer be the launch vehicle. The B-52 launch limited the vehicle to something like Mach 7 to 8. If we used a hydrogen-oxygen propulsion system, the vehicle would have a fairly high cross-sectional area. It would end up with a larger volume-to-weight ratio than on the X-15. Even looking at methane, RP or JP fuel, or alcohol, we were confined to something below Mach 8.

We all had been involved in the Shuttle carrier aircraft (SCA) and had seen a 160,000-pound vehicle could be launched off the top of a 747. That gave us a much larger target weight to work with than if we were confined to the B-52. We also knew that the SCA could carry a maximum payload of about 250,000 pounds. At the time, we also had proposals for mating wing to wing two subsonic aircraft, C-5s or 747s, and putting the experimental vehicle under the center wing between the fuselages.

All of those areas were under continual discussion. Should we be advocating a new air-launch vehicle? Or could we use the SCA? This second option was eliminated, as the SCA was considered a national resource. It was an absolute requirement for getting the Shuttle back to the launch point. There was no way that we were going to be using the SCA, because it might jeopardize the ability to return the Shuttle to Cape Canaveral from Edwards.

We talked about building our own SCA, an initial cost figure that got added into what we might do. We also looked at strengthening the B-52 wing—how big and how heavy a vehicle could we carry? We convinced ourselves we could get into the 80,000-pound region with a redesign of the B-52 wing and adapter. However, we then started to be constrained in terms of size because the vehicle would be below the B-52 wing. Even though the B-52 is a high-wing vehicle, by the time we add the adapter, there is a limit to the size of the vehicle we could carry and still safely take off and land. My primary focus at that time was that we probably could use a 747 in an SCA configuration to launch the yet-to-be-determined hypersonic research vehicle.

With the approval and encouragement of the Ames Research Center and the

Dryden facility managers, we formalized our proposal and, as much as possible, collaborated with Ames on it. Ames had very good hypersonic capability, both in terms of testing and in terms of the credentials of many of the Ames people. Many had spent their careers in trying to understand hypersonic aerodynamics and aerothermodynamics.

In late 1983, we formed a team. We investigated promising hypersonic projects whose data we might profit from as we built up our own competence. [87] Many at the time talked about a Shuttle rescue vehicle capable of fitting into the Orbiter's payload bay and bringing astronauts home in an emergency. Another project offered particular appeal, but after pursuing it for about two years, I ran into an obstacle I couldn't surmount, despite my optimism about the project. A major hypersonic initiative was just getting underway, and I was asked to curtail my advocacy of the more limited project. I was told that I was hurting another, far more ambitious, far more well-thought-out-hypersonic program. The person who made these arguments to me did not make the request on a personal basis, but out of respect for the things that he and I had supported over the years, I felt that he probably had a good reason for asking. I didn't want to hurt a program that was actually going to go out and do things related to what I was advocating.

The Start of NASP

The big project was the National Aerospace Plane (NASP). Dryden Center Director Ken Szalai was fully involved with this new program, and he asked that I also be briefed so that I could become a supporter and a contributor. I agreed and it now looked like I would be entering a new phase of working on exciting things—hypersonic flight with an air-breather that apparently had technologies that I was unaware of. This included a scramjet engine able to take the vehicle all the way to orbit and an airframe able to manage the incredible amount of heat generated while flying at high dynamic pressures between Mach 6 and 28.

However, I didn't know what kind of materials they were going to use. Most scramjet designs require a sharp inlet. A lip on the inlet cowl is very hard to protect, because anything that gets very hot tends to melt. Keeping a sharp edge on an inlet, even if it is made of very high temperature material, had never been done. But I was told they had solved those problems, and they were in the process of designing and building a vehicle. The original numbers they gave me were unbelievable. The vehicle was in the weight class of the X-15A-2 with tanks on, and it was going to taxi to the runway, take off, and make it into orbit.

As time went on, the NASP design weight increased as did the time and cost for development, but it seemed to me that there had to be some new physics—things that I was totally unaware of—for this thing to work. Partly because of the request that had suggested I was hurting this other program and partly because I knew I didn't know everything there was to know, I was optimistic that I was going to be involved in this program that finally would take us from runway to orbit.

As I worked on the NASP, I kept looking for the new technologies that were going to enable us to do something so far beyond what I had been hoping to do as the next step. I was told by people in the NASP program that there were new scramjet combined-cycle engines that would do all of the things that we needed to do, and there were materials available that had properties far beyond anything I had seen. But as I poked around and tried to find out what these were, I was disappointed. It seemed to me that this mostly was a great extrapolation of what they had been doing in ground facilities in the scramjet area at Langley, G.E., and various other places. While these extrapolations were arguably acceptable for a briefing, it seemed to me that they would be very difficult to deliver in an engineering sense.

I think that's the big disconnect between what I had been advocating and what the NASP was selling. I was claiming that we needed the validation of a scramjet actually flying before we could put a lot of faith in our ability to go out and build an operational vehicle. In doing that, the scramjet not only needed to be refined and developed from the data we had from ground tests, but it also needed to be given a great deal of robustness. Even if it worked as it did in the ground tests, we would still have variations in the angle of attack, angle of sideslip, and various vehicle rotation rates, plus all of the changes in the atmospheric characteristics, such as wind sheers, and density and temperature gradients. However, the NASP people were promising that, with enough money and enough talent, they would design something that would taxi to the runway and go into orbit.

There was, in this case, too much of a "can do" attitude. It's hard for me to say that, because I am sometimes accused of being a little overly optimistic. If you asked honest questions and posed your own doubts on some of these technologies, you were not considered a team player and were told to "get with the program, and let's make this thing work!" So I did the best that I could, but I had a lot of difficulty believing that any of this was going to pay off. However, assuming that there were other people, buried more deeply in the project than I was, who might have some insights and actual validations, perhaps from some other classified program, perhaps there was reason for some hope.

One of the things that Dryden did for the NASP program was to develop a very high-quality simulation for a vehicle with an air-breathing engine. It had come from our own Shuttle simulator. It had all of the uncertainties that can be put into a simulation of flying from take-off, through the acceleration in the supersonic and the hypersonic flight regimes and to the return to Edwards. It also modeled, as well as we knew, with what the various heating rates on the vehicle would be. The simulator also had the best atmospheric model existing at the time. This gave us the tools, even though we didn't know quite how all of these problems, in reality, were going to be solved, to simulate a mission.

If a NASP were to take off from Edwards, where would it be when it got to Mach 1, Mach 6, Mach 10, Mach 12, or to whatever speed we were using in the simulator, and then turn around and come back to land at Edwards? At high Mach numbers, a turn covers several Western states. (Fig. 1, adapted from reference 88) Just from an operational point of view, we could start to see some of those issues and propose solutions in terms of tracking and having the flight planned to be over less populated areas.

Dryden was enthusiastically committed to supporting NASP. Even though a lot of us didn't see how these technologies were going to actually come about, we took the position that the technical leadership within NASA and within the DoD would be taking care of those issues. Within the scenario of which we were part, our doubting whether they were achievable added nothing. We would have our discussions privately on the reality of what we were working on, but we did our very best to be contributors to the overall program. I spent a couple of years adding what I could in terms of what types of instrumentation were required and what types of uncertainties should be put on some of the wind-tunnel values they had for various NASP shapes. However, it was all simulated, of course, because we had no aircraft to base this on.

The Flaws of NASP

My own feeling is that NASP was dead not at birth, but before birth. The objectives were far too grandiose. I liken it to the Wright brothers trying to design an SR-71 as their first flight vehicle. The SR-71 never would have flown if the intermediate steps in the history of flight between 1903 and the early 1960s had not been taken. The question I asked myself is, how did the technical leadership of the country let themselves be sucked into something that was so beyond what the laws of physics told us we could do? I'd still make the same statement today, that we still don't have the technologies in materials, in propulsion, in understanding the boundary-layer conditions, in dealing with excess heat, and the other kinds of issues.

I have concluded that the NASP program was approved so far beyond any

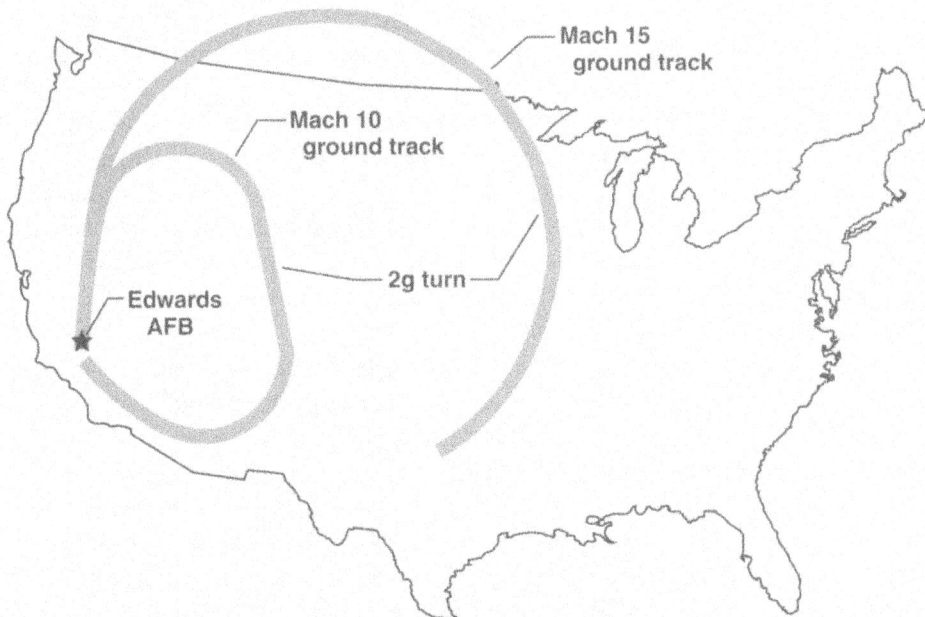

Fig. 1. Ground tracks required for turning at hypersonic speeds.

reasonably attainable objective because the Apollo program had shown that if we put enough money and talent into something, we can accomplish great things in a relatively short period of time. The time from Sputnik to the first man walking on the Moon was less than 12 years. You had to have lived through that time to realize how rapidly we overtook the Soviets after Sputnik. It's incomprehensible in today's world, and I claim that it was done partly because the world was simpler in those days. We also had an enemy, so we could marshal public opinion, and the fiscal and techno-logical resources for a program such as Apollo. Apollo had a very focused objective. It was really a matter of taking things that we had already done and making them bigger—making bigger engines for the Saturn rocket, making the turbopumps larger than what we'd already done, and putting them together to make a very large booster. With that amount of throw weight, then we had enough capability to get to the Moon and back.

The Apollo program was the crowning achievement of the early NASA and perhaps of NASA for all time. It happened because we understood the physics of that program, and we had all the tools needed in terms of materials, fuels, electronics, and a very primitive computer technology for pulling it all off. Apollo was very much an exercise in increasing the reliability of all systems required for going to the Moon and back. However, the objective of Apollo was only to go to the Moon and back. It was not to have man explore the rest of the solar system.

If that had been the goal, I believe Apollo probably would have occurred 40 to 50 years later if we had let technology lead us to a means of getting to the Moon, establishing outposts on the Moon, and then doing the same with Mars. Once we had done that, we would end up with a permanent presence. I think we eventually will end up with a permanent presence on both the Moon and Mars, but it's far, far in the future. Now, because we've been to the Moon and back, it's hard to have a near-term objective of returning to the Moon, but that's really the next thing that we need to do with manned exploration.

I think that lack of a realization for what it really took to go to the Moon and back made it look as if we could do anything we wanted if enough resources, money, and talent are put into a program. A little bit of that way of thinking was why the Shuttle was rushed into being made an operational vehicle. Building the Shuttle was actually a very hard problem. There was just enough technology in the 1970s to design and build a vehicle that was essentially reusable, but incredibly complex, and which requires an incredible amount of manpower to operate.

However, the Shuttle also, in my opinion, is a great success story. Once again, it looked as if we could start with virtually no intermediate steps and go directly to an operational vehicle. The people who believed that weren't involved and didn't realize how many razor-edge decisions and technologies came into existence just in time. For example, if the development of the Shuttle's main engines or the ceramic thermal protection system hadn't worked, there wouldn't have been a Shuttle flying in 1981. Once we had those technologies, however, then we could take traditional materials, avionics, and rocketry—both solid and liquid—and just barely close the design. That is not really the way we wanted to build an operational vehicle. The complexity and the possible failures show that the Shuttle doesn't have a lot of excess margin in terms of being a successful operational vehicle.

In the mid-1980s, the "visionaries" in the country, Congress, NASA leadership, and the associated bureaucracies believed that we really needed an air-breather to go from runway to orbit, which I certainly believed. However, because we had solved huge problems and been successful on two very, very difficult engineering projects— Apollo and the Shuttle—I think that they felt that with enough money and enough technologies, the NASP could be built and flown. They bought into it, and NASP became the only focus of hypersonic research from the mid-1980s until the early 1990s. The NASP failed to build a vehicle of any kind to demonstrate the technologies that had been projected and extrapolated from ground tests and with the various computational techniques. As a result, all that we got from the NASP was little more than a huge stack of view graphs, a lot of disappointed people, and a substantial loss of credibility by NASA.

At the working level, I have talked to few people who were involved with NASP in materials, aerodynamics, aerothermodynamics, catalytic heating, and scramjet technologies who felt that they had failed because the technology wasn't available to solve the NASP problem. They felt this was their one chance to make a very substantial contribution, so they all had worked really hard. There are some other people, however, who think that we actually failed because NASP lost funding, or because the program didn't get the support that it needed. I don't feel that way as a technologist. I believe NASP lasted far too long. They had failed because the objective was far beyond anything that should have ever been approved by the leadership within the country.

I remember two presentations at the AIAA Aerospace Engineering Conference and Show held in Los Angeles in February 1987, about a year after the Challenger accident. The event was mostly a collection of displays and an opportunity for people to get together, but it did have a few papers. The first session included a presentation given by the head of the NASP program. My recollection is that he said that for a few billion dollars, within several years they would demonstrate a NASP taking off from a runway and going into orbit. People in attendance were really enthused to hear this, because NASP had been reported in the media, but a lot of the details hadn't been released. Here was the announcement by the leader of the effort saying that these things were going to happen. I'm thinking to myself at the time that this would be the most complicated thing I could think of to do! I don't know how to solve one of his problems, much less the thousands or tens of thousands that really face this program.

Reality set in, I thought, when Dick Truly gave a presentation right after lunch. He was trying to get the Shuttle back on track after the loss of the Challenger. His briefing was very honest. As I recall, he was asking Congress to give him a little over $2 billion over three years to build an additional Shuttle. The cost was going to be higher because it was a low production kind of a vehicle. But he was advocating building an additional copy of a vehicle that had already been designed and successfully flown many times. He followed the chief of NASP, who had said NASP was going to be less complicated and far lighter than the Shuttle, that it was going to do all of the things that we needed to do. It would make the Shuttle irrelevant, and he was going to do all of that in several years for a few billion dollars!

I thought that after the technologists at the meeting heard those two briefings, all of them would tell me, "Boy, talk about unreality and reality!" But I didn't hear that. Most seemed to think NASP was the way to go. The nation really wanted to have another success like Apollo or the Shuttle, and it thought NASP was going to be that. It took the country and even the engineers quite a while to realize that NASP wasn't going to work.

There were always, of course, engineers who pooh-pooh everything. But I try to be a positive engineer, and even though I don't think something is the best idea, once it's funded and it's going to happen, I do my level best to make it happen. I was doing that on NASP, but I was wondering where were all of these key technologies that were "thoroughly understood"? Where was that information? When was it going to become available? I'm really astounded that NASP took as many years as it did to finally fall from favor. It really had a good ride from the mid-1980s until the early 1990s, when the majority of people who had informed opinions were finally saying, "I don't think this is really going to happen."

Hypersonic Projects After NASP

After two or three years of working on NASP, I teamed with a Dryden propulsion engineer named Henry Arnaiz and started advocating a hypersonic test vehicle. It had to be something that was going to be near-term to help NASP. Of course the NASP people were claiming they're going to go to orbit in a few years, it would be hard to help the program. We had to advocate programs on the basis of if NASP should fall behind or should fall short of its objectives, here's what we need to do. All of the money in the hypersonic field was going to NASP, and the only people we could try to sell a supporting hypersonic vehicle to were the people who were involved in NASP.

I started with the working level people at Ames and Langley. I was trying to come up with a critical mass of people who could advocate a hypersonic boundary-layer test series for trying to understand exactly what goes on. This was important not only for the aerodynamics and aerothermodynamics, but also for the air-breather, because the inlet is very much an aerothermodynamic issue. We needed to make sure that our codes were properly predicting what was going to go on in the inlet and also what was going to go on after the fuel was burned and the exhaust was expanded out the rear end.

The Pegasus air-launched rocket seemed promising. It would soon be available, and a lot of people at Dryden liked it because it could be launched from a B-52 at Edwards AFB. The air-launched Pegasus would give plenty of performance for whatever vehicles we chose to launch. It could also accelerate a larger vehicles with deployable scramjet experiments up to the Mach 10 to 15 region. With a two-flight program, we felt that we could study the essentials: boundary layer and shock impingement heating. Again, the most important factor was that the Pegasus was air-launched from the B-52. (Fig. 2) Air-launch was something that Dryden had done for over 40 years. As far as the payload, there were a lot of proposals. But our costs were far less than NASP, and we did find more and more support as time went on. All of this seemed awfully conservative compared to what NASP was claiming. Just think, here at the start of the 21st century, we still don't have a significant amount of new validated data from any kind of vehicle for aerodynamic and aerothermodynamic work. We've lost over a decade in obtaining hypersonic information.

During the X-15 program we had seen proposals for multistage rockets that would be launched from the B-52. The first stage would be quite similar in shape to the X-15 from the wing station to the back of the vehicle. There then would be upper stages for air-launching a satellite into orbit. Those, I think, never got seriously considered, but we did realize that the possibility was there with a 40,000-or 50,000-pound rocket that we could actually orbit a satellite from the B-52. The idea, as I remember, was that the first stage would have been recoverable, whether it was solid rocket or liquid rocket, because it has the biggest and most expensive engine. However, those ideas never got very far.

Then, in the mid-1980s, NASA noticed that we were running out of the Scout boosters. We needed to come up with an alternative for launching small satellites or for other types of missions that required a smaller rocket than either the Atlas or Titan booster. Orbital Sciences Corporation (OSC) came up with a proposal for a multistaged, winged, air-launched vehicle that initially would be launched from the B-52. Since Dryden didn't want to be in the business of launching satellites, the plan, once the system was perfected, was that the vehicle would be launched from an OSC-owned airplane, later defined as a modified Lockheed L-1011 airliner.

When the first Pegasus flew in 1990, it managed to put its payload into a peculiar orbit, but, nevertheless, Pegasus did orbit something on its first flight. I've always kept in my mind that the launch of the first Pegasus was the first mission that actually went runway to orbit. It was unmanned, it was totally expendable (except for the Dryden B-52 launch aircraft), but it did achieve orbit. Considering all the technologies and things that went on, it was a significant happening. On the first few flights, they did intentional stability-and-control maneuvers that I analyzed in collaboration with the people at OSC and other places. I'd also been involved in some of the initial discussions about the launch dynamics, energy management, and the launch of a winged vehicle from the B-52.

By the time the program was ready for its first flight, it was very much driven by OSC, with Dryden as the primary provider of the B-52, the test range, the communications, the tracking, and the safety of the flight. Dryden played a major role, but it wasn't really a research mission. It was planned to be an operational vehicle, and relatively inexpensive satellites were the first payloads. On the sixth flight, they went to a different configuration that had more performance. It was called the XL, and it failed on its first flight. I was involved in the accident review, and I participated in analyzing the aerodynamic data.

Henry Arnaiz died unexpectedly in the middle of our advocacy of Pegasus-launched hypersonic vehicles. His death took the wind out of my sails for six months

Fig. 2. An early B-52-launched Pegasus shown shortly after rocket ignition, along with an F-18 chase plane. (EC91 348-4)

or a year. I tried to figure out how to make this project happen without Henry's very able support and very optimistic approach. He had been hard to discourage. I missed him and his optimism. He also had been one of my best friends. It was a tragic loss for me, and I'm still not sure I've gotten over it. After Henry died, I realized how much I had relied on him to take the detailed propulsion issues and either arrange for further analysis or consult with his cronies to give the ideas more credibility.

I had to learn a lot more about scramjets in terms of the inlet geometry, what the inlet was actually doing, what kinds of ignition would be used, how to mix the fuel with the incoming air and burn it, how long the combustion chamber had to be, and how to expand combustion products in the exhaust and exit stage to get the full benefit of the scramjet. I had to carry more of the burden of responsibility than I had before and never, of course, could I carry it as ably as Henry had before me.

In planning for Pegasus launched vehicles, we realized an additional benefit: we could get more data from the Pegasus itself as we proceeded to the hypersonic flights, because its first stage actually went to Mach 8. We were advocating a wing glove on the Pegasus to obtain aerodynamic information up to the burnout of the first stage. That experiment would be mostly to define boundary-layer conditions. We did work on some early Pegasus flights and obtained data on the wing. Later, in 1998, we flew a Pegasus that had boundary-layer instrumentation on a specially-made glove integrated with the wing that was called the Pegasus hypersonic experiment (PHYSX). The single PHYSX flight gave us some very high-quality boundary-layer data up to Mach 8 and altitudes of about 200,000 feet. [89]

Before NASP, we had also talked about another possibility, of putting another stage on a rocket and actually lift a vehicle up to the Mach 20 region. The test vehicle would need an improved thermal protection system. Those were the proposals until NASP clouded the horizon for any real research in the hypersonic area. Then, as it started to look like the NASP was not going to come close to meeting any of its promises, people started to step back, thinking it probably wasn't going to happen.

At that occurred, around 1988, we started seriously putting together proposals to try to interest either the technologists or the NASP program in the concepts we'd been talking about. Those concepts included thermal protection systems that would allow the boundary layers to be studied free of ablation or outgassing. These concepts would allow us to get into the real gas and the catalytic effects that only occur above Mach 12 or 15 and to provide data on what the materials really do when they're exposed to loads, temperatures, and dynamic pressures that really can't be simulated on the ground.

Those were the proposals initially, and we did get some significant interest from the technologists, primarily people at Langley and Ames. We tried to come up with a consensus among the three centers—Ames, Dryden, and Langley. Of course, Dryden was part of Ames at that time. The proposals would come up with something that the technologists thought would solve their problem, so they could validate their codes and have greater confidence in their understanding of the phenomenology. Then when they designed something more complex, they would be successful.

The HALO Manned Research Vehicle

During this period, around 1990, Milt Thompson and I and others had also been looking at a B-52-dropped manned research vehicle to see if the air-breather could be worked into something that was manned. All of those studies resulted in a vehicle that was either too small to put a man in, or not able to attain speeds above about Mach

4.5 or so.

We went on to the next best thing, which we'd also looked at back in the mid-1980s, of air-launching a larger vehicle from the top of the 747 Shuttle carrier aircraft (SCA). There were two versions of that larger vehicle. One was to have a NASP shape in front of an expendable rocket. The NASP-shaped front would be recovered and might even be manned. The other was a quite large NASP shape that used rockets to accelerate to the conditions we needed. Both were lofted launches. We didn't look at any depressed trajectories—that is, where the vehicle was always in high dynamic pressures for aerodynamic control—because we would be faced with solving the same heating problems as with NASP. If the whole mission is flown at high dynamic pressure, the vehicle incurs a huge heat budget that has to be dissipated somewhere, as well as having possible localized hot spots on the vehicle to worry about.

I remember Mach 10 was the favored speed. Getting higher than Mach 10 with a manned vehicle was going to be complicated, because we didn't have proven materials that would take the extreme heat. Everything would also have to be actively cooled, and that would have made things more complicated. Different from some of our earlier advocacies, these vehicles were all NASP-like shapes because that seemed to be the only game in town.

About late 1991, we also started looking at launching a vehicle from the top of the SR-71. Earlier, in 1990 and 1991, Dryden had acquired three SR-71s. (Fig. 3) We could tell that if the hypersonic vehicle was launched at Mach 3.2 and 80,000 feet, it could be a very small vehicle compared to what would have been launched off the 747 SCA. We did quite a few trade-off studies on that.

Steve Ishmael was one of our SR-71 pilots. I had known him for 15 years or so at Dryden, and I worked with him quite a bit on several programs, including the X-29. Steve and I developed a good relationship, and we had a lot of discussions of what could be done in terms of launching a manned airplane off the top of an SR-71.

Over the years at Dryden, pilots have done what we call "flyovers." They usually fly it in afterburner, if available, coming in a few feet over the dry lakebed and straight at the main Dryden buildings at near-sonic speed. Each aircraft is fairly quiet as it approaches, getting rapidly larger at those high speeds. Just as the aircraft gets near the main building, it starts a slow climb to clear the building with adequate margin. As the aircraft comes overhead, the observers get the full ear-shattering noise that an aircraft in afterburner makes when the observer is in close proximity, as we would be. As it passes in an instant overhead, we turn our heads to watch it fly away. This was the moment of maximum intensity, when we got the full benefit of the sound and light show with the maximum noise and the exceedingly bright flame coming from the afterburners.

All of the high-performance aircraft are unforgettable, but the very best and most unforgettable flyovers were made by the SR-71s flown by Steve Ishmael. They were fast, close, exceedingly loud, and had the very brightest flame, especially after he learned to dump TEB (tri-ethyl borate) behind the engine just as he flew over the building. The flame coming out of the SR-71 could be greatly extended (more than the length of the SR-71) by turning on the afterburner, then turning it off before it ignites, but after the TEB is ejected, then turning the afterburner on again and leaving it on. The first cloud of TEB is ignited by exposure to air and is then joined to the second flame from the activation of the afterburner, resulting in a very long and bright flame.

As you might guess, the SR-71-launch vehicle was the most appealing to everybody. First, the SR-71 (Fig. 4) is, in my opinion, the prettiest, most impressive airplane ever made. We believed that we could get the vehicle up to the Mach 3 and 70,000 foot region, then launch a vehicle off it that had been mounted on the top of the SR-71. All three of the proposals that Steve and I had been working on were eventually called the hypersonic air-launch option (HALO), for lack of a better name. The option studied by far the most was the one for the SR-71-launched vehicle.

We were not making a very big step, because the D-21 had been launched from the M-21. The D-21 weighed about 12,000 or 13,000 pounds. The D-21 ramjet was made by Marquardt and was based on the Bomarc engine. For our constraint on the vehicle, we used the size and the cross-sectional area of the D-21. We discovered that with a 17,000-or 18,000-pound vehicle with a cross-section only slightly larger than a D-21, we could get the SR-71 up to 70,000 feet and Mach 3 for the launch. Then, we would be able to get up to Mach 10 with the manned HALO vehicle. We'd use an RL-10 rocket, a restartable rocket with a good track record.

Fig. 3. SR-71B flying over Mount Whitney during a December 8, 1994 flight. (EC94 42883-4)

The HALO vehicle was quite a mature design in terms of the codes available at the time for designing vehicles. We were using the Ames HAVOC code that optimized the overall vehicle design, including estimated aerodynamics, aerothermodynamics, shape, thermal protective system, and structure. The hypersonic groups at Ames and Dryden had long discussed getting hypersonic data either with or without a NASP. We had detailed conversations with them from our earlier advocacy. They were among the most supportive of the technologists. They had the nation's leading expertise in catalytic effects. They had the people who had interest in the boundary layer, the boundary-layer heating, shock-impingement heating, and, to a lesser extent, the scramjet technologies. The primary work on scramjets had been defined by Langley, however, and the designs that Langley was advocating were much closer to those proposed for the NASP and HALO.

We were looking at three launch possibilities-from a B-52, the 747 SCA, or an SR-71. Depending on what Mach number we wanted to get to, the subsonic launches on the B-52 and the 747 SCA would get quite heavy. Speeds in the neighborhood of Mach 5 could be reached with a vehicle hanging on the B-52. However, the vehicle need to reach Mach 10 to 12 for the scramjet demonstration, while the people studying catalytic effects wanted something in the region of Mach 14.

We also looked at a vehicle launched from the 747 SCA. That vehicle, we figured, could reach Mach 12 to 14 and weigh in the neighborhood of 80,000 to 100,000 pounds, depending on what experiments it carried and what kind of cooling was proposed. However, the one that we pushed the hardest was the SR-71 launch that mimicked what the D-21 had done. The launch would be done at Mach 3 and 70,000 feet. The additional drag of the HALO vehicle would slow the acceleration of the SR-71 somewhat. The amount of that drag could only be found in flight tests with

Fig. 4. SR-71B shortly after takeoff showing the shock diamonds in the "twisted" exhaust plumes entrained in the strong aerodynamic vortex shed from the aircraft. (EC92 1284-01)

a dummy HALO vehicle mounted on top of the SR-71.

Using the Ames HAVOC code, Jeff Bowles at Ames could turn out a complete conceptual design of the HALO very quickly in response to each issue. In general, what he came up with was a manned 17,000- to 18,000-pound vehicle with about a 45 percent mass fraction. It would be about 60 feet long with a cross-sectional area of about 40 square feet and a wing span of about 20 feet. It was basically a NASP-like vehicle, but perhaps a little more slender than NASP to keep the cross-sectional area down. As far as the 2-D inlets and the scramjet was concerned, the vehicle was more than adequate for what we needed. It really was a mini-NASP vehicle in an unclassified sense. We also had from the HAVOC code the component weights, the highest heating rate areas, and drag and lift calculations for the HALO mated to the SR-71. (Fig. 5)

One concern was exactly how we were going to get the SR-71 supersonic, for the SR-71 is marginal going through the Mach 1 region. The SR-71 pilots normally do this with a moderate dive maneuver. We talked to the Lockheed people who had been working on the SR-71 for 30 years, and they were quite confident that we could boost the thrust on the SR-71 to accelerate through Mach 1, even with the additional drag that we would have from the HALO vehicle. In addition, there's also interference drag between two mated vehicles, as with the HALO perched on top of the SR-71. Ed Saltzman of Dryden did some independent estimates based on other flight experiments that we had done to see what he thought the transonic drag would be that we would need to overcome.

Fig. 5. Three view drawings of the proposed HALO vehicle mated to the SR-71.

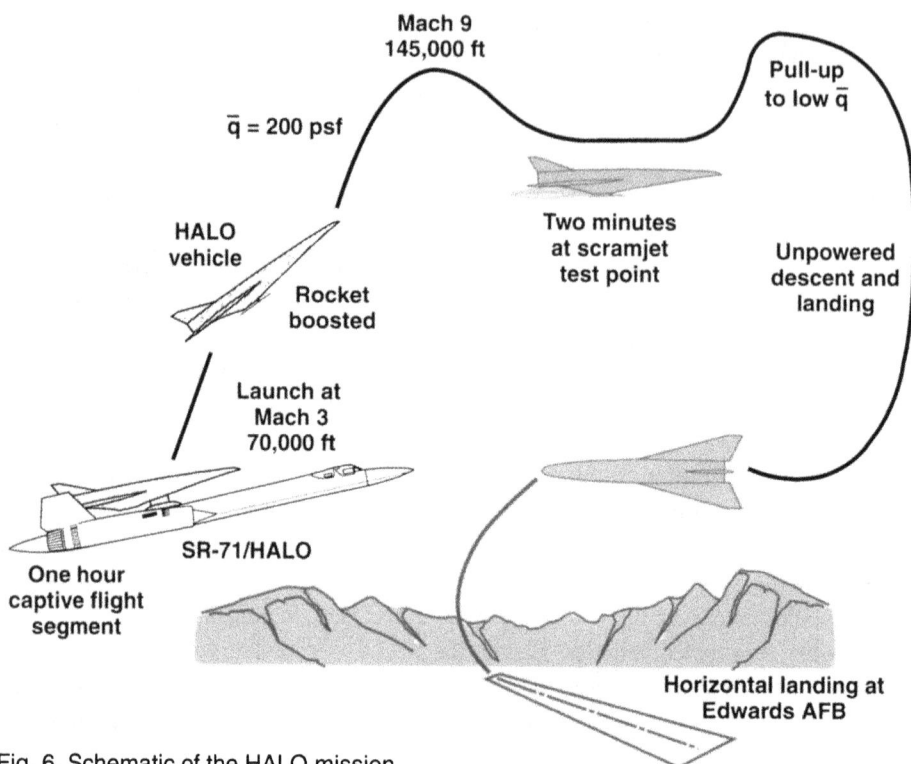

Fig. 6. Schematic of the HALO mission.

We convinced ourselves that we might actually be able to increase the thrust by tinkering with the engine operating parameters. This shortens the life of the engine, because it is outside its normal operating range, but it will put out more thrust. Perhaps a better idea was nitrous oxide injection, much as used in drag racing, just prior to the acceleration through Mach 1. That would give added performance to get supersonic with the two mated vehicles.

The payload on the HALO was going to have about 500 pounds of instrumentation and various other experiments that it would carry in addition to the pilot and the scramjet. It would be powered by an RL-10 rocket, which has a heritage back to the 1960s. It's considered one of the most reliable rocket engines that has ever been made. It's appeared on many vehicles, and my recollection is that in close to 300 flights there had only been a few failures. The RL-10 would later be used on the McDonnell-Douglas DC-X program and be very successful, so it would have been a good choice for the HALO.

After the air-launch of the manned HALO from the SR-71, its RL-10 hydrogen-powered engine would ignite and it would go into a fairly steep climb, reminiscent of X-15 altitude flights, and get out of the atmosphere to fly at low dynamic pressures. (Fig. 6) Then, when the HALO would get near the Mach number that we wanted tested, the onboard pilot would lower the altitude with a dip maneuver and bring the vehicle level for a speed run. For Mach 10, Mach 8, or Mach 6, we could get two minutes of test data using the RL-10 (if necessary) to maintain the flight condition,

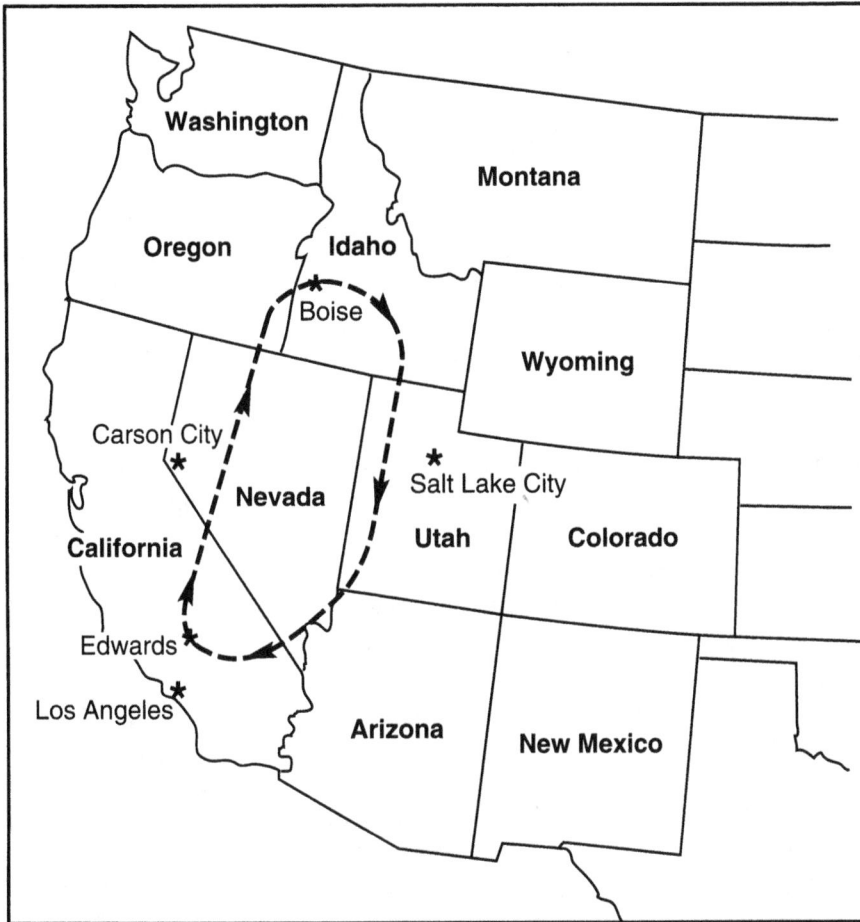

Fig. 7. Ground track of a typical HALO Mach 10 mission.

which primarily would be to test the scramjet. We could try various ways to start the inlet and also do variations at angle of attack and angle of sideslip, pitch rate, roll rate, and yaw rate, to get a feel for how robust that particular design was. On subsequent flights, we would have other modular scramjets on the bottom of the HALO.

We could take off with the SR-71 from Edwards, then go up to Boise or Mountain Home in Idaho to launch the HALO. (Fig. 7) It would climb, finish the turn while it still had adequate dynamic pressure, and head south. It would be at hypersonic velocities fairly promptly and come back over Utah. After performing the dip maneuver to test the scramjet engine, the HALO would decelerate by climbing back up to low dynamic pressures until it was no longer in the high hypersonic heating Mach numbers. As Mach number got down into the Mach 5 to 6 region, it would start back into the denser atmosphere. The HALO would then return to Edwards AFB with the high key/low key maneuver, coming in for a manned landing like the X-15. The SR-71 also would soon arrive back, because it also would be going Mach 3 at the time of the launch.

It was fairly well thought out. We had ways of getting maneuvers. We looked at diving into the atmosphere from Mach 4 up to Mach 14 with a little larger vehicle, although it would not have been able to be accommodated on the SR-71. I think the Mach 12 vehicle on the SR-71 looked like a sure thing with the HAVOC code. When we looked at reasonable safety factors and margins that all designers build in, it probably could fly in the Mach 10 area and could get two minutes of test time at Mach 10, or a lower speed if that was desired. If the scramjet was putting out enough thrust, of course, the HALO could maintain that Mach number with the air-breathing scramjet. Once the HALO vehicle got through the dip maneuver, the rocket engine would be shut off, because we would no longer need it. Although the RL-10 could be restarted in flight, I don't think we ever planned to do that, but it did leave other options.

We had a full temperature model of all the hot spots including the cowl lip on the scramjet engine and the leading edge of the wings. We would be flying something that, for all practical purposes, would look like a NASP. It would not, as the NASP had planned, taxi to the runway and take off. Since we would be using hydrogen, we also would carry some hydrogen on board the SR-71 to top the HALO off just prior to launch, to take care of boil off. That problem was a fact of physics that couldn't be avoided. We could plan to accept the loss of hydrogen, but, of course, that would reduce the total performance.

Steve Ishmael was quite willing to fly the HALO. There would be plenty of room for a cockpit with an ejection seat that could be used in an emergency prior to launch or after the vehicle had slowed to a speed and altitude where ejection would be reasonable. I felt our design was more mature than the NASP, even though we had a lot fewer people working on it. The vehicle was designed to get hypersonic data at up to a dynamic pressure of 2,000 pounds per square foot (psf) in the dip maneuver. (Fig. 8) Most scramjet people like to think of flying at dynamic pressures between 800 and 2,000 psf. This higher range of dynamic pressures improves the performance, but of course the heating gets horrendous, especially on the wing leading edge and the cowl lip at the higher Mach numbers.

There were other experiments that we would try to accommodate. With the Mach 10 vehicle, we probably would have had some difficulty getting very meaningful data for the specialists in catalytic heating, but experiments involving materials, boundary-layer studies, or shock impingements at high dynamic pressures and high Mach numbers could be done on the HALO vehicle.

The fall-back position, if we couldn't convince people to use the SR-71, was to build a 60,000- or 80,000-pound vehicle that would do much the same thing at Mach 10 from the modified 747 SCA. It would be a larger vehicle. From some points of view, that was good; from some others, that was bad. But we thought we had covered a complete set of characteristics.

By the time we got to the HALO briefings, the NASP program wanted to try to salvage something less than the whole enchilada they had planned. They had their own way of wanting to do what I believe the HALO would have done for them. Not surprisingly, we didn't get a lot of official support from the NASP program on the

HALO. I still believe the HALO is the very best thing that can be done if substantial research data is needed in the Mach 10 to 12 region for an air-breather. A larger vehicle on the 747 SCA, perhaps an 80,000- to 100,000-pound vehicle, could probably get up to Mach 14 and really start covering the catalytic effects, but I think that Mach 10 would be a good place to start. Then, if there were lots of successes there and those problems were understood, we could go to a larger vehicle launch from the 747 SCA or perhaps some new launch vehicle. Nevertheless, the HALO was, to me, the culmination of the search that we had started on in 1983 for a new X-vehicle.

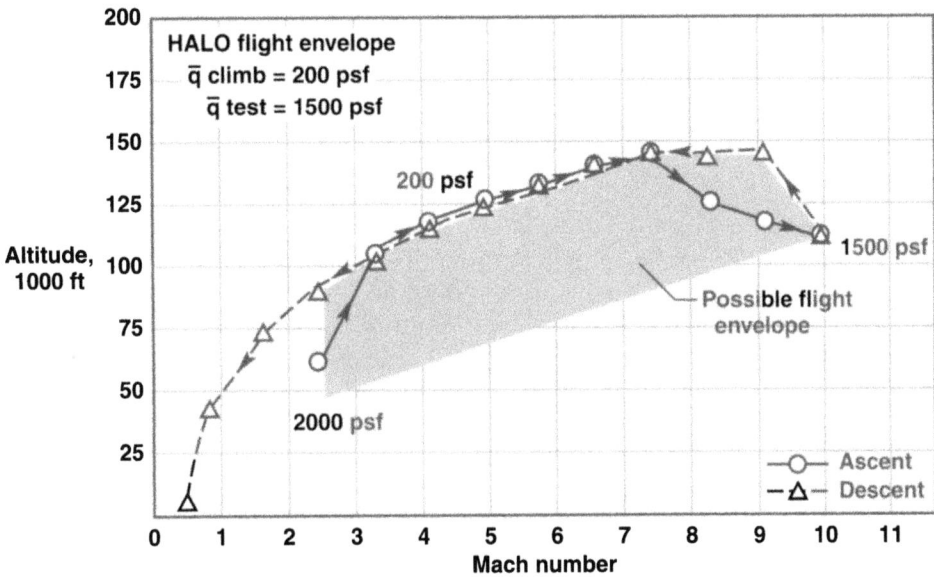

Fig. 8. Flight envelope showing a typical Mach 10 HALO mission.

XVI

Hypersonics in the 1990s

The programs we were proposing to fly with the Pegasus from 1988 to 1990 were presented to the NASP people and we got some favorable response. However, we hadn't found anyone to support the funding of the program. We also were trying to make sure that we maintained our partnerships with the technical community. We briefed the best technical experts within NASA on our proposed experiments in the fields of boundary-layer characteristics, aerothermodynamics, catalytic heating, shock impingement heating, and the experts in the basic physics of inlets and scramjets.

In the process, I noted a dichotomy that was quite baffling to me as a researcher and technologist. On the one hand, the NASP community was saying that we were proposing things that they thought were fairly well understood and that the Pegasus-launched vehicles would add nothing. On the other hand, we had the expert technologists telling us that so little was known that they needed high quality data so they could predict aerodynamic and aerothermodynamic characteristics. The technologists would support the Pegasus vehicle efforts, but we had to guarantee them that we were going to get their experiment exactly to the angle of attack, angle of sideslip, Mach number, and altitude that they specified. If we couldn't achieve such precision, the tests would not provide the data they needed. I and some others like me appeared to be the only middle ground. I don't want to make it sound as if I'm the only one who had figured this out, but I think I gave the bulk of the presentations. I started explaining that in my presentations, telling them that there was a dichotomy between the true experts and the people who were signed up to build the NASP vehicle. However, they never could see that what I was trying to do was help both of them.

HALO

In late 1991, we started additional studies and advocacies, going back to some of the things that we'd looked at earlier. We knew of the two-stage vehicles-the first-stage air-breather and the second-stage rocket-such as the Beta vehicle and the German Sänger, and we were starting to go back to see if we could design something that would get scramjet information in the Mach 12 region. This eventually became the SR-71-launched HALO concept.

Center management felt that I needed to include a cost estimate for the HALO. I thought it was in the neighborhood of a few hundred million dollars, because it used a proven propulsion system (the RL-10), and most of the systems (actuators, landing gear, avionics, and ejection seat) were off-the-shelf items. All that remained was the design, building, and integration of the systems into the airframe. The airframe would be fairly expensive, because of the high heating environment that it would fly in. With the small size and low weight, I believed that $200 million to $300 million should cover it. For a vehicle capable of 50 to 100 flights, that amount seemed very cost-effective. Although I had to admit that I was no expert in cost estimates, I really hadn't seen much correlation in the past between estimated costs made by others and the actual costs for aerospace vehicles.

To give HALO more credibility, our management felt we needed an independent cost estimate. Ben Rich, of U-2 and SR-71 fame, had recently retired as the head of the Lockheed Skunk Works. We had worked with him in the past, and from my point of view, Ben had always been far easier to deal with than Kelly Johnson. He was very experienced and more patient and casual than Kelly. We contacted Ben and arranged to hire him as a consultant to review the HALO proposal. I gave him a presentation on HALO with far more detail than I had normally given. He asked a lot of fairly detailed questions and seemed to think that it technically was a fairly sound approach, including using the SR-71. We had asked him for a cost estimate, not including the cost of the scramjet modules.

I was astonished at his cost estimate of one billion dollars. He said that in the environment of the 1990s, anything worth doing always seemed to eat up a billion dollars, mostly in satisfying the bureaucracy. In addition to that, the NASP people were telling us that the HALO scramjet modules would also cost a billion dollars. The scramjet modules were totally fixed geometry, in terms of the injection, the inlet, and the exhaust. I thought it would be hard to spend more than $10 million on a module. Now I had to carry along with my HALO presentation a cost estimate of two billion dollars for an 18,000-pound vehicle. Whenever I could, I left out the cost estimate, saying it was undetermined. Cost estimates are one means that opponents of concepts use to kill them. I think Ben's estimate was an unfortunate round number, but in the case of the scramjet cost estimate, I think the NASP people wanted to kill any alternative proposals. This was one more lesson in not asking questions that you don't want answered.

After about six months of working on the technical design and honing the presentation, we combined the HALO program with the Pegasus-launched vehicles. The idea was that the whole hypersonic area could be covered. The vehicle would fly at dynamic pressures from essentially zero pounds per square foot up to 2,000 pounds per square foot. All of the proposed designs for aircraft in the hypersonic region fit into that particular flight envelope. This included the thermal protection system for obtaining the real gas and the catalytic and the boundary-layer-effect experiments at the very high Mach numbers.

The HALO was a vehicle designed to learn how to fly a scramjet—not just to understand the physics, but also to give it enough robustness that it could maneuver, and that it would not cause problems with unstarts or have flight conditions where it wouldn't be reliable. The forward portion of the vehicle was a detachable nose so that we could change the ramp angle of the inlet. The lower portion of the vehicle was a scramjet module that would allow us to use the same HALO configuration forebody and change the details in the scramjet itself with modules that fit the mold lines of the HALO. The engines themselves would be different in terms of injection, the Mach number they were designed for, or to demonstrate scramjet thrust while maneuvering. Several attempts could be made at a given Mach number with different scramjet designs to find out which ones were better at some conditions than at others.

The HALO program had been briefed to Langley, to NASP, to DoD, and to Ames management, since we were still part of Ames in 1992. On September 3, 1992, Dan Goldin, the NASA administrator at the time, was coming to Dryden to find out what we did and let us know what he expected of us. He'd become the administrator in April of 1992, so he'd been on the job for about five months when he came out to Dryden. The hypersonic proposals were among the things that we were going to present to him. The agenda was controlled by Ames management at Moffett, but they very much wanted the administrator briefed on the hypersonic program as we currently saw it.

I was asked to put together a half-hour presentation in which I combined the HALO with the Pegasus-launched vehicle programs. There were a couple of other things that I also wanted to tell the administrator about. As it turned out, briefing Dan Goldin wasn't like briefing other administrators. I don't think I got to the second chart before he started challenging me on what I was saying. He got me off the subject. We digressed for an hour and a half, debating space debris and single-staged reusable vehicles and rockets vs. air-breathers and all of those things, and what were the good ideas, and what were the bad ideas, and how strongly did I feel about it. After about an hour and a half, it was mentioned that we needed to break for lunch.

I asked the administrator if he would mind if I finished one of several briefings that I'd planned. I did finish the hypersonic briefing, and I think it was well received. He liked it, and he started going around talking about it after he left Dryden. A program that we were trying to keep within NASA suddenly became visible, and there were multiple newspaper stories. I was directed to talk to the Aviation Week staff so they could do an article. The people at Langley and Headquarters were very upset, because they thought this definitely was a breach of our agreements on doing things jointly. I understood their feeling, but I also thought that the NASP was dying. It was clear that the administrator did not like NASP and that he was looking for an off-ramp. HALO looked like the best idea around. I don't regret having briefed it. I certainly did not do it without management's approval, at least within Ames. We started a joint project with Langley on trying to refine the HALO. Unfortunately, they didn't like the SR-71; they didn't like HALO being manned. They didn't like much about HALO, and it was not going to be jointly advocated in anything like its present form.

Ames, in reaction to Langley's criticisms, redesigned the HALO as an unmanned smaller vehicle to fit on top of the SR-71, addressing most of the concerns that Langley had expressed. It was called the HALO II, which was about where that effort ended. We continued to advocate scramjet-testing with some of the engine manufacturers. However, funding wasn't there. Without the support of Headquarters and Langley, there was no way that any of these things were going to be funded.

The Goldin Years

My first association with Dan Goldin was several months earlier when he became the NASA administrator. He believed NASA was outdated, fat with bureaucracy, and irrelevant. He wanted to change all of that, so he formed red and blue teams on all of the major efforts that NASA was involved in. I, along with Ken Szalai, who chaired the red team, and Bill Dana of Dryden, served on the Moon-Mars red team, which was supposed to be briefed on all proposals, plans, and funding of manned missions to Mars. I sat on that committee for five or six months. Dan Goldin was very interested in that particular topic. I think, like a lot of us, he really did want to go to Mars, but he wanted to make sure that we had a good program before we tried to get support from Congress.

The first President George Bush had started the Space Initiative, but it was never picked up seriously by Congress. However, the Mars program started at that time to focus in on what they wanted to do. As a result, it was an official high-priority program within NASA. That was a very interesting study, and I was a strong supporter of wanting to do it. I was impressed with all of the hard work that had been done by various contractors, led by Johnson Space Center. At the time we were doing the red team study, the Manned Mars Mission was being led by Doug Cooke, who'd been to Dryden in 1975 and been trained in the parameter identification techniques to use on the Shuttle ALT program. He worked on the first dozen or so orbital flights of the Shuttle with us. Then, he had gone on to other jobs, some within the space station program.

The red team study was an intensive effort. We traveled all over the country and heard briefings for up to 10 or 12 hours. We tried to absorb and respond to those briefings and come up with the pros and cons of various aspects of the program. It was a very difficult, but intensely interesting, task. We did a lot of our work in Washington, D.C., and Dan Goldin knew where we were. The man was very busy, but occasionally he'd drop in and ask how we were doing. I was very impressed with Dan Goldin's energy and enthusiasm that seemed to start before six o'clock in the morning and go till ten o'clock at night. We gave him briefings at both of those times in various places around the country, and he was just as energetic and antagonistic at those times as he was at any other time. I admired his energy and his desire to make NASA all it could be and to start doing exciting things both in space and in aeronautics.

Having had some interaction with him before on the Moon-Mars program, I found that the interaction I had with him at the generic hypersonics briefing at Dryden in September left me unaware of how long we had been off the subject. I think a number of people at Dryden were very uncomfortable that I was arguing with the administrator when he was there to evaluate our work. Nevertheless, to finish the HALO briefing, he told me he wanted me to go to Washington at some near date and give him a full briefing on the advantages of reusable vehicles, rockets, and air-breathers, and he wanted me to be prepared to defend their use.

That was not my specialty, but I think he asked me to do it partly because of the association with the Moon-Mars team and my willingness to debate him on the issues off the subject. He said, "You tell me when you'll be ready to give the briefing—one month, two months, or three months." Although I think I picked three months, somehow I gave him the briefing in two months. But I've never worked harder or traveled more than I did during that time. It was a field I was familiar with, but in no way was I an expert in it. I had to pick brains around the country so that I would be in a position to give the administrator a very important briefing on the advantages of reusability. I went to all of the credible people that I knew. Some of them I knew from aeronautics, and some of them I knew from working on NASP. I knew them for their honesty, their expertise in their field, and their willingness to cooperate and to contribute whatever they could. I also had lots of contacts at Ames, Langley, McDonnell Douglas, Rockwell, and Lockheed, and with DoD people. I ended up with a stack several feet tall of documentation of what various people from the military and space thought about reusability and how we ought to demonstrate each technology.

I also picked the brains of the people around Dryden I had been working with on both the HALO and NASP programs. I used the NASP simulator to help me understand the pros and cons of some of these reusability concepts. On October 30, 1992, I gave Dan Goldin what was supposed to be a half-hour briefing, but it turned into an hour-and-a-half briefing. We stayed on the subject this time, but he had a lot of questions, and it took a long time to answer them. He's very engaged when you give him a briefing. Some people don't like that, but I love it. I felt I was talking to somebody who cared about what I was talking about. He may not have been agreeing with what I was saying, but he was at least agreeing that it was a topic worth hearing and debating. My basic bottom line was that we don't build any transport system to throw it away. Why would we build a vehicle to put things in space that we would throw away? We don't do that with airplanes, cars, or shoes. Why do we need to do that with space vehicles?

In my opinion, the outcome of that meeting was that Dan Goldin decided a major thrust within NASA should be in demonstrating the technology of the reusable launch vehicle (RLV) and in putting less attention on the expendable technology, such as the National Launch System (NLS) that was then a primary focus of NASA. This resulted in the formation of the Access to Space Committee. I was asked to be on that team. It was composed of some very good people from around the nation, mostly within NASA. It was an opportunity that was hard for me to turn down, but I had really burned myself out between HALO and justifying the advantages of reusable vehicles versus expendable vehicles.

Steve Ishmael was picked from Dryden to sit on the Access to Space Committee. He was a better representative than I would have been for many reasons. One of them was that he was a research pilot that I'd actually worked with in the simulator, looking at some of the reusable vehicles in the HALO program. Another reason was that in addition to understanding the key issues on orbiting vehicles, he had extensive experience in the operational aspects of exceedingly complex aerospace vehicles. Steve and I had a good relationship. He spent a lot of hours working on the Access to Space Committee over a six-month period. They were looking at single-stage-to-orbit-rockets, two-stage-to-orbit rockets, and a first-stage-air-breather/second stage rocket. They also were discussing what NASA should do so that we would be looking at the right technology and the right kind of program. As an ex-officio member, I got to participate in that discussion.

The X-33 and the Return of the Lifting Bodies

The outcome of that discussion, which I agreed with at the time and still agree with, was the X-33 program. When the program was announced, we knew there were three companies interested in it: Rockwell in Downey, California, McDonnell Douglas in St. Louis, Missouri, and Lockheed Martin at the Skunk Works in Palmdale, California. Within a phase one, they had a certain amount of time within a certain budget to flesh out their concepts. (Fig. 1)

Fig. 1. Three industry proposed SSTO vehicles on which they based their X-33 designs. From left to right: The Rockwell wing-body vehicle, the McDonnell-Douglas vertical lander, and the Lockheed lifting body. (EC95 43320-1)

The Rockwell people were looking at something like the X-2000 simulation that I was familiar with. The X-2000 simulator was the official stalking horse for the Access to Space Committee. The X-2000 was a wing-body configuration using the shape, aerodynamic, and aerothermodynamic databases from previous Langley studies with a hydrogen-oxygen engine model from Marshall Space Center. The Rockwell proposal was really a 1990s wing-body version of the Space Shuttle as a single-stage-to-orbit fully reusable vehicle. The X-33, of course, was a demonstrator, and it wasn't planned to go into orbit.

The McDonnell Douglas group was looking at a vertical lander. It was basically an advanced version of their DC-X, which was, in my opinion, the most successful program that I've seen in the last 20 years. The DC-X program was done on budget and on time. They did a difficult problem, and it all worked. However, it failed in the end after it had met most of its goals because one of their inexpensive parts broke. I had worked with McDonnell Douglas throughout the DC-X program which had been led by one of my favorite astronauts, Pete Conrad. I worked with the people at McDonnell Douglas on the analysis of their DC-X flight data and on planning the 180-degree rotation maneuver. They were really a good group of guys to work with.

We worked with all three contractors, assisting and evaluating anything that they cared to share with us on the X-33, primarily the aerothermodynamic and aerodynamic databases that included the mass properties and the propulsion system. We'd get them on the simulator and design a simple control system for them. Steve Ishmael would fly them, John Bresina did all of the simulator implementation, and I was helping to evaluate them. I was trying to pass on to Steve many of the things I had learned working on the Shuttle and how different that was than flying high-performance aircraft.

It was a lot of fun as well as a lot of work. I didn't actually favor any one contractor, but the one I found was easiest to work with was McDonnell Douglas. Peter Hoag, who had been one of the Air Force lifting-body pilots that I had worked with in the 1960s, was the primary manager I worked with at McDonnell Douglas. The people at McDonnell Douglas challenged us to critique what they were doing and give them feedback, which we did. They gave us responses to every criticism we had. They were a technically excellent group of people, very well organized and very intent on their proposal.

My biggest problem with their proposal, then and now, is that it requires the pilot to turn the engine on at precisely the right moment after it's been in space and through a reentry. That engine must start as it nears the ground. It can't start too early or the vehicle has to carry too much weight in fuel. However, if it doesn't start right when it should, the vehicle sink rate will be too high when the vehicle gets to the ground, resulting in a crash—the worst possible failure at the worst possible time.

Even though I did not let that bias affect me in any way, it still remained my largest concern. A configuration change near landing—whether it's deployment of a wing or a rocket or starting an engine—seems like a bad idea when a dead-stick landing is an easy thing to do without a configuration change, from both the physics and the piloting point of view.

We worked a lot with the people at McDonnell Douglas. From what we could tell, they were running way ahead of the other two contractors in terms of their activity and refinement of their design. Rockwell also shared their databases with us, but their design was much like the X-2000. They were basically doing a 1990s design of a refined Shuttle using a single Space Shuttle main engine (SSME). This design was easier than the others to do in the conceptual competition, because they were starting with the Shuttle, and we already knew what it would and wouldn't do. We worked with them quite a bit, too.

The one we worked with the least was Lockheed Martin. We knew the people at Lockheed Martin were working on a lifting-body configuration, but I don't think that they really wanted us to help initially. It was the Skunks Works doing it, and the way the Skunks Works succeeds is by doing the work without a lot of interaction with either management or outside evaluations. However, I think they knew that, to have a viable proposal, it needed to be on an equal footing in terms of evaluation. Consequently, they also supplied us with databases.

We did the same thing with all three configurations. We used our simulator to take them aloft and to orbit; we also brought them back and landed them. We reacted with each of the contractors as to the pros and cons of our evaluation of what they were doing. They made modifications to correct some of the things pointed out in the evaluations, while with other things, they felt we were not right in our evaluation. One of the toughest things for me was to make sure that I didn't share anything that I had learned from talking to one of the three competitors with any of the others. As a long-time NASA employee, I've been in that position many times. Sometimes I don't give them an answer right away, because I have to think, "Where was I working on this particular problem last, and was it with one of their competitors?" If it was, I can't tell them about it. I can only react as if I hadn't had that experience before.

Once that early conceptual phase ended, each contractor put in a proposal for what it wanted to do with the X-33 and what its single-stage-to-orbit vehicle would be. I was asked to be an evaluator in several areas, but I refused because I was contaminated from having worked with all three of the competitors. I had a lot of biases by then in regard to which proposals I liked and which companies I thought had the best approach to making the X-33 a success.

Eventually, the Lockheed Martin proposal was selected. I thought it was picked for all of the right reasons. I guess the Lockheed people had a very good business plan, which wasn't something I was very interested in, but they were doing something that I thought, if they succeeded, would give us the best set of options. They proposed a metallic thermal protection system much like a very advanced X-15. This system has a lot of advantages in terms of operability, maintenance, and resistance to damage. They picked the aerospike engine, which had never been flown, although the design concept had been around for 30 years. They also picked the lifting-body shape. Therefore, after the X-33 had flown and validated the design, the person or the company to actually propose a money-making single-stage-to-orbit operational vehicle would have two choices for the thermal protection system—ceramic tiles like the Shuttle had or metallic; two choices for the type of engine configuration—nozzle or aerospike; and two choices of aerodynamic configuration—wing-body or lifting body. In all three of those cases, a good argument for either choice can be made. They were all viable concepts.

One of the advantages of an aerospike that doesn't get mentioned much is that it's very good at distributing the loads and carrying them throughout the configuration, especially with a lifting-body design. A lot of extra stiffening and structural weight doesn't get wasted with the aerospike integrated with a lifting-body configuration. I found this to be an opportunity to return to where I had been in 1962, only working this time on a lifting-body configuration with Lockheed Martin. I was very active early in the program. I was, however, very disappointed in the schedule and the money given for the X-33 program. The NASP program never had a chance of success because the physics and the attending complications were far beyond our capability to resolve. Any of the X-33 concepts were certain to fly, if given adequate time and budget.

I thought the X-33 was at least a five-year program requiring probably four or five billion dollars to find out if the technologies were good, there being lots of issues. All of the proposals relied heavily on composite materials, and we didn't have much experience with them in these flight regimes. I also thought the program was very underfunded, the budget way too small to do anything anywhere nearly as complicated as this program. I think the Rockwell proposal perhaps could have accomplished the intended goals, because their design was really a second-generation wing-body and they were going to use the SSME rocket. The Rockwell proposal was one that might have actually succeeded in the three-year period.

It was very exciting to get back to work on the lifting body, using new materials and a new shape. This one really had quite high volumetric efficiency, but suffered from the same problem that the other lifting bodies had in the 1960s and 1970s. That problem was the packaging of all the engines, systems, fuel tanks, and payload into a configuration that does not have its center of gravity too far aft. The people at Lockheed Martin struggled with that problem. From the aerodynamic and aerothermodynamic point of view, they closed the design. It didn't have the fuel fraction (the fraction of the total weight of the vehicle that was the hydrogen and oxygen) that they wanted, and they didn't have time to do a couple of iterations to get maturity in the structural and the aerodynamic design. They had to finalize the mold lines and the weight long before they had a chance to get the design down to where it was going to be a viable program to fly Mach 15. Dryden, in an attempt to assist Lockheed with the X-33, modified an SR-71 [90] to fly an X-33 shape to predict the aerospike characteristics. (Fig. 2) This proposal, called the linear aerospike SR experiment (LASRE), was flown and was partially successful.

Fig. 2. The SR-71 carrying the LASRE experiment to investigate the in-flight characteristics of the aerospike engine. (EC98 44509-15)

The people at Lockheed Martin struggled with several technologies, but the one that killed the program, even though it should not have killed the program, was the failure of the composite hydrogen tank. That was probably a big mistake in the program objectives in wanting a composite tank, because the time was so short, and the technologists designing large cryogenic tanks made out of composites had absolutely zero experience with large composite structures. To expect people to solve that problem and fly it successfully in a short period was totally unrealistic. However, the vehicle still could have been flown with a hydrogen tank fabricated from aluminum rather than from composite material. Of course, it would not have demonstrated the cryogenic capability of the reusable composite hydrogen tank. However, I'm not sure that a few flights would have made the composite hydrogen tank a good enough option that someone trying to build a single-stage-to-orbit RLV would have used it anyway.

Even though I give the people at Lockheed Martin high marks for advancing many technologies simultaneously, I think the composite tank decision probably was a time bomb in the program. NASA was not willing to take the hit both monetarily and in scheduling to let them do the vehicle with an aluminum tank. For the most part, the X-33 was built and in the hangar, waiting for the hydrogen tank, scheduled to fly six months after delivery. Although there were still things to be done before it could fly, it essentially was built, as opposed to the NASP which was never even completely designed. I think the Lockheed-Martin version of the X-33 would have been very successful, and we would have learned a lot about the lifting body shape, the aerospike, and the metallic thermal protection system. We didn't get to learn about those because it was decided to terminate the program after the failure of the hydrogen tank.

It may seem as if a single-stage-to-orbit rocket vehicle is a digression from my commitment to going from runway to orbit, but it really isn't. We can take the technologies developed on a single-stage-to-orbit rocket-based RLV and use that as the second stage on the air-breather. All of the technologies that we'd been looking at on the X-2000, the various X-33 proposals, and the Lockheed Martin orbital version of the X-33, the VentureStar, were applicable as well on the second stage of a runway-to-orbit vehicle. The people at Lockheed Martin were not solving the scramjet problem, but they were looking at more concepts for what that second stage could look like, and it could have looked like a wing-body or a lifting body. In working with Lockheed for four or five years, I found a lot of the problems we'd had with lifting bodies in the 1960s and 1970s reappeared, but Lockheed managed to solve many of them. I think it's regrettable that we didn't get to see how that configuration did through a full hypersonic test to be able to understand the design.

Another return to lifting-body technology that occurred in the late 1980s and the early 1990s was getting to work a little with Langley on a proposed space station crew rescue vehicle, the HL-20. The HL-20 [91, 92] had some very nice characteristics which Langley had refined somewhat and showed, at least in the wind tunnel, the ability to land it horizontally with an L/D max in the area of three and a half to four. It also had good hypersonic characteristics. Langley was proposing it for two possible functions: first as the rescue vehicle, then called the assured crew rescue vehicle (ACRV), and second, as a shuttle for human pay-loads. The ACRV would function as an emergency vehicle that would leave a space station as a life boat to get the crew back to Earth. If there was a fire, an explosion, or a collision of some kind on the space station, the crew would be able to hop into the HL-20, get away from the space station, deorbit at the proper time, and land virtually any place on Earth within a day or so.

Langley worked jointly with the Johnson Space Center, and I thought it was a really good idea. It was another opportunity to return to looking at yet another lifting body. It shared a lot of characteristics with the X-24A, both aerodynamically and in its lack of attractiveness, but they were challenged on their design, because for a true rescue vehicle, we needed to get back much faster than a day or two. Langley looked at using the HL-20 to get the cross range that we needed, but once it was subsonic, it would come down on parachutes. Langley showed that it would be able to land virtually any time on any orbit. That was the essence of the assured crew rescue vehicle. I'm not sure what the reasons were for it dropping from favor in the early 1990s. However, Langley was unable to get the funding to go further. Of course, if Langley had gone ahead, there probably would have been a lifting-body test program—very similar to the M2, HL-10,

Fig. 3. The X-38 vehicle shown shortly after launch from the B-52. (EC99 45080-24)

and X-24 programs that we'd done in the 1960s and 1970s-to check out the landing characteristics and the aerodynamics. There also probably would have been a hypersonic unmanned program to show that the HL-20 could return from orbit much as the X-23 had. At any rate, it didn't happen. Langley also proposed to sit it on top of a rocket booster with a crew in it and use that to transport crews to and from the space station. Those were good proposals, and the HL-20 looked like a pretty good lifting body, especially from the performance and volumetric efficiency point of view. The people at Langley worked hard on it, and they wrote some technical reports and papers on it. But it's sitting there on the shelf waiting for a user to come along and come up with a program for it.

In about 1992, we were contacted by the Johnson Space Center, which was also advocating a crew vehicle, either for transporting people back and forth to the space station or as a rescue vehicle. The people at Johnson had been jointly advocating the HL-20 ACRV in the late 1980s with Langley. They were looking at lifting bodies and wanted our input as to what they should use. We had a telecon with the Johnson people who were proposing this program. There were a few of us in the room. The ones I remember being there with me were Ken Szalai, Dale Reed, and Milt Thompson. The people at Johnson wanted our reaction to their proposal to use a lifting-body configuration as a rescue vehicle and to bring it in on a parafoil. A lifting body flies fairly well once it gets subsonic, so I didn't think that the addition of a parafoil was the best idea, but that wasn't the question they were asking us. They were asking us what lifting bodies, based on our experience, we would recommend. All of us in the room picked either the HL-20 or the X-24A, primarily because of their hypersonic and subsonic L/D and their volumetric efficiency. The X-23 had flown from above Mach 20 down to Mach 2 successfully, and the X-24A had flown from the B-52 up to Mach 1.6 and back to land successfully. Based on this demonstrated capability, they would probably want to choose the X-24A. That's more or less what they did in picking the X-38 shape.

I continued to work off and on with the X-38 team. They've modified the configuration somewhat from the X-24A. An unmanned version of the X-38 was launched from the B-52 several times, where the technologies of deploying the largest parafoil ever made was demonstrated. The X-38 looks like an X-24A except for the very back end of the vehicle. Following its launch from the B-52 (Fig. 3), the X-38 briefly flew subsonically, then it deployed the parafoil (Fig. 4) and demonstrated the landing.

Fig. 4. The X-38 gliding under the parafoil as it maneuvers in preparation for landing. (E01 204-70)

335

My contribution to the X-38 has been very small compared to my contribution to the other lifting bodies or the Shuttle, but it was fun to relive my youth by once again participating in a flying lifting-body program. I've worked with some of the people at Dryden who were involved in the X-38. In fact, after each of their flights, I have made my best attempt to extract stability-and-control and performance information. I've also been involved in several independent reviews when they make a major modification or try to resolve an issue on the X-38. However, they didn't fly very often, and the flights were fairly short. I think the amount of analysis that I've done and fed back into the project has been appropriate. The new generation of engineers both at Johnson and Dryden have tried to learn a lot of these things themselves.

Late in my career at Dryden, I've managed to return to lifting body work with the HL-20 that was designed and not built, the X-33 that was built but never flown, and the X-24A redesigned somewhat and redesignated as the X-38 that was designed, built, and flown. It's fun to watch other people working on these things and learning some of the same lessons that we learned in the 1960s. Its also fun to watch them demonstrate new technologies such as the parafoil. At the end of your career, it is nice to return to your beginnings. I'm very happy to have been able to do that.

Reflections

That brings me to the point of looking at where we have gotten in my hope of going from runway to orbit in my career. We've demonstrated on various lifting-body shapes from Mach 2 on down that they're a good reentry vehicle, based on the technologies of the 1960s. We've demonstrated on the X-15 that a winged vehicle can reenter and obtain hypersonic data and data essentially outside of the atmosphere with the reaction controls and also at quite high dynamic pressures of over 2,000 pounds per square foot. We demonstrated the Space Shuttle going from Mach 28 down to doing a precise landing at Edwards or Kennedy. We've solved the shape problem. We've had one shape successfully look at the whole flight envelope—the Shuttle. We have every reason to believe that the X-24A-type shape will work as well, and, based on predictions, so should the X-33 shape.

The thing we still haven't gotten in my career is the first piece of scramjet data, and I've been working for NASA for over 40 years. One of the first things I worked on was preparing to look at various ramjet/scramjet technologies on the X-15. Some were based on plans of flying a delta-wing X-15 configuration that was powered by a scramjet engine and would be able to maintain air-breathing flight at Mach 5 or so for some extended period of time. Neither of those things have happened. The dummy ramjet that burned off in 1967 marked the termination of all such programs. From all of the other proposals for air-breathers—the X-24C/NHFRF proposals, the Pegasus proposals, the HALO proposals, and all of the work that had been done on the NASP— we still do not have one single data point on a scramjet engine. The scramjet would deliver the required propulsion needed to get the air-breathing vehicle up to the Mach 8 to 10 region to then deploy a second-stage rocket that would be either a lifting body or a wing-body, but the piece we haven't done yet is the scramjet.

Since there has been no NASA flight demonstration of a scramjet, it will probably not be relevant to any near-term attempts to go from runway to orbit. In my opinion, it will take at least 10 years, and probably longer, from the time of a meaningful flight test of a fully functioning scramjet before it can be included in any RLV concept. To have any benefit from a scramjet, in addition to getting the required thrust, we must learn how to deal with the tremendous amounts of heat that will be generated over the entire vehicle, and especially over the sharp inlet cowl lip at the Mach numbers and dynamic pressures required to produce sufficient thrust to accelerate the vehicle. This extreme heating environment will challenge the designer in structural concepts and in developing viable new materials. Any meaningful near-term concept will need to rely on turbojets or ramjets for any air-breathing concept to go from runway to orbit.

Epilogue

Stairway To Heaven

The programs on which I worked during the 40 years of my career at Dryden used two different paradigms. The old paradigm started with the with the air-launch of the X-1. That paradigm continued with the D-558, the X-2, the X-15 rocket airplanes. Each mothership—the B-29, the B-50, or the B-52—took the rocket vehicle to a launch point such that it would be able to land safely, whether or not the rocket worked. The initial envelope expansion was done by starting out unpowered and doing drop tests to see how the vehicle would perform. Then, we moved to a low-speed rocket-powered flight, followed by analysis of the data from that low-speed regime. We then resolved any issues and identified the areas where we were less certain of what might happen. Once they were understood, we went to flights at a higher speed or altitude with launch points farther out. Little by little, in this way, we expanded the envelope where the rocket aircraft was capable of flying. (The exception was the Shuttle ALT program, which made only glide flights from the 747 SCA.)

Once we understood the trouble-free parts versus the marginal parts of the flight envelope and identified the areas where we did not want to fly, we proceeded with the research programs to understand the aerodynamics, the aerothermodynamics, the structural characteristics, the materials characteristics, and the flying qualities of the vehicle, including the flight dynamics. That was the paradigm I was used to when I came to Dryden in 1962, and that paradigm had been very successful.

The New Paradigm

The other paradigm really started in 1961 with Yuri Gagarin and the first manned spaceflight by the Soviet Union as well as the first manned flight to orbit the Earth. This flight resulted in a manned space race between the United States and the Soviet Union. This new paradigm has been very successful, but it is very different from the old paradigm. The new paradigm is more for achieving a near-term goal than it is for achieving goals in scientific or engineering research. However, the new paradigm may have included research goals, as with today's International Space Station and some of the Shuttle flights. The ultimate test and demonstration of the new paradigm occurred after President John F. Kennedy declared in 1961 that the United States would go to the Moon and return safely within the decade.

As I've pointed out earlier, the process of getting to the Moon was really one of making things much more robust and of building larger rockets including ones that used hydrogen and oxygen. However, developing those engines and the components that went with them was analogous to what we'd done before. The design had many very competent people studying the various tradeoffs. I think that the way they chose to go to the Moon, with lunar rendezvous and returning the command module to the Earth, probably was the optimal way to do what Kennedy had challenged the nation to do. Many of those ideas and concepts had been floating around for several years by then. But it was Kennedy's statement that focused us on getting it done, to making the decisions that needed to be made in a timely fashion to meet his goal of going to the Moon and returning to the Earth safely within the decade. The technologists that needed to work on this space project had been involved in the missile programs that started in the 1950s. Of course, we also used the heritage of the V-2 rocket that the Germans had developed, as well as some German engineers and scientists headed by Wernher von Braun. (Fig. 1) We expanded beyond the Atlas and Titan ballistic missiles into the Saturn V booster. In addition to making bigger rockets, we also had to make the rockets and the spacecraft exceedingly reliable. We had no way rescuing anybody on the way to the Moon or on the way back, so we needed to improve and document system reliability. However, the basic physics of going to the Moon and back had been understood for some time. Of course, this was

Fig. 1. Wernher von Braun in an X-15 during the early 1960s. (E-7608)

enhanced by computer technology, even though that technology was still very primitive in the 1960s.

With the great success of Apollo, which was probably the crowning achievement of the engineering world for the twentieth century, many people felt that with enough money and enough skilled people, there were few things that we couldn't do. The new paradigm was continued in the development of the Space Shuttle, which used some technologies that hadn't been proven before. The process was quite different from the step-by-step old paradigm. To build the Shuttle, we needed to do a great deal of testing in wind tunnels and to a lesser extent in those days, studies with computational techniques. The experience of all of the people who had been involved with rockets, hypersonic aircraft such as the X-15, and the space capsules were put together to design a space vehicle that was reusable, launched from the Earth, and return to land horizontally.

The old paradigm was broken between 1975 and 1981, and it was broken for good reason. If the objective of the United States was to put crews and payload into Earth orbit with a vehicle that would return and land and then be used again for future flights, there was no way we were going to reach that goal in a few years with the old paradigm. It would have taken two or three times as long to do

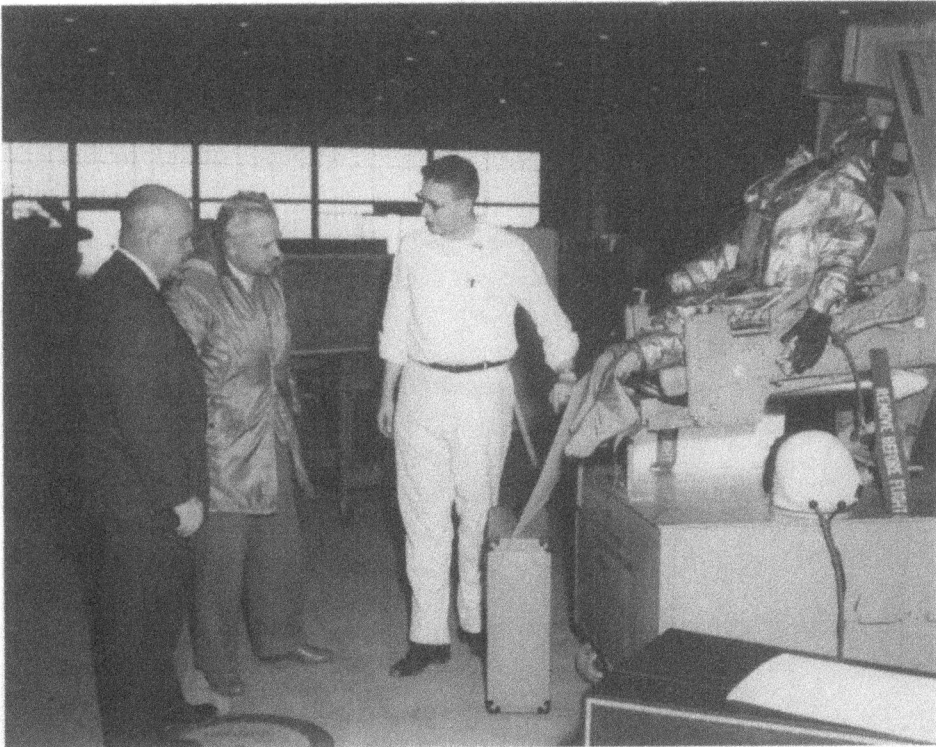

Fig. 2. Dryden Director Paul Bikle, German rocket pioneer Eügen Sänger, and engineer Garry Layton study the X-15 ejection seat during discussions at the time of the X-15 and X-20 programs. Sänger's proposals in the 1940s led to the concept of the X-20 Dyna-Soar. (E-6524)

that with the old paradigm, and it may or may not have been more expensive; it would be difficult to determine at this point whether one way might have worked better than the other. Nevertheless, there were hints that the old paradigm was going to be broken anyway, even for aeronautical research. The X-20 Dyna-Soar (Fig. 2) was to be the follow-on to the X-15, and it would provide aerodynamic and aerothermodynamic data up to orbital speeds. However, there was no known way, at the time it was being proposed in the late 1950s and early 1960s, that we could have conceived of a way of building a fully reusable booster. The X-20 thermal protection system, the engines, and the aerodynamic shape forced the program to use a vertical rocket launch with a Titan III booster. Even though that never occurred, it did show that, independent of the capsule world, the old paradigm was starting to run out of steam by the 1960s.

I believe that because the Apollo program was driven by the manned space race between the two giant world powers as a demonstration of military might and technical capability, going to the Moon happened forty or fifty years sooner than it would have happened had we continued with the old paradigm of going beyond the X-15. With the old paradigm, we would have learned first how to build a fully reusable vehicle to get to orbit and back, and then we would have used it to build the lunar vehicle in space. If we had used the old paradigm, I believe that we probably would still be waiting to go to the Moon. We might have

Fig. 3. The Gossamer Albatross preparing for a sunrise flight at Dryden in April 1980. Shortly after the manpowered Albatross successfully flew across the English Channel, it was brought to Dryden for an instrumented flight test program to evaluate its lift, drag, stability and control, and thrust characteristics. (ECN-14339)

Fig. 4. Helios vehicle being flight tested at Dryden in preparation for its record setting altitude flight in Hawaii. (EC99 45285-5)

succeeded in landing a man on the Moon by 2010 or 2020, using the old paradigm. That's not based on any detailed study, but knowing how slowly the old paradigm proceeds in terms of far-out objectives, I believe that it would have been in that neighborhood. On the other hand, had we kept to the old paradigm and taken that long to proceed, we also would have learned a great deal just in getting there. We would have learned more about materials, about operational considerations, and about reducing the amount of oversight that needed to be done on the vehicle. We also would have continued to make the flights progressively cheaper, safer, and more trouble-free.

The old paradigm is more analogous to the way aviation proceeded from the era of the Wright brothers up to the B-2, the F-22, the Boeing-777, and the Airbus 380. It proceeded based mostly on commercial or military utility, tweaking each concept and improving our capability with occasional quantum jumps. The old paradigm is not dead within NASA as a whole or Dryden in particular. I think a good example of that is the work done by Aerovironment and related groups with the Gossamer Condor, the Gossamer Albatross (Fig. 3), and then the various other vehicles up to the Helios (Fig. 4) that reached nearly 100,000 feet on solar power alone. That work by Aerovironment is a good example of the old paradigm of "build a little, test a little." Probably seven or eight aircraft were needed to learn how to build the Helios so that it stood a chance of reaching nearly 100,000 feet. Although the old paradigm is still with us, it's very slow-paced. Programs under the old paradigm don't have nearly the staff or the oversight that the more recent bigger programs (such as the NASP and the X-33) have that haven't done so well. The old paradigm is a series of evolving stepping stones, each based on the increase in knowledge provided by the vehicles in previous steps.

The old paradigm meant that we ended up developing a continuum within each of the technologies and in all of the disciplines that I consider part of building an aerospace vehicle. These would include—but not be limited to—aerodynamics, aerothermodynamics, propulsion concepts, flight dynamics, pilot/vehicle interface and handling qualities, and properties of materials. That is the lifetime of materials, how they fail, and how we make vehicles lighter by using new materials, and structural designs all being focused on getting the lightest, most efficient structure for a given concept. The structural design for a 747 is quite a different concept than the design for a reusable launch vehicle like the Shuttle. We use different techniques, both internally and externally in the Shuttle. Finally, the design gets involved in the types of control surfaces, including use of the reaction-control systems, the avionics, and the other systems. These will include the software, the computer management of systems, and the use of the computer as the primary interface with the pilot for the vehicle to fly from launch to Mach 28, then down to touchdown and landing.

Losing Our Way

In the first 20 years of my career, to about 1981 or 1982, many new configurations—some good, some bad-were being developed, tested, and evaluated. Decisions were made as to which of those configurations should and shouldn't be used. However, we have basically stopped working on a runway-to-orbit concept, since 1981. I claim that the paradigm was broken in 1981 by the first flight of the Space Shuttle. After that, we essentially didn't develop anything involving a new vehicle. The one exception is the Pegasus, first launched in 1990. That effort demonstrated that if the B-52 took off and launched the Pegasus at an altitude of about 40,000 feet, the multistage winged vehicle could put several hundred pounds into orbit. It has continued to do that, even after most Pegasus flights were switched to being launched from the L-1011.

There have been changes, but if I had to pick a date when I believe we started to lose our way, it would be 1981. I suspect that those engineers 20 years older than I am would pick an earlier date. I believe that the reason that the new paradigm replaced that old paradigm was the giant technological gaps needed to be bridged in a relatively short time for Apollo and the Shuttle. Their schedule did not allow time for the old paradigm to be a viable option. The new paradigm would put a lot of resources and money into taking a large step from previous experience. However, some problems are too difficult to solve, no matter what resources are available, such as controlled nuclear fusion or a cure for cancer.

The old paradigm was to build-a-little, test-a-little. Even though the old paradigm resulted in the major new airframes and rocket systems of the X-1, D-558, X-2, X-15, and the various lifting bodies, it involved working in a very conservative fashion. Because the air-launch gave us a full-flight abort capability, if anything was out of tolerance or not functioning properly, we could abort the drop with the vehicle still attached to the mothership, return to Edwards, fix it, then go back and try again. This was also true at any point after the vehicle was launched from the mothership. The pilot could dump fuel if the engine didn't start and always find a safe place to land, because that was a requirement of the old paradigm. Consequently, there were never any serious issues of losing the vehicle due to being too far from the parts of the flight envelope previously explored, analyzed, and understood. If a vehicle problem caused an abort, we would fix it. If it's due to data that we don't understand, we can analyze that information until we believe we understand it enough to mitigate the risk for the next flight. That is the crux of the old build-a-little, test-a-little paradigm.

When it came time to do the Shuttle, the funding was available to do it. We could not analyze related flight data, because none existed. We had only predictions on which to put tolerances and variations in convincing ourselves that there was a high probability that the Shuttle would be successful. However, it was more risk than we would have had with the old paradigm. It was also very expensive to meet the short schedule, for it involved about 100,000 people. We needed to have sufficient confidence that—even with all of these new flight regimes and a new type of vehicle—that there was a high probability that we would succeed in not only getting the crew back safely-but, almost as important, the vehicle as well. We could fix the problems identified on the first flight and return, for a time, to more of the build-a-little, test-a-little approach. We then could expand the envelope of the vehicle and improve the center-of-gravity placard limitations that were placed on payloads so that it would be of greater utility. A little bit of that occurred after the first flight in the Shuttle program. There was not much of the "build-a-little," but there was some "test-a-little," so that we could convince ourselves that we had a vehicle that was a viable concept.

Of course, the Shuttle has now been flying for over 20 years, but it is still a very complex system and a very complex way of getting people and payloads to and from orbit. It's quite a remarkable accomplishment. As a technologist I found that although the Shuttle was not as exciting as the Apollo program, it was more complicated to do. There's a lot of risk, and although I think we understand most of the uncertainties better than we did in 1981, the Shuttle is still exceedingly complex and its design only barely closed so that we could meet the goal of putting 50,000 pounds in orbit.

It is my belief that, as a result of the Shuttle, the old paradigm was mothballed for hypersonic vehicles, and not used again. We're still having problems achieving the next goal of getting low-cost payloads into Earth orbit because we have abandoned the old paradigm. In this context, low cost per pound is in the range of several hundred dollars per pound. The real goal is for space to be of more utility so that we can become a spacefaring civilization.

I don't mean to imply that we haven't had proposals for returning to the old paradigm. There were many approaches that would have returned us to the old paradigm, but none of them have gotten to the point where they were serious contenders for something that actually might be approved, funded, designed, built, and tested. The Pegasus-launched vehicles and HALO proposals were aimed at returning to the old paradigm. The X-33, as implemented, was being done with the new paradigm. The success of the Shuttle has moved the engineering decision-making from what used to be the technical leadership of the country into more of the political arena. And this is a sign of the times in many things other than just aerospace.

Early in my career, I had the opportunity to see the way technical leadership worked within this country, even though I was far too junior, of course, to be involved in it. The technical leadership was made up of people who had a strong feeling for technology as well as outstanding capabilities in management. (See Fig. 1, 2, 5) These people had risen through the aircraft companies, the military, or through the NACA and NASA. They had been technologists early in their

Fig. 5 Three key technical leaders shown next to the X-15 during 1964. From left to right: Center Director Paul Bikle, NASA Deputy Adminstrator Hugh Dryden, and former center director Walt Williams. (E-11146)

careers. They had worked their way up through management. They kept their keen appreciation for what could and couldn't be done. There were a large number of such people when I started in the 1960s. I didn't know most of them because I didn't need to know most of them as various ideas such as the X-15s and the lifting bodies got fleshed out, discussed, modified, and evaluated by them. By the time an idea had become a program that Congress needed to approve, it was well thought out, and a good program of significant potential.

Their conclusions were valuable; they mitigated risk, both technical risk and physical risk to the crew and vehicle. Their conclusions were well thought through, including a consideration of whether the step was too large technically. They also considered whether the step was too big to reasonably mitigate the safety of the crew and the vehicle. Those matters were well understood by what I referred to as the technical leadership. I used to see and read about that leadership. It was within NASA. It was within my own organization at Dryden. It was at North American Aviation, at Lockheed, at Northrop, at Grumman, at Edwards AFB, and at Wright-Patterson AFB. I saw that this kind of technical leadership was available wherever I looked.

When I talk about these technical leaders, I don't think that they, in themselves, had all of the information they needed because technology is too broad for most people to ever attain that. However, I think that as they rose in management through their organizations, they recognized who was good to get an opinion from and who wasn't. They remembered who knew what was going on in terms of technical and program risk, and who didn't, who understood the technology, and who understood the real details, the day-to-day things that are required for making a proper assessment.

By having such individuals for contact by visit, telephone call, or letter, they were able to form assessments of what the next step should be that far exceeded their own personal capabilities. However, they really did know the technology. They were smart, they knew the key technical people, and they had a good idea of what we should and shouldn't be doing. They also had their cohorts from whom they could request independent assessments. The people who were leading these efforts tapped into the heartbeat of technology and what was doable through contacts with very competent people throughout the industry, NASA, and the DoD.

I think that process enhanced their vision. Their vision was usually fairly ponderous and conservative, I think, because they knew that there weren't a lot of opportunities to advance technology and they needed to consider carefully what program would be next. On the X-15, for instance, proposals were given to a committee that basically drove the advocacy and the approval until there were reasonable estimates of cost, time, objectives, and the manpower requirements. I don't mean to imply that the technical leaders were never wrong or that they occasionally didn't discard some really good ideas. I just mean that a technical person who is proposing something prefers that disagreements are based on technical points that can either be won or lost. Sometimes you lose even if you are right, but at least you had the opportunity to state your case. If it had been done well enough, there was a chance that the program might happen. I think that after the Apollo and Shuttle successes, the process involving the technical leaders was largely bypassed because of the short-term concerns for public approval and nation's prestige.

The old paradigm's technical leaders were tremendously competent managers. I believe they were people who had vision as well as a good technical grasp of many areas. They may not have been experts, but they knew how to ask the right questions. I have given briefings to people in that category early in the Shuttle program and I always have been impressed by how they could ask a critical question even though they may not have understood completely what we were talking about. I think that is a talent. I doubt if it can be taught. I suspect it is something one is born with. One of my favorite examples is George Low. I don't think he'd ever worked in parameter identification and maybe he had never even heard of it before I gave him a briefing, but he definitely understood the concept. He drew me further into the mathematics and into the confidence in the answers that we had gotten. I don't remember many other managers asking me these kinds of questions since. Such visionaries were around. Some of them were NASA employees. Some of them were generals or DoD people. Some of them had spent their careers at various Air Force labs or in various aerospace companies.

For the last two decades or so, if we wanted to start a program, it had to be big and expensive, usually with fairly near-term results. It needed to be showy, and we needed to go right to the top for political approval. However, for real technology advancement, near-term results are usually not what we are interested in while doing a program. It's the longer-term result we're usually interested in. I think part of the reason that we get into emphasizing shorter-term results is that politicians are elected every two to six years, and they want to see results that they can claim credit for to help them get reelected. That's their job so I don't blame them for that.

In the current system, there's no credibility for how doable something is. Is it really going to happen? Or is it something that we will invest three or four years in, only to discover that it's not going to work? The other problem is that, since politicians are directing the effort, political whims over several years end up redirecting program goals and objectives. What we may have already built becomes inappropriate under the newest plan. It's very difficult to do something that's really hard and really complicated if we don't maintain the same goal. It's not that I believe that goals can't be changed, but we need to know what the impact of changing the goals will be. Frequently, changing goals is merely a consequence of a technical problem that we've encountered, or perhaps new technology has been developed and the way we were proposing to do something is no longer a good idea.

The current system, by its very nature, eliminates the establishment of informed and competent technical leaders and leadership simply because there's insufficient reward for being a competent technologist. Currently, decision makers don't necessarily have to have a technical grasp of what they are making decisions on. If we haven't been using the tried and true technical assessment of each step and developing the people who are capable of real technical leadership, they're not going to be there when we need them. I see a paucity of such individuals. Today, I don't know very many of them. In the 1960s and 1970s, I could have made a list of a hundred people that I thought would be good to assess large projects. Now I know of only a few people like that. However, that's because we haven't succeeded in doing anything difficult. We've tried and failed on several occasions to build and fly something.

With the political insistence on both short-term return on investment and unrealistic goals, I think that there is a perception that people are failing because they don't know what they're doing. Actually, we are failing because the job is too difficult; it is too big a step, and it's going to be impossible to deliver it on schedule for the given money. I think the bureaucrats and the politicians seem to be unaware of the extreme length of time required to build a flight vehicle for some of the programs that have been approved. Some of these programs keep going, such as the International Space Station, but with a great deal of rancor because the politicians feel that they were lied to. And they may have been lied to by people with wishful thinking and short-term goals who didn't have the technical vision.

Seeing projects fail, the bureaucrats and the politicians tend to eliminate or neuter irreplaceable laboratories and organizations, not realizing that it takes decades to build such laboratories, which are, in the end, concentrations of technical people. Eliminating or neutering such a group has irreversible consequences because once the people are scattered, the capability disappears. This practice always reminds me of the story of Napoleon's artillery forces shooting the nose off the Sphinx for target practice. It seemed like a good idea at the time to someone, but those of us looking at the Sphinx today really wish they had realized that the irreversible consequences of their target practice would disfigure the Sphinx forever. I think the country will eventually come to regret greatly the work of the politicians and bureaucrats who have been getting rid of or reducing the scope of the national laboratories in the last ten years. I won't claim that all of those laboratories have been completely destroyed, but a great deal of their value has been lost and can only be redeemed over long periods of time dedicated to recreating the capabilities.

A government laboratory is much more than just a collection of buildings and equipment. Rather, I believe, the primary component is the specialized team of people who work there, with their qualifications and varying levels of experience and knowledge. This team, the laboratory's major asset, will have many members who have spent most of their careers working on a wide range of engineering and scientific tasks. Because of this depth of experience, this team embodies the most important quality of such an enterprise, the corporate or institutional memory.

Although the term "corporate memory" probably originated in private industry, its presence there is limited to a corporation's experience and held, primarily, by only the inner core of the corporation's technical staff. When the term "corporate memory" or, more appropriately, "institutional memory," is applied to a government research laboratory, however, the term becomes much broader in scope, being more global than it can be in private industry. This global nature accrues from a greater variety of experience. It benefits from a greater interaction with other agencies, research centers, universities, and private industry, and it relies on the longer tenure of the technical staff. Perhaps most importantly, this institutional corporate memory extends throughout the entire technical staff, from the most elite research groups (the inner core of researchers and technicians) to the outer reaches of the technical staff (the typical journeyman researchers and craft specialists). This difference between the limited holding of corporate memory in industry and the global holding of institutional memory in government laboratories is the greatest strength of the government laboratories.

I believe that this institutional memory is fostered when employees in government laboratories conduct research and projects in partnership with other government laboratories, universities, and private industry. Over a career under such conditions, a government laboratory employee will gain extensive experience from working on various projects with a wide variety of partners. It's this accrued experience, the global nature of institutional memory, that is limited to government laboratories.

The greatest benefit of a government laboratory is that the specialized team, inculcated with institutional knowledge, is constantly performing research—pure research, applied research—tracking down interesting results and generally producing useful technical knowledge. This is the purpose of every government laboratory. Of course, knowledge unshared is knowledge denied, and government laboratories have the vital role of sharing technical knowledge. Because the laboratory has no proprietary ownership of the knowledge, this knowledge is made available to everyone on an even-handed basis. All of industry benefits, not just one company. It is not uncommon for a technique developed at a government laboratory to become the gold standard of the industry, used universally.

The institutional memory and research gave government laboratories great influence within the political environment. Because of the competency of the laboratories, the politicians and bureaucrats relied on them for assistance in deciding the direction of long-term research programs, including aerospace technologies. This reliance by the decision makers on the laboratories is another and higher form of the technical leadership that I used to see throughout the aerospace industry.

There's really no mechanism that I know of today within aerospace for starting a long-term, relatively high-cost incremental technology development program. I see it happening in some other high tech areas where there's investment in nano-devices and microelectromechanical systems (MEMs) devices, in lasers, in chip technology, or in molecular biology. Those are all long-term projects, but they work their way through the system, much as the people working on the Human Genome Project have done. As technology improved, the work schedule on the Human Genome Project shortened, not lengthened, because the people working on the project developed new computer and algorithmic capabilities that made it much faster to completely map the human genome. That can happen with other long-term incremental programs, too. When I criticize elected people for making technical decisions with the advice of organizations in the DoD and NASA, along with the technical staffs, I don't want to make it sound as if I don't think they're trying to get the job done. I simply don't think they know that there's a better way to go about it than the way that's been used for the last 20 years. They simply don't have the long-term benefit of knowing the way that projects were done before the early 1980s.

Going from runway to orbit, to me, would have to be done in an incremental fashion, eventually ending up with a very viable and robust capability of low-cost access to Earth orbit. We're not going to do it in one step, but I think the political reality is that they won't let us do it in any way other than in one step. Since we continue to fail, it's more or less a subject that's not even discussed anymore, because the politicians and the bureaucrats believe that we're unable to deliver. We are unable to deliver on the short-term, high-cost programs with unrealistic objectives that they have been approving. Some people within NASA and the DoD have been so overly optimistic that we really have lost our credibility with the politicians. The politicians, I'm sure, would like to do something that's very good for the country as well as good for them, but we don't give them much encouragement that that's going to happen, based on the last 20 years of failures.

One of the consequences of our failure on quite a number of high-visibility, high-cost programs is that we also have a work force that lacks experience in testing and flying real things. They have experience in advocacies, in conceptual design, and in exercising ground-test facilities in terms of materials, wind tunnels, or computational fluid dynamics or computational structural approaches, but they really haven't had the opportunity to take all of these designs and all of these predictions that their information says will succeed, then go out and try to fly it. They don't get to discover that things are not always as predicted or find out what they need to do to fix them. Engineers learn a great deal from little failures. They go back over them to understand, document, and write papers about them. Later, as they are training new engineers coming into the work force, they can explain some of these issues. This is what Hugh L. Dryden (Fig.5) was talking about when he said that flight research separated "the real from the imagined."

Another aspect is that failures have always made the news. It seems as if the news is always more bad than it is good, no matter what the subject may be. On the other hand, earlier, in the 1940s through the 1970s, I think that we were allowed to take both technical and personal risks and go do things. That doesn't mean that we were necessarily cavalier about it. It meant we studied it until we had included all of the information that we knew and all of the things that we understood, and then we tried to estimate the likelihood of success. Now, when we fail, we know we're going to get negative news coverage. That's just the way life has become.

In my view, something that's changed in the last 20 years is that news has become a form of entertainment. A lot of reporters are out trying to make a name for themselves. They do an investigation of a failure, then give a very superficial assessment of it. The result usually makes it look like somebody didn't do a good job. I don't think that this was as true of the media 20 years ago. What this change has done is to make people in management very averse to risk. They don't want to see their program, their name, or their organization on the front page of a newspaper or in a news broadcast with media coverage having only a superficial assessment of a failure. Management people have no recourse with the media for arguing what they did to mitigate the risks, or what the benefits were versus the costs. Management doesn't get to argue those issues with the media. People in management also don't want to see any of those negative media reports, so they tend to be very conservative. This turns into a no-win situation. If they're doing something that's not very challenging, they'll get criticized for not being very useful. But if they fail trying to do something harder, they get criticized that they don't know what they're doing. They're damned if they do and damned if they don't. I think that's too bad. I'm a big believer in freedom of the media, and I think that the media can provide an honest, independent assessment of what's going on put in terms that the average individual can understand. I believe that's a very good thing, because there are people connected to programs and technologies that would love to have no scrutiny whatsoever, and that is unhealthy. Right now, however, I think that media presentation is weighed too much on the entertainment side. If we do something well with even a little failure, only the small failure gets highlighted in the news.

The programs that I've discussed earlier in this book have had a significant amount of risk, both to the vehicle and to the pilot. However, the risk was mitigated by understanding what the hazards were and making a decision based on understanding those hazards and what their probability of occurrence was. In an incremental program using the old paradigm, the next test is only a short step from where we have been and does not require a large study of the hazards to proceed. Only when the hazards for the small step were assessed did they go ahead and fly the test mission. That's a very healthy way to do it. I think we always want to have safety as a top priority, but it isn't the only thing to consider. If we do only those things that are extremely safe, we're going to advance at such a slow pace that I don't think it'll even be noticed that we are advancing at all. Risk is part of the assessment, but it's informed risk that's used to mitigate the overall risk in a program, and informed risk is important. Once risk is discussed and assessed, once we know what the risks are and that there's some small probability of a failure, we go ahead and do it. Then if there is a mishap, we don't pillory the person who approved the test, because if we do that routinely, soon nobody will approve a test. Then, nobody will do anything. I would call it irresponsible for the media to try to be entertaining by keeping things shallow so that there's no real discussion of the technical issues. Media irresponsibility, coupled with the lack of technical leadership and the uninformed bureaucracy and the politicians making the decisions make up a major discouragement. It's not an optimistic model of the future for a country that hopes to continue to be the technological leader of the world.

Finding the Stairway to Heaven

My reason for discussing these issues is that we need to get beyond them, so that we can get back to doing interesting things and make a good assessment of the most economical and reliable ways of access to space and what its benefits might be. The best way to do that is to demonstrate various approaches and have the environment available for the capitalists, the innovators, and the visionaries to evaluate the results. I think we'll find that it's all very worthwhile, whether it's space tourism or whether it's manufacturing things in space that we can't manufacture on the Earth, perhaps learning to manufacture them on Earth after we've manufactured them in space. Usually if we eliminate a constraint on manufacturing, we find there are new products we can make without that constraint. In the case of space in a microgravity environment, as in the space station, there are things that we can make there that are very difficult, if not impossible to make with current technologies on Earth. There's not a long list of those, because scientists and technologists haven't had the time to experiment to see the kinds of things that might be produced in space. There are short lists of a dozen or so ideas that people have had in terms of medicines or materials that they'd like to manufacture in space.

Nevertheless, I believe that once we're permanently in space, there will be a huge explosion of ideas, concepts, and technologies that will be developed and perhaps manufactured only in space. We also might learn what the crystalline structure of the material is, then learn how to duplicate it on the ground. Space offers an exciting environment for developing new technologies without the constraint of Earth's gravity. In addition to that, of course, there's the manufacturing of things that will be used only in space. A lot of science fiction has been written about manufacturing in space or building spaceships in space to go somewhere else, where people spend years away from their home planet. Once we attain some critical mass, I think we will be manufacturing satellites and assembling spaceships to go to Mars or to the moons of Saturn. This is yet 50 years in the future, not something attainable in the near term. I think those are things that we need to do and we need to learn about doing, along with how we can attain reliable low-cost access to space.

I think the low cost and low risk is the step we're ready for. We know how to get into space. We've been capable of putting humans in Earth orbit for over 40 years. We've sent people to the Moon and back. We've built several space stations such as Skylab and the various Soviet space stations. Now we're building the International Space Station, which is still a fairly tiny step compared to what we need to do in order to fully utilize space. I also don't think there should be a huge rush to do this. I think it's more important to learn how to do it in a cost-effective and in an environmentally-sensitive manner than it is to do it in a hurry.

With that said, I can return to the topic at hand, which is figuring out a way to take off from a runway and end up orbiting a significant amount of mass, then returning and landing with a significant amount of mass. We've done little bits and pieces of that. My idea of how to do all of that is not new. I believe we need to go back to the old paradigm that I referred to earlier, incremental testing. To me it is sobering to realize that it has been over 35 years since Bill Dana flew the last mission of a fully reusable vehicle with the October 24, 1968 X-15 flight to 255,000 feet and a Mach number of 5.4.

We need to once again build a series of evolving vehicles that make us experienced and a lot smarter about how to do each next step. That means going back to the "build a little, test a little" old paradigm. It won't be cheap, but I'm not sure it'll cost any more than what we have spent in the last ten or fifteen years on programs that never flew, such as the NASP and the X-33. I'm guessing that to do what I'm about to suggest would be in the neighborhood of a billion dollars per year over a ten-year period. That is a lot of money, but I think that kind of money can be found.

Ways and Means for Low-Cost Spaceflight

I'd like to examine for a moment the two basic ways of getting into space—expendable launch vehicles (ELVs) and reusable launch vehicles (RLVs). Expendable boosters are the most compact way of getting a payload into orbit. There are three problems with ELVs, however. The first is that we lose all of the expensive equipment we have put in the expendable vehicle. The second is that, in most cases, debris is spread into Earth orbit that both satellites and manned vehicles must avoid. The third is that ELVs leave us without an easy means to return payloads from orbit to Earth.

However, there are some cases where I believe expendables are the right answer. I think Apollo was one of those cases, because developing a reusable capability to reach the Moon would have taken many years. I also would propose going to Mars in the next decade or two with expendables, much as Robert Zubrin has proposed in The Case For Mars. For lifting really big near-term payloads that cannot be done by the Shuttle, Ariane 5, or Titan or Atlas rockets, I think we need to develop a large expendable booster that will do that. The Russians have the Energiya with a significant payload capability, but it was launched only twice over a decade ago.

I'm not anti-ELVs. In fact, I believe that they're a stepping stone to where we want to go. ELVs have allowed us to go out and learn about the space environment, about orbiting, about reliability, and about the effects of spaceflight on human beings. They've allowed us to explore the rest of the solar system with probes. I don't think any of that could have been accomplished in the near-term without the expendable vehicles.

My proposal is basically to take a longer term view of how we might develop RLV technology, which would then be used to replace the expendable vehicles. I believe that will be cheaper and safer, and I also think that it is just the way that transportation systems develop. Transportation systems frequently start out with fairly hokey initial designs and sometimes even derided concepts that eventually develop into such things as the Wright Flyer, horseless carriages, and the early steamships. Nevertheless, those primitive beginnings were essential. I do believe that for the next 50 years there will be a need for the best ELVs that we can have in terms of reliability, cost, and safety. After that, we should be able to manufacture anything in space with supplies brought up by reusable vehicles, no longer needing ELVs.

Basically what NASP claimed it was going to do was what the X-33 later was approved to do: to investigate the viability in today's technology of a single-stage-to-orbit RLV. The X-33 program was also looking at technologies that could be used for reusable launch vehicles. The same technologies also can include a reusable vehicle on top of an ELV stack to carry either cargo or people from Earth to orbit and back. There was an HL-20 proposal to do precisely that. Only the vehicle launched from the Titan III booster would have been reusable. Some of the proposals for air-launch vehicles are derivatives of a Pegasus approach where the first stage-the most expensive one-would be reuseable. So it's not a matter of RLV or ELVs, rather it's a matter of combining the two. It's really a continuum from a single-stage-to-orbit, fully reusable vehicle to a reusable module on an expendable vehicle, and then to multiple-stage vehicles where each stage is fully reusable.

I think the result of the X-33 program showed that the single-stage-to-orbit chemical rocket (nonnuclear) RLV was going to be a minimum of three to four million pounds, of which 88 or 89 percent would have been the fuel. There was hope that the lighter-weight composite materials, the efficiency of the lifting body as an airframe, and the aerospike engines might produce a vehicle with a launch weight of two and a half million pounds or even as low as two million pounds. However, I think the X-33 also showed that if we used mature technologies to make sure that the vehicle was reliable and cheap, a three- to four-million pound vehicle is probably more likely. Probably the launch weight would be more than that, because we've never done it. The Shuttle is the closest that we've come to it, and at launch it weighs more (4.5 million pounds), and it's not fully reusable.

It's interesting to go back to some of the older studies done on reusable vehicles in the 1940s and early 1950s that used hydrogen and oxygen propellants. The number that kept coming up was that it would take about a three- to four-million pound vehicle to reach orbit and come back to Earth. Of course, that was before much was known about lifting bodies or space environments. However, since then, that number hasn't changed very much, which surprises me. It would seem that their weight estimates would be lower than our weight estimates, because they knew so many fewer things than we know now. However, it could be just a coincidence that those numbers remained much the same. There's really no technical reason why we can't build a three- or four-million pound vehicle. It's just that it's going to be a very large vehicle that uses only hydrogen and oxygen for fuel. Transporting it to the launch site would be very complicated. If it landed anywhere other than the launch site, it would be hard to get it back to the launch site. The vehicle would have large dimensions because of the liquid hydrogen fuel, and with a 90-percent fuel fraction, the empty vehicle would weigh at least 300,000 to 400,000 pounds. A single vehicle in the three- to four-million pound category is probably beyond what would be reasonable to do. It's not impossible to do, obviously. It could be done, but it would be very expensive, and it would be a huge vehicle.

The X-33 showed me that some of the weight savings with the extreme use of composite materials, including the hydrogen tank, may have been pushing technology beyond our experience, and we may not want an operational vehicle to use these lightweight composites until they have been demonstrated multiple times in an RLV environment. We just don't have much experience with composites in the space environment and in the launch and reentry environment. I believe that we still don't have the experience to use composites in the load-bearing parts of the vehicle, whether they are fuel tanks or the structure. I don't believe that would be commercially viable with the state of knowledge of composites, and I don't think we would realize the goal of several hundred dollars per pound to orbit that most of us would like to see for orbiting payloads. I may not be right, but that's my opinion, based on my experience in advocating quite a few concepts and working on some of them.

I believe that any practical RLV concepts in the foreseeable future will use multiple-stage-to-orbit vehicles. I probably can't come up with an exhaustive list of what those possibilities are, but the first stage could be an expendable stage, it could be a fully reusable and flyable stage, or it could be a stage that's recoverable by parachutes or parafoils. It also could be one of these more exotic magnetic levitation (maglev) accelerated rocket systems. I'm not saying they're good or they're bad, but I think there are simpler ways to investigate some of the longer poles in the tent than using more complex techniques. I believe a two-stage-to-orbit RLV is within our ability to accomplish, at least for modest payloads.

I would like to examine for a moment just two RLV concepts: one where both the first and second stages are rockets, and the other where the first-stage is an air breather, and the second stage is a rocket. A two-stage-to-orbit RLV, where both stages are rockets, has a couple of advantages. The primary one is that the two rocket stages would take the very lowest amount of energy and the least amount of time to attain orbit. It's just a matter of physics that getting the second stage to orbit would be done more quickly if both stages were rocket. This was what the original Shuttle was to have been, with both stages being flyable. We could argue that the first stage would be recovered with parachutes or a parafoil, and those are certainly possibilities.

There are three disadvantages to such a design, however, the first is not as difficult as the other two. Mating the two vehicles and placing them vertically for launch is more difficult than a horizontal launch. There have been many attempts and studies over the years to show that a horizontal launch would be possible with a rocket, but for the most part it's not a very practical or efficient thing to do.

The second disadvantage is the abort scenario. That is, once the first stage has been committed to launch, the second stage is also committed to launch. Any failure on the first stage probably means the loss of usefulness for both stages. That doesn't mean that they'll blow up, but it does mean that neither of those stages is going to get where they were originally intended to be. With an abort of the second stage, we still have the problem of where it comes down, and how we will get it back to the launch site.

The third disadvantage is that the second-stage separation usually takes place with both vehicles at a fairly steep angle with respect to the ground. Optimally, most of the second-stage rocket separations are in the neighborhood of Mach 3 to 5 with altitudes of 100,000 to 200,000 feet. Once the second stage separates, the first stage is burned out but flyable. It becomes a glider coasting up into the rare atmosphere. It must perform some sort of a reentry, and then it must try to return to the launch site. That's the hard part, because the first stage had all the momentum heading away from the launch site. In addition, it needs to have enough energy to glide back to the launch point. Putting engines on it, either jet or more rocket engines, makes the design of the first-stage less desirable. Unless the first stage is aerodynamically efficient—that is, a high L/D max supersonically and subsonically—the stage separation would need to be at Mach 2 or below to return to the launch site with any margin.

The other approach, which is an extension of the old X-1 to X-15 paradigm, is an air-launch with the rocket-powered orbiting stage mated to an air-breather stage. Obviously as opposed to the two-stage rocket just described, it is less efficient in terms of energy. However, the horizontal mating, the all-flight abort capability, and the recovery of both stages are advantages. In addition, the air-breather provides many more launch opportunities or longer launch windows by flying the air-breather under the orbit that you want to attain. With a vertical rocket first stage, you have to wait for the orbit you want to be over the launch site to have a launch opportunity. In the case of the X-15, the air-breather stage was the B-52. We know we can make larger launch vehicles like the 747/SCA, which launched the Shuttle ALT program. It has the capability of launching 250,000 pounds.

I wanted to state what the next two reasonable approaches would be for reducing the cost of orbiting a payload by using a two-stage vehicle—either the two-stage rocket or the air-breather/rocket combination. But I think it's now time to return to the old paradigm and see if we can successfully do an envelope expansion and get the technological continuum needed to define a commercial money-making project for low-cost access to space. This would be regardless of whether the first stage is a rocket or an air breather. I believe my approach is based on the historical development of demonstrated aerospace technologies. I think this approach is every bit as valid today as it was before, if not more valid.

My preference, of course, is to return to the old paradigm and use the approach of envelope expansion of build a little and test a little with a series of air-launched rocket spacecraft with ever-increasing velocities and fuel fractions until we know the strengths and weaknesses of a wide variety of configurations. Then we will be able to make informed proposals with a high likelihood of realizing low cost per pound to orbit. In addition, the same approach would allow the opportunity to try to mature the scramjet, both the combustion technology and the extremely hostile thermal environment, to see if it will offer a more attractive first-stage air-launch vehicle in the more distant future. However, we don't need a scramjet for the first stage air breather concept to be viable.

The detailed description of how best to attain low-cost payloads to orbit—and, ultimately, to further manned exploration of the solar system—is beyond the scope of this narrative. However, such an approach can easily be laid out by using the old paradigm. It is a characteristic of the old paradigm to provide a steady incremental progression of aeronautical research vehicles. Their success has been based on a determined understanding of the vehicle configuration characteristics, both external and internal, and the flow regimes in which they perform. This formula has been backed up by a knowledgeable leadership and technical staff that was careful, confident, and on guard against unsubstantiated claims and unwarranted enthusiasm. To mimic the past, we should proceed with a long-term project to conceive, design, build, and flight-test a wide variety of promising second stage rocket powered RLV concepts that contribute to low-cost access to orbit. This should be done with air-launched vehicles in an incremental envelope expansion that results in separating the "real from the imagined." As we proceed through the flight validation, the successes and failures of the various concepts will determine the next set of concepts to be flight-validated until it is known which RLV architecture will provide a viable, profitable approach for low-cost access to space. This approach will only work with a national commitment to a long-term research program.

After fleshing out and incrementally flight-validating the concepts—including the shape, propulsion system, materials, structural concepts, payload type, and size—and therefore knowing the valid concepts, we can return to the question we have been examining. That is, which one of the known and by then proven concepts—a single-stage-to-orbit system, an all-rocket two-stage-to-orbit vehicle, or a air-breather-first-stage/rocket-second-stage concept—is preferable? At that point, the choice and implementation of the concept becomes another incremental step in defining a routine low-cost-to-orbit architecture. In addition to having successfully flight-validated the various configurations, we once again will have created an experienced workforce of engineers, technicians, and managers to venture into and beyond any of the ideas I have talked about in this book.

Of course, my hope is that when this process is finally completed, the result will be the old paradigm's vision of going from runway to orbit. That would initially be a two-stage vehicle with an air-breathing first stage and a rocket-powered second stage for carrying the people and cargo into orbit. The first stage would be the descendant of the B-29s, B-50s, and B-52s used for the first generation of rocket planes, and of the SR-71 or 747 we had hoped to use to launch the HALO. The second stage could be either a lifting body, the descendent of the different concepts I worked with over the past four decades, or a wing-body concept, an improved version of the Shuttle and such later concepts as the X-2000 or the Rockwell proposal for the X-33 program. Ultimately, farther in the future, we would like to learn how to design and fly a combined-cycle rocket/air-breather, single-stage-to-orbit vehicle all the way to orbit. That was also NASP's goal, but in order to achieve it, we must devote as much build-a-little, test-a-little effort as needed to succeed finally in the ultimate demonstration of going from runway to orbit.

Whatever technologies emerged during this development process, we would know at the completion of this process how to handle all of the technical and safety issues. Solving these issues would be hard, but the country needs to keep doing things that are hard! If we don't keep doing things that are hard, soon all that any of us will be able to do is make presentations and listen to presentations, and we won't know how to do anything. Then some other nation will take up the idea of going higher and faster, and that nation will capture the low-cost commercial launch market as well.

Appendices

Appendix A: Various Terms Used For Pilots

Throughout the text I have used the terms "pilot," "test pilot," "research test pilot," and "research pilot." All of the Dryden pilots are research test pilots, who during envelope expansion are test pilots, which require skills of anticipating and reacting to unexpected and sometimes treacherous aircraft characteristics. Once the vehicle and its flight envelope is understood, then the pilot uses a different set of skills that require obtaining high-quality data by performing precise flying at the required conditions. In this case, I usually refer to them as research test pilots or research pilots.

Appendix B: Names of the NASA Dryden Flight Research Center

The NASA Dryden Flight Research Center is the current name of the facility that is at Edwards Air Force Base. This facility has been known by various names throughout its history. It was originally called the NACA Muroc Unit from September 1946 until September 1947, then was renamed the Muroc Flight Test Unit, and was operated as a division of NACA Langley Laboratory. In November 1949, the facility was renamed the NACA High-Speed Flight Research Station, but was still part of Langley. It then gained its independence from Langley in 1954, and it was named the High-Speed Flight Station until 1959, when it became the NASA Flight Research Center. This name was used until 1976, when it then became the NASA Dryden Flight Research Center. In 1981, due to a NASA reorganization, it then became part of the NASA Ames Research Center, and was known as the Dryden Flight Research Facility. This lasted until 1993, when it regained its independence and was again called the Dryden Flight Research Center, which is the current name. In order to avoid confusing the reader, the facility is referred to as Dryden throughout the text, even though it was then known by one of the other name when many of the events described occurred.

When referring to Ames Research Center, I always mean the Ames Research Center that is at Moffett Field, California. I always refer to Dryden as the organization that is at Edwards Air Force Base, California even when it was an organizational unit under Ames Research Center. When I refer to NASA that includes the NACA, which was the predecessor of NASA. The NACA existed from 1915 until October 1958.

Appendix C: Three Views of Vehicles

In the text six lifting bodies, the X-15, and the Space Shuttle Orbiter are discussed that have flown at Dryden. The following three views show all eight of these vehicles to help the interested reader better understand some of the discussions in the text. (Reproduced from Reference 93)

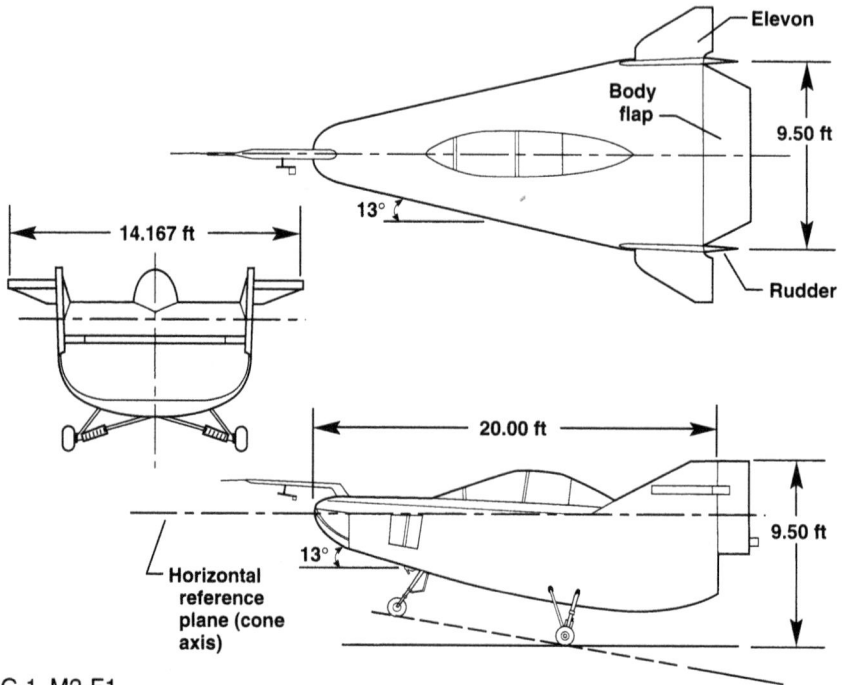

Fig. C-1. M2-F1

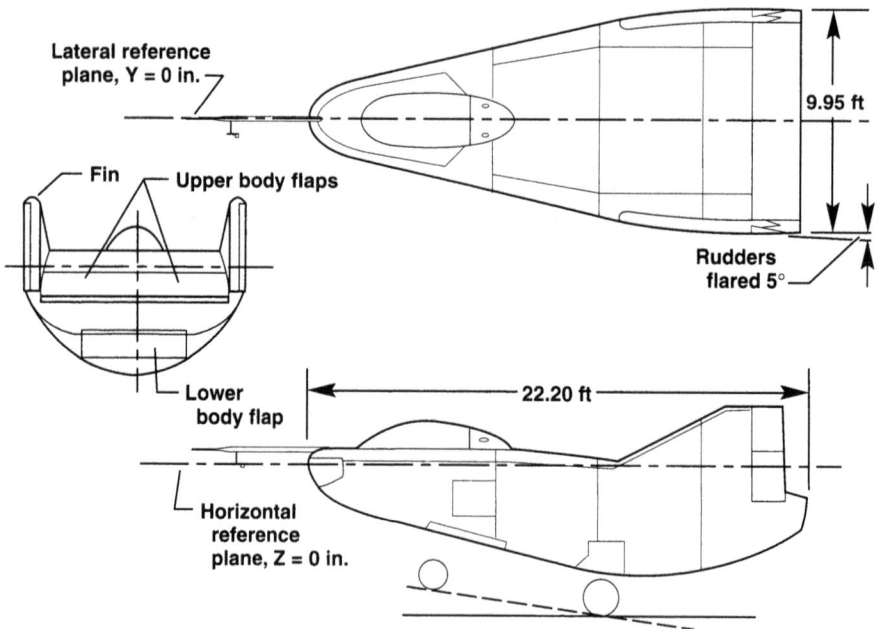

Fig. C-2. M2-F2

Fig. C-3. M2-F3

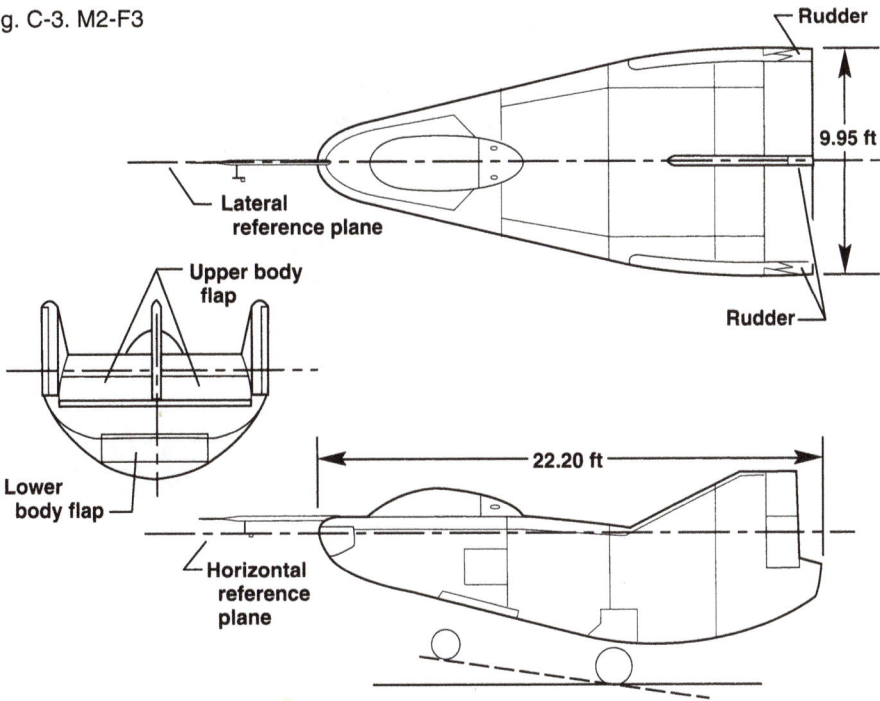

9.95 ft

Rudder

Lateral
reference plane

Upper body
flap

Rudder

Lower
body flap

22.20 ft

Horizontal
reference
plane

Fig. C-4. HL-10

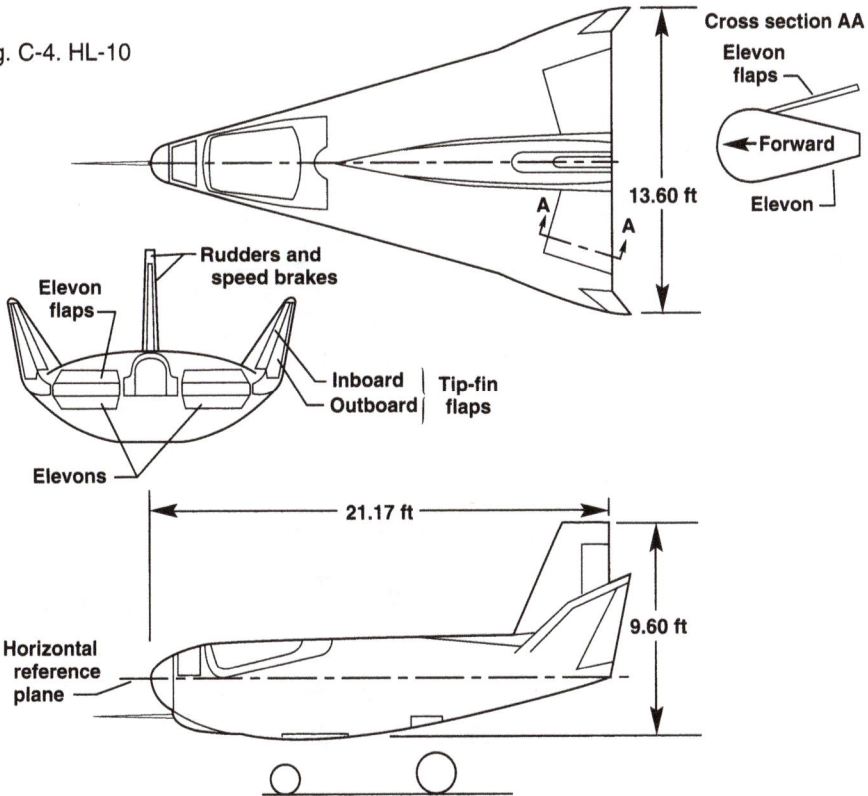

Cross section AA

Elevon
flaps

←Forward

Elevon

13.60 ft

Rudders and
speed brakes

Elevon
flaps

Inboard
Outboard

Tip-fin
flaps

Elevons

A

A

21.17 ft

9.60 ft

Horizontal
reference
plane

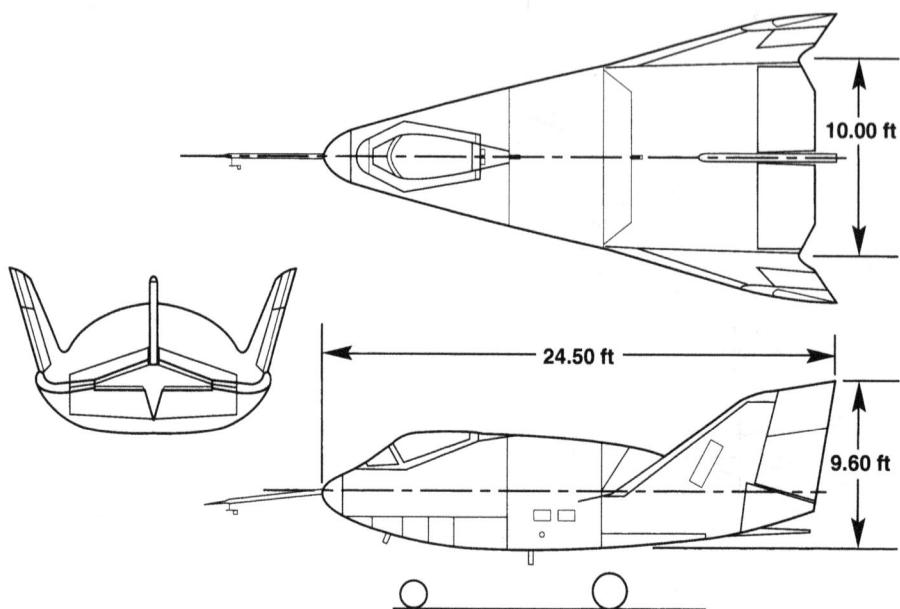

10.00 ft

24.50 ft

9.60 ft

Fig. C-5. X-24A

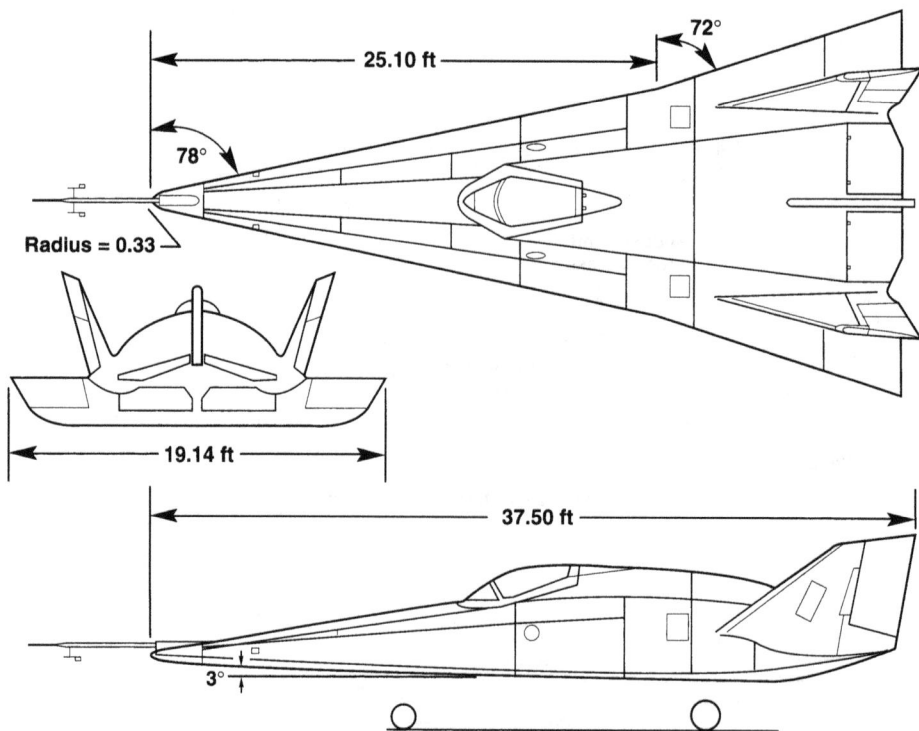

72°

25.10 ft

78°

Radius = 0.33

19.14 ft

37.50 ft

3°

Fig. C-6. X-24B

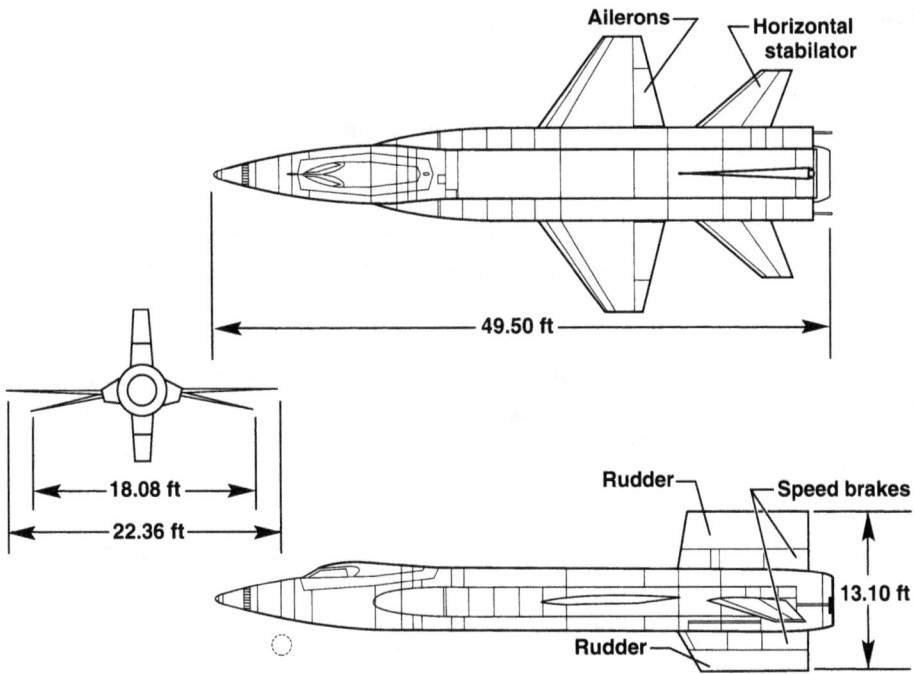

Ailerons —

— Horizontal
stabilator

⟵ 49.50 ft ⟶

⟵ 18.08 ft ⟶
⟵ 22.36 ft ⟶

Rudder —

— Speed brakes

13.10 ft

Rudder —

Fig. C-7. X-15

OMS pods —

Elevons

Tailcone (used
only in earliest
drop tests)

⟵ 78.07 ft ⟶

⟵ 122.25 ft ⟶

Rudders —

46.33 ft

56.58 ft

Body
flap

⟵ 107.53 ft ⟶

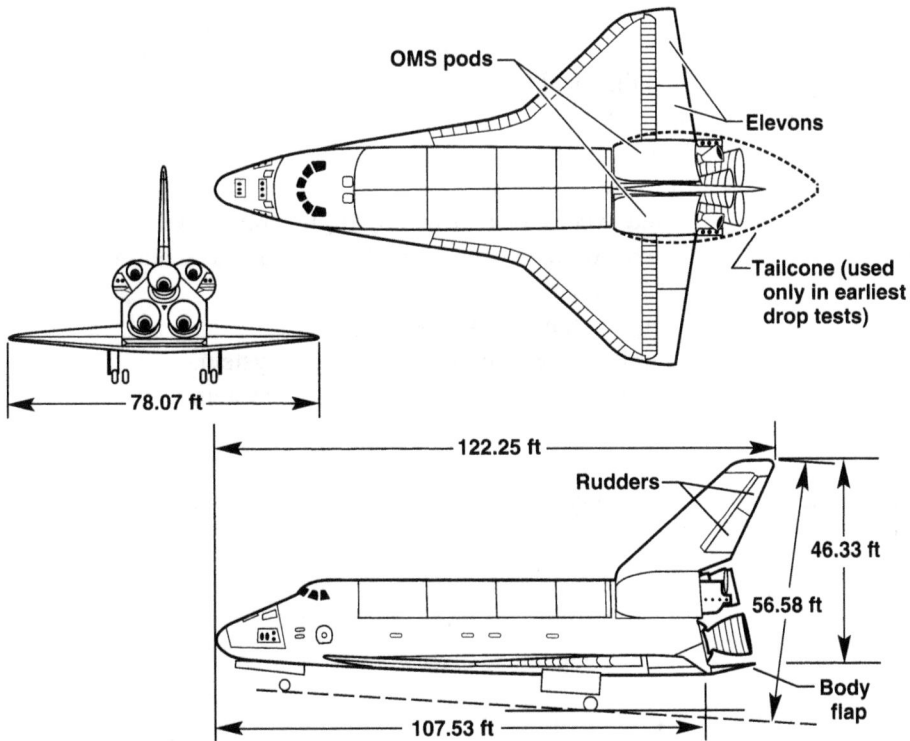

Fig. C-8. Space Shuttle Orbiter

Appendix D: Aircraft Nomenclature

This appendix shows four views of an aircraft and the variables that engineers use to describe aircraft flight. It is hoped that the reader interested in learning more about aircraft flight will find geometric definitions useful.

ALPHA:	Also α, also angle of attack, the angle between the relative wind and the longitudinal axis of the body, x-z plane, deg.
AN:	Acceleration normal to the body axis, positive upwards, g.
AX:	Acceleration along the body axis, positive forward, g.
BETA:	Also β, also angle of sideslip, angle between the relative wind and the longitudinal axis in the x-y plane, deg.
Flight Path Angle:	See GAMMA
g:	Gravitational acceleration.
GAMMA:	Also γ, angle between the flight path and the local horizon, deg.
L:	Rolling moment, ft - lbf.
Local Horizon:	Local "level" reference plane.
M:	Pitching moment, ft - lbf.
N:	Yawing moment, ft - lbf.
p:	Roll rate, deg/sec.
PHI:	Also ϕ, roll attitude angle rotation, angle between the wings and the local horizon, deg.
Pitch Attitude:	See THETA.
q:	Pitch rate, deg/sec.
r:	Yaw rate, deg/sec.
Relative Wind:	"Wind" vector established by flight path vector.
THETA:	Also θ, Angle between longitudinal body axis and the local horizon, deg.
x:	Along the longitudinal body axis.
y:	Lateral with respect to the body axis system.
z:	Vertical with respect to the body axis system.

NOTE: The wings are in the x-y plane, the vertical tail center plane is in the x-z plane.

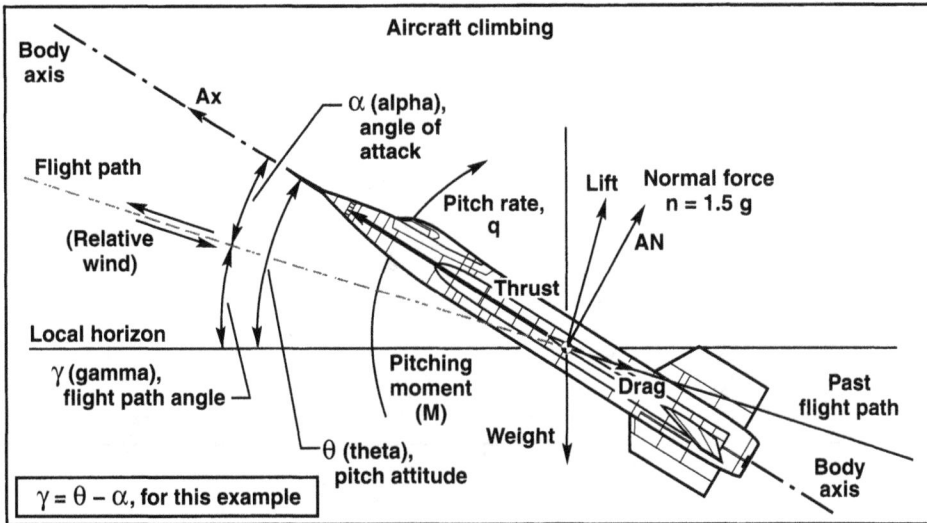

Fig. D-1. Diagram showing longitudinal terms with aircraft climbing.

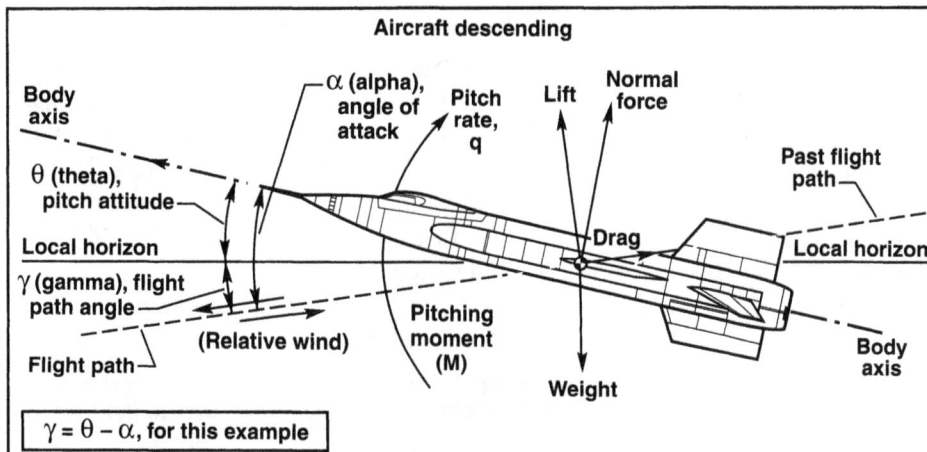

Fig. D-2. Diagram showing longitudinal terms with aircraft descending.

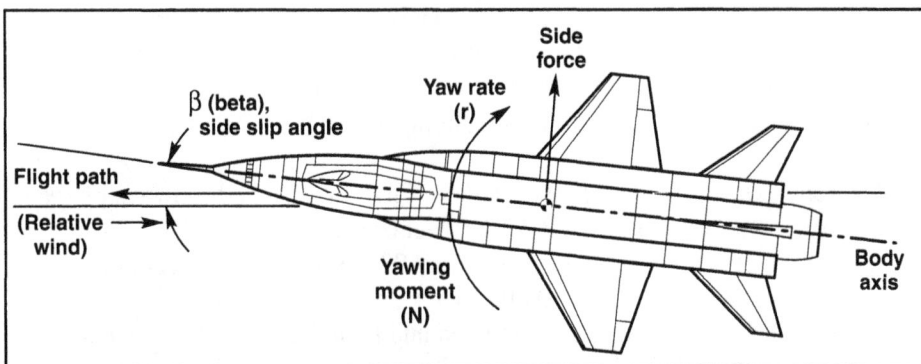

Fig. D-3. Diagram showing directional terms.

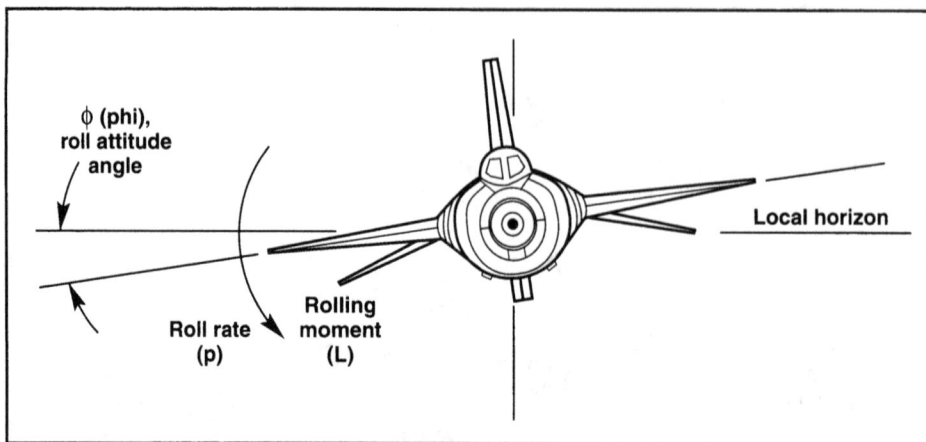

Fig. D-4. Diagram showing lateral terms.

Appendix E: Runways at Edwards AFB

Throughout this narration I've talked about runway numbers and their proximity to Dryden. Those runways are designated by the magnetic direction that they are aligned with and the last digit is dropped. The magnetic deviation at Edwards is 13.6 degrees east of true north. But, since the runways were laid out when only magnetic compasses were used they're all magnetic headings. Therefore runway 18 is really 180 degrees south of magnetic north, so runway 18 is going south. If you were landing north on that same runway it would be 36 or 360. The three lakebed runways up near Dryden and the compass rose are runway 18, runway 23 and runway 15. If you go the opposite direction landing to the east, runway 23 becomes runway 05 which I mentioned once in the STS-44 lakebed runway landing. If you go to the southeast it is runway 15, if to the northwest it is runway 33 which I only refer to in connection with the landing of STS-37.

The other lakebed runway that I referred to is runway 17, which is out in the middle of the lake bed. If you land the opposite direction on that you would be landing on runway 35 which I don't refer to anywhere in the narrative. The last runway is the concrete runway, sometimes called the main runway, and it's a 15,000 foot runway with a 1,000 foot overrun. If you're landing towards the southwest it's runway 22, and if you're landing towards the northeast it's runway 04.

The runways vary widely in their length. First, the lakebed runways by the compass rose. Runway 15 is about 32,000 feet or six miles long. Runway 18 is about 22,000 feet or four miles long. Runway 23 is about 26,000 feet or about five miles long. Runway 17 is a 40,000 foot runway, or about seven plus miles.

I'll describe the closest point to Dryden and then the most obvious touch down point. Runway 15's closest point to Dryden is about three quarters of a mile, and the normal touch down point is about one mile from Dryden. The

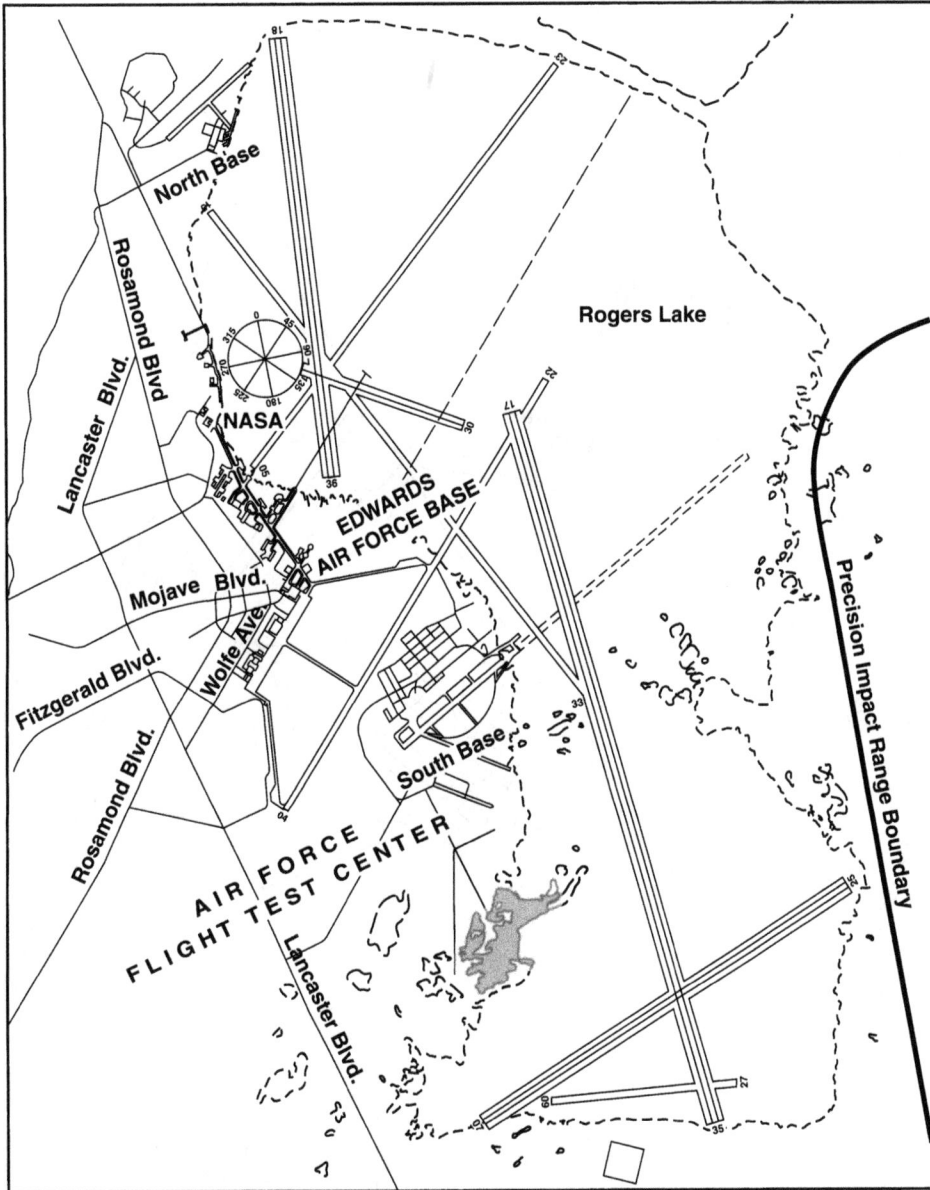

Fig. E-1. Rogers Dry Lake runways.

North Base

Rogers Lake

Lancaster Blvd.

Rosamond Blvd.

NASA

EDWARDS
AIR FORCE BASE

Mojave Blvd.

Wolfe Ave.

Fitzgerald Blvd.

Rosamond Blvd.

South Base

Precision Impact Range Boundary

AIR FORCE
FLIGHT TEST CENTER

Lancaster Blvd.

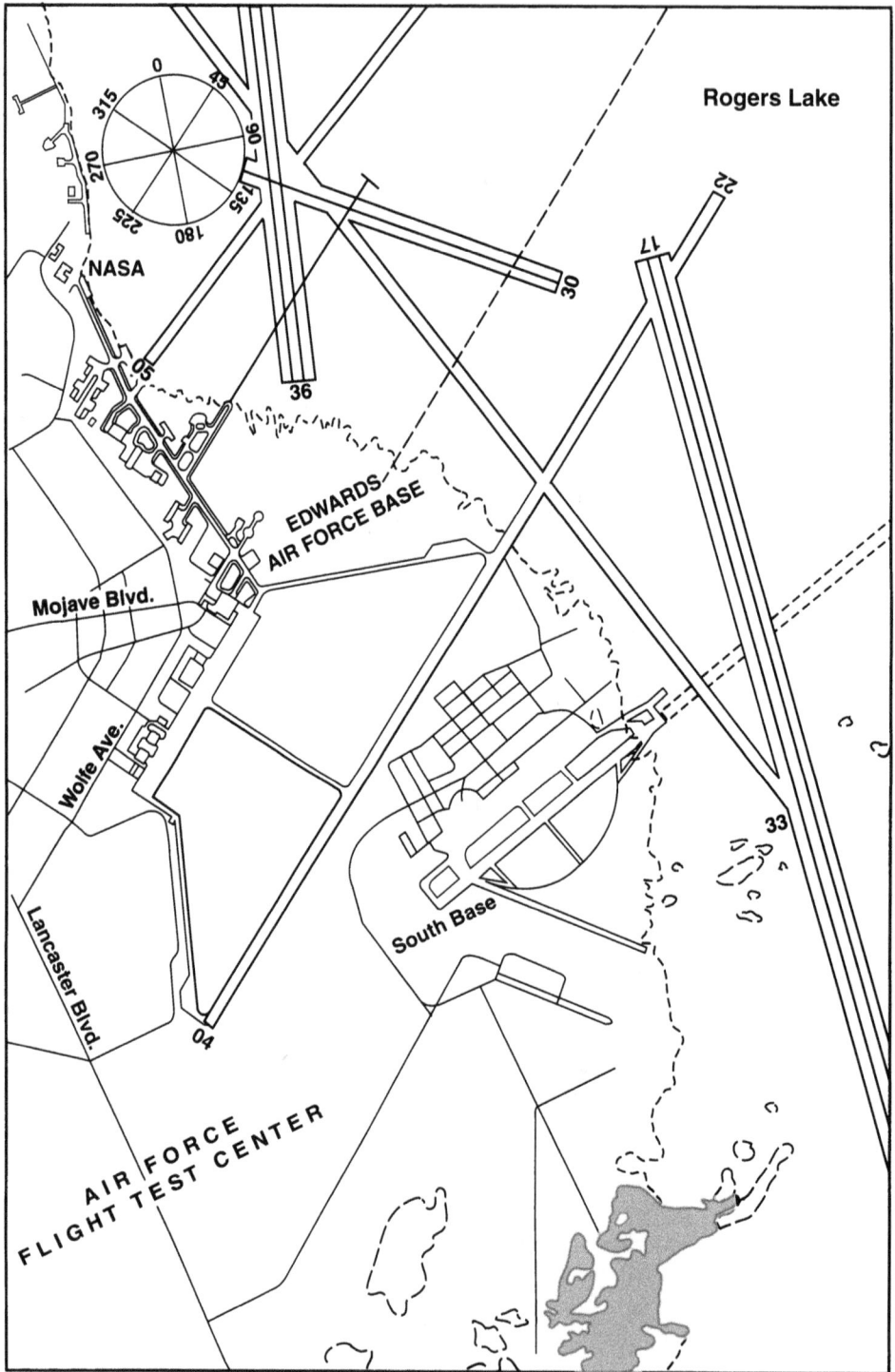

Fig. E-2. South Rogers Dry Lake runways.

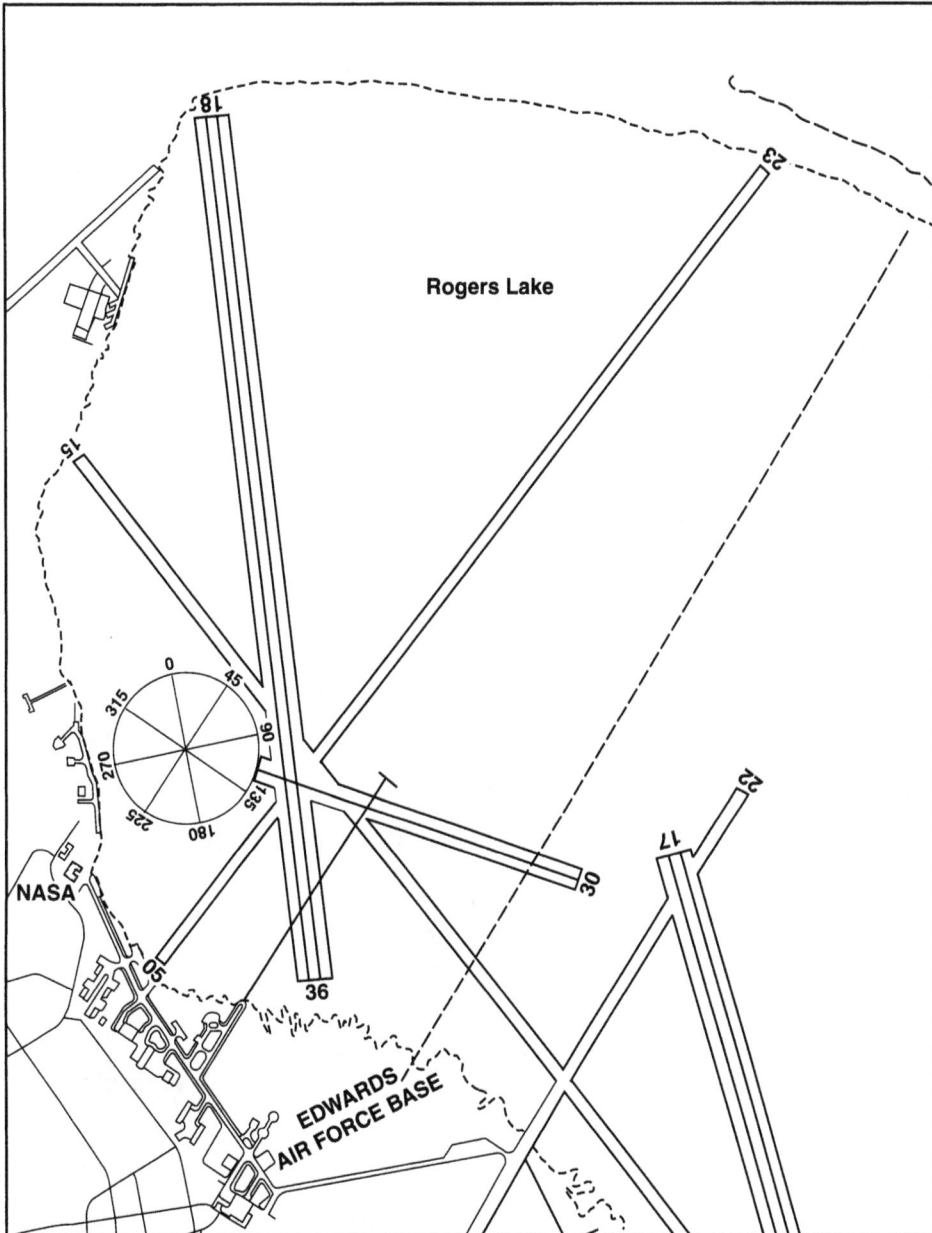

Fig. E-3. North Rogers Dry Lake runways.

closest point of runway 17 is two and a half miles from the Dryden ramp, and the more or less normal touch down point is about three and a half miles from Dryden. Runway 18 is about three quarters of a mile from Dryden. The normal touch down point is about two miles from Dryden. Runway 22 or 04 is a 15,000 foot concrete runway with an overrun onto the lakebed and into the desert. But we don't usually include that because it's normally not used. Its closest point to Dryden is about two and a half miles, and the normal touch down point is about two and a half miles away. Runway 23 is about a half mile from Dryden at its closest point, and the normal touch down point is coming more or less at Dryden about three miles out. The one landing I referred to on runway 05—the actual touch down point was probably one half to one mile from Dryden. The closest approach that it made to Dryden was about 1,500 or 2,000 feet.

Appendix F: STS-1

The first flight of the Space Shuttle Columbia (STS-1) is discussed in some detail the text. The time histories of some of the key flight parameters are shown and the events are labeled to help the reader get a better understanding of the events that are discussed in the text. The figures are adapted from Reference 58.

STS-1 Flight Events
1. Entry interface
2. Control Activation Oscillation 330,000 feet, Mach 24
3. Bank maneuver to the right: ~ 70° bank resulting in large 50 second oscillation
4. Excessive body flap deflection, Mach = 25 to 15
5. First bank reversal; roll to the left to ~70° bank angle
6. 10% Longitudinal Acceleration (AX) jump for turbulent boundary layer transition wedge formation
7. Second bank reversal, roll to the right to ~ 50° bank angle
8. Third bank reversal, roll to the left to ~40° bank angle
9. Fourth bank reversal roll to the right to ~30° bank angle
10. Quarter hertz oscillation during transonic buffet
11. Heading Alignment Circle (HAC)
12. 280 to 300 knots indicated approach with negative flight path angle of 22 to 25 degrees

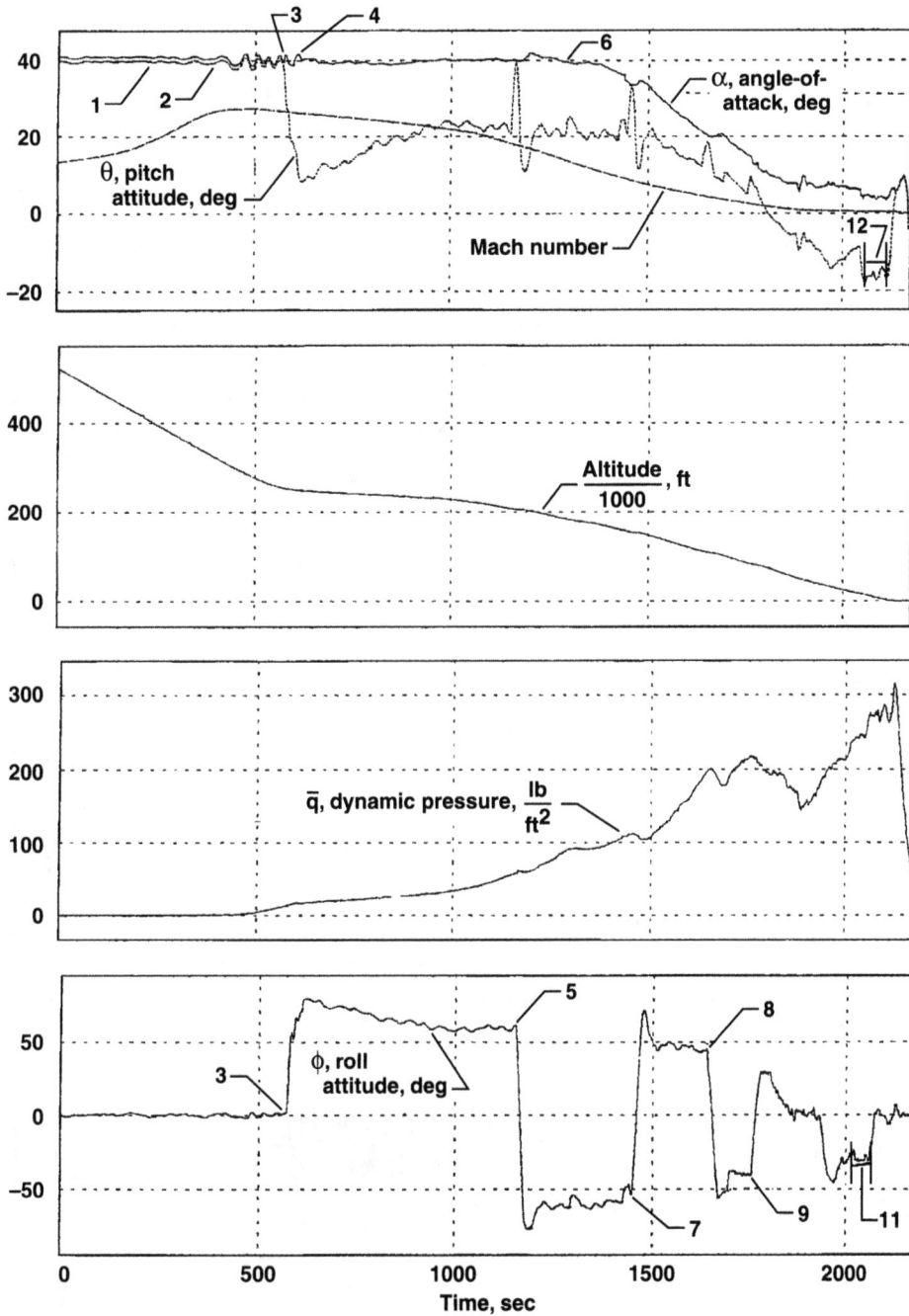

Fig. F-1. Time histories, including altitude and dynamic pressure, of events during STS-1.

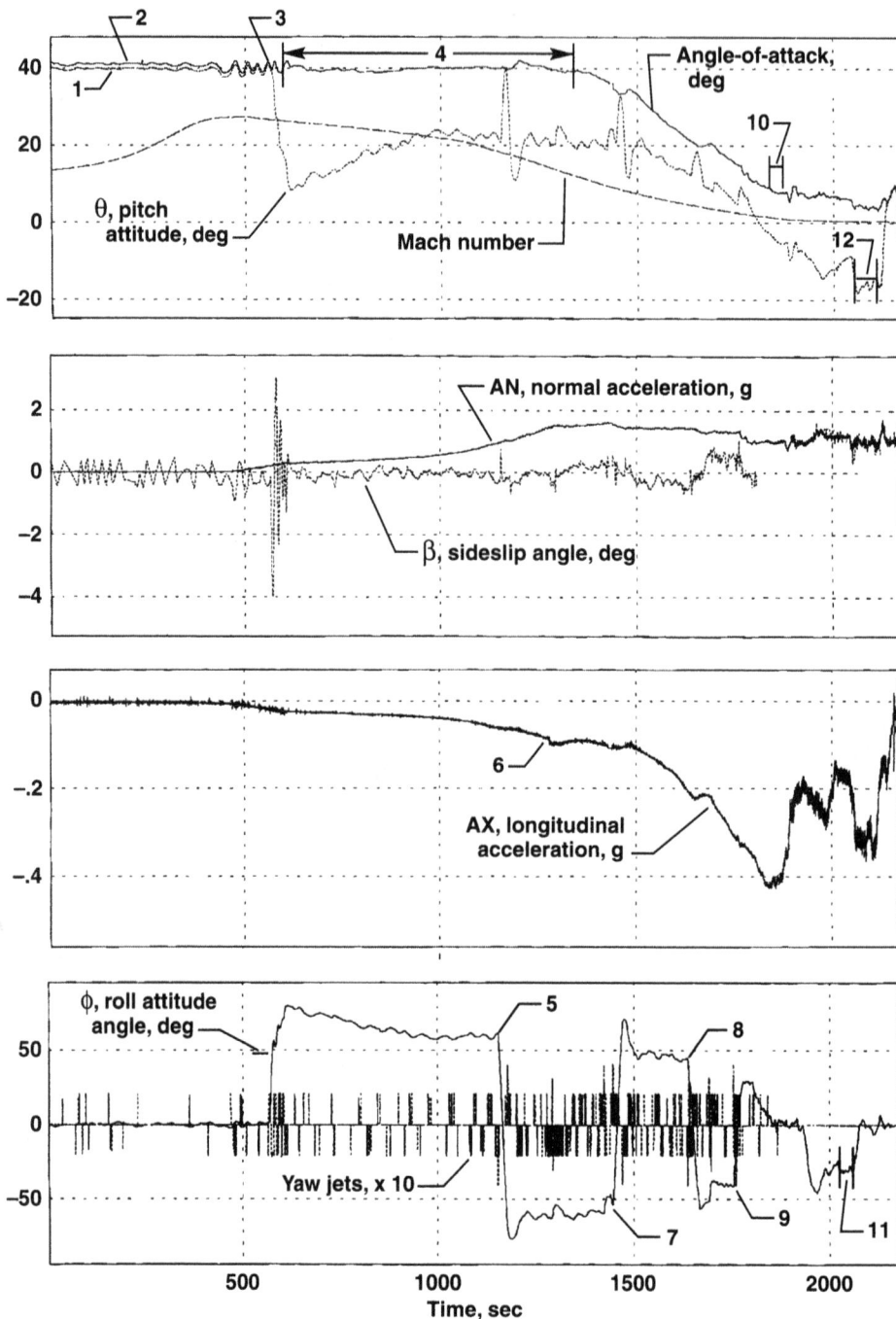

Fig. F-2 Time histories, including acceleration and yaw jet firings, of events during STS-1.

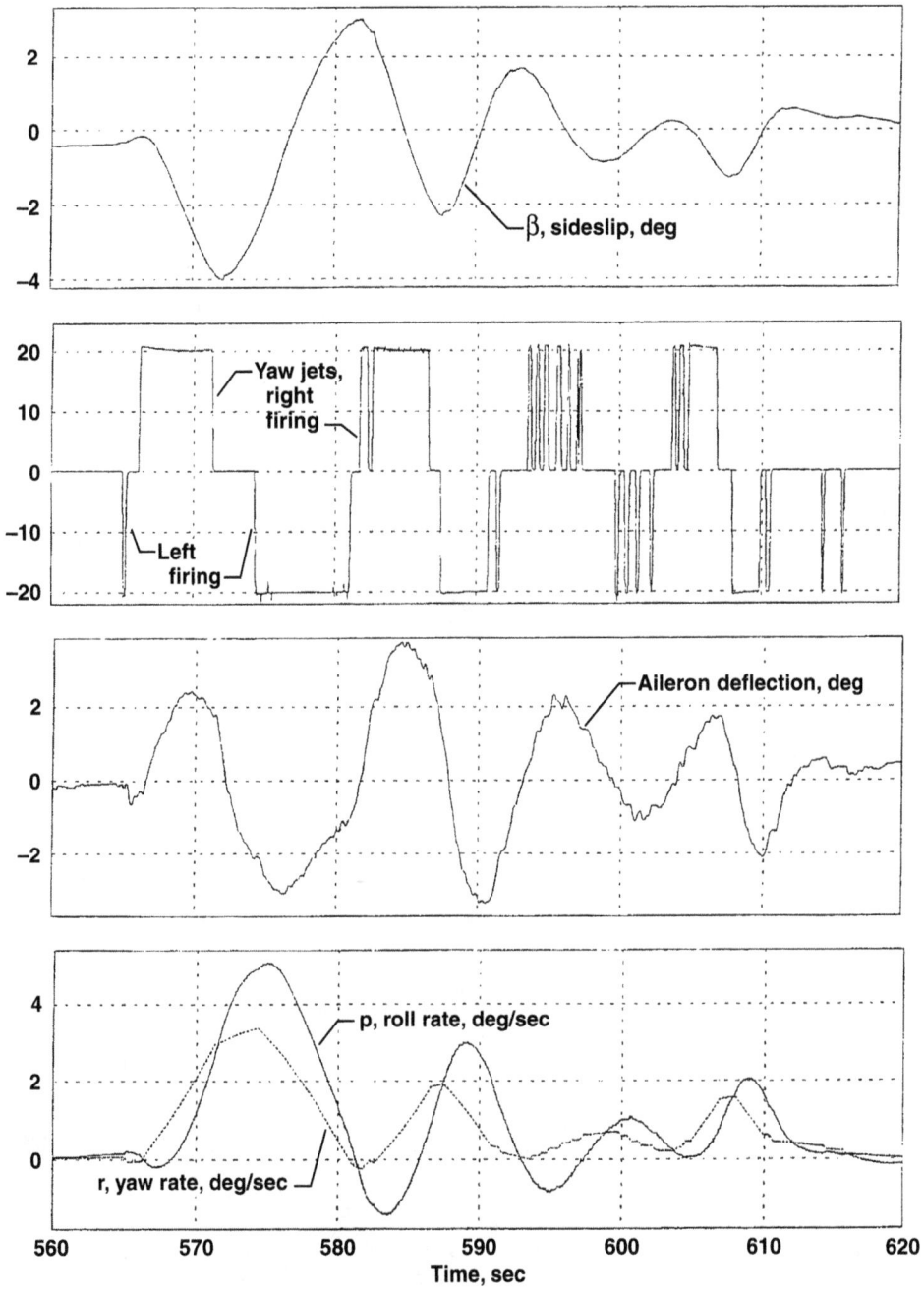

Fig. F-3. Time histories of the first bank maneuver on STS-1.

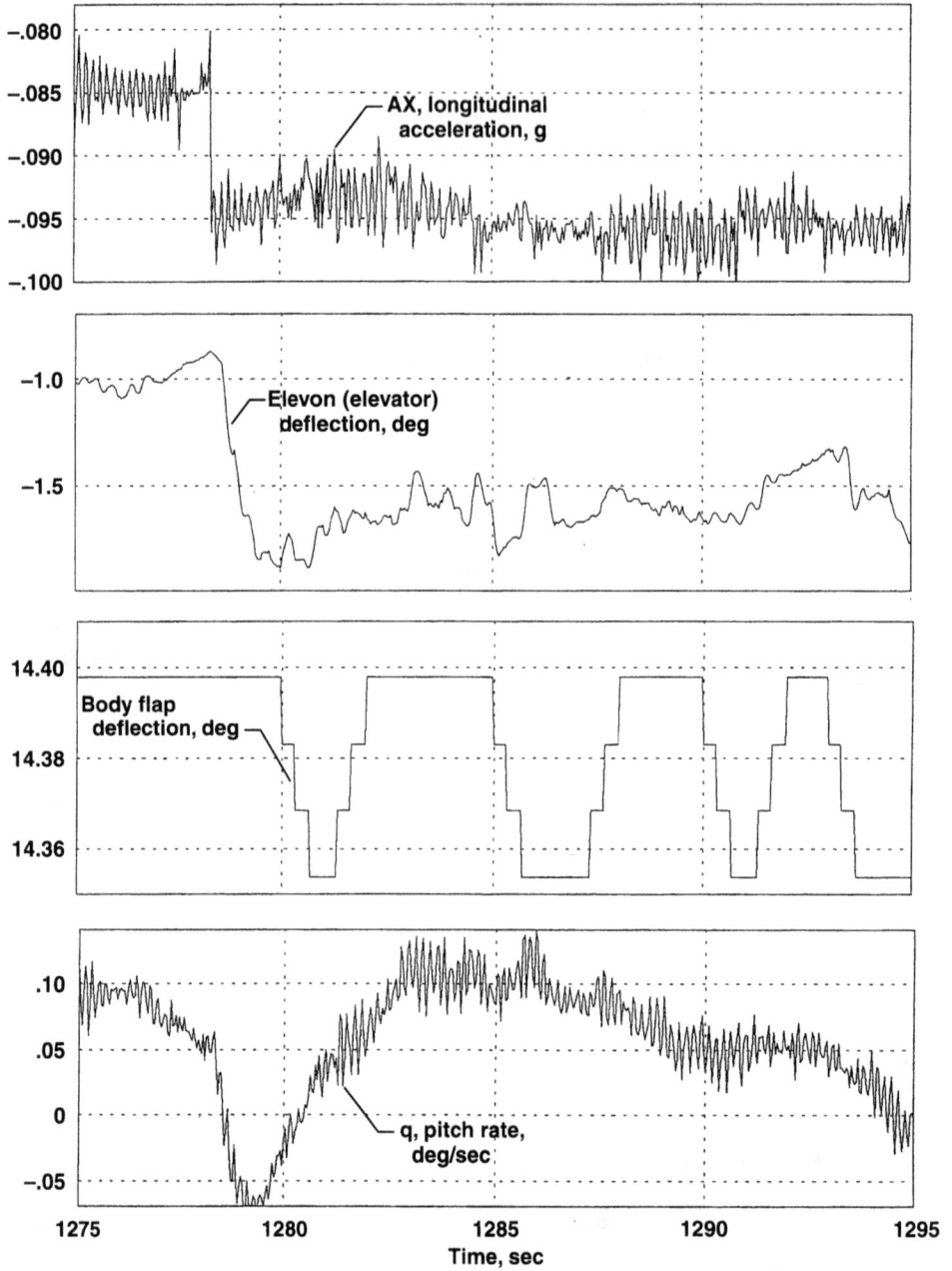

Fig. F-4. Time histories showing boundary layer transition on STS-1.

Appendix G: STS-2

The maneuvers done by the commander on STS-2 that are discussed in the text are shown in the following time histories. These six maneuvers identified are believed by the author to be the record setting maneuvers for flights with a winged vehicles for both altitude and speed. The figures are adapted from Reference 59.

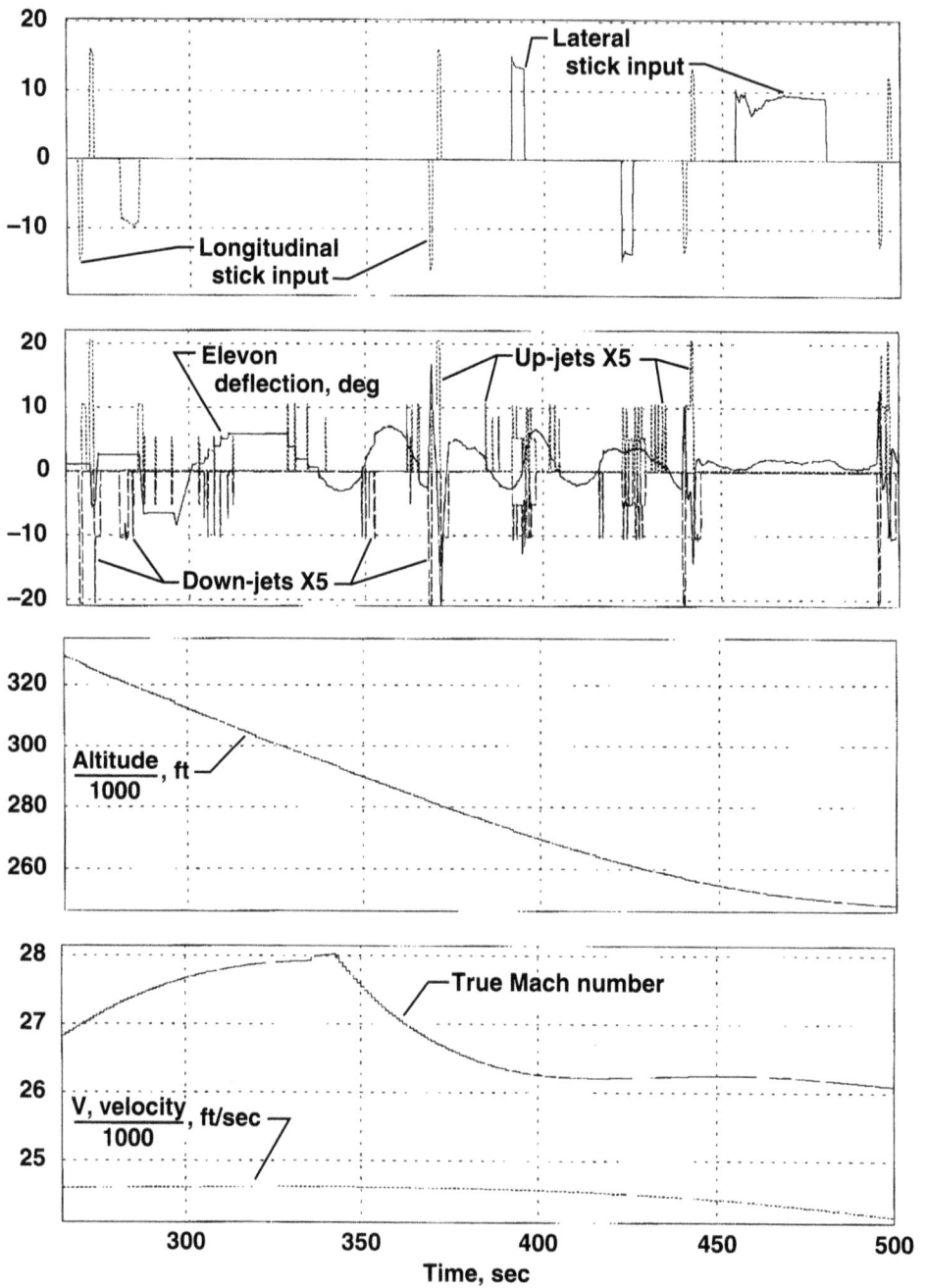

Fig. G-1. Time histories of high speed, high altitude manual maneuvers on STS-2.

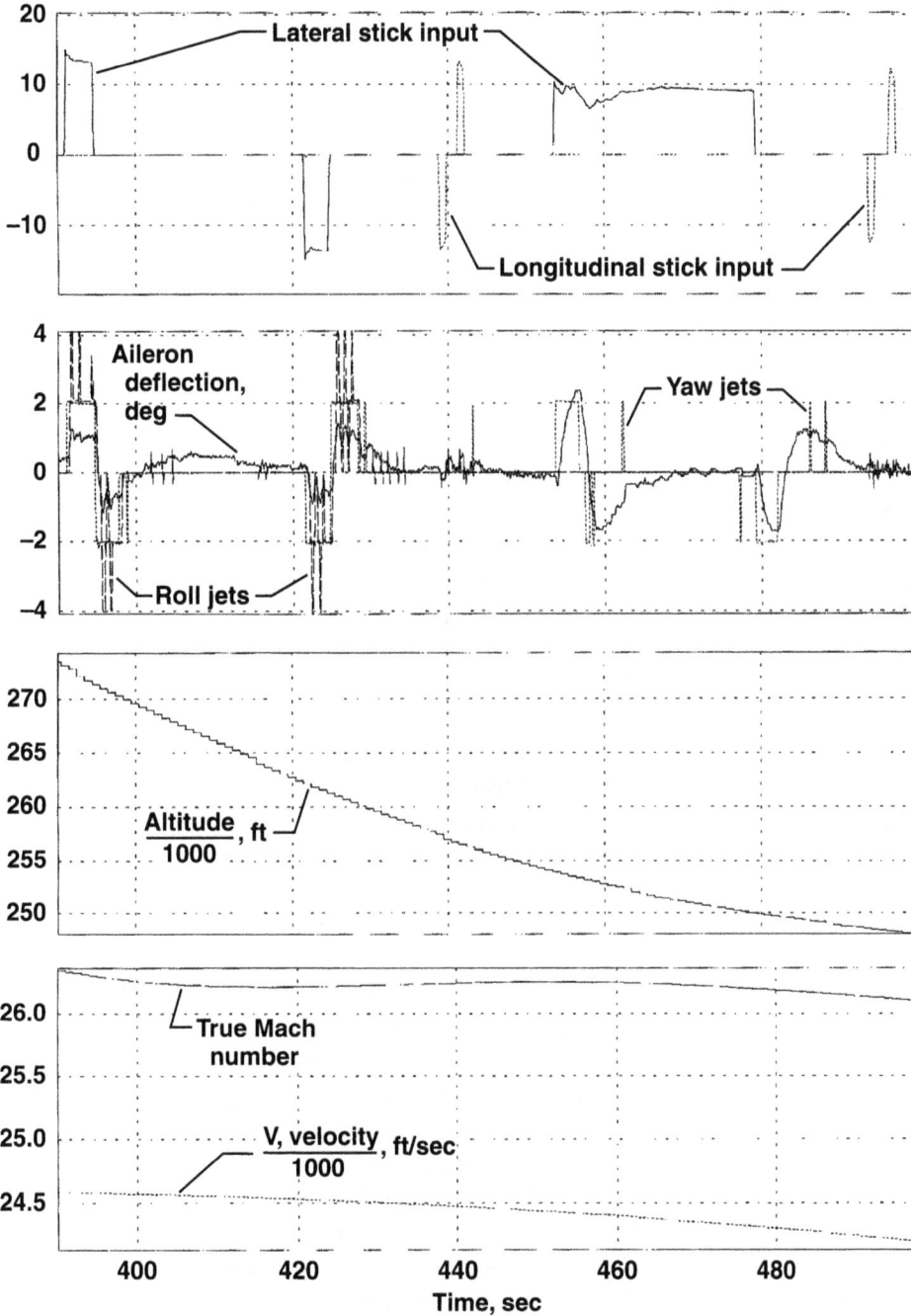

Fig. G-2. Time histories of high speed, high altitude manual maneuvers on STS-2 with a more sensitive time scale.

Acronyms

ACIP	Aerodynamic Coefficient Identification
ACRV	Assured Crew Rescue Vehicle
ADDB	Aerodynamic Data Book
AFFDL	Air Force Flight Dynamics Laboratory
AFFTC	Air Force Flight Test Center
AFRSI	Advanced Flexible Reusable Surface Insulation
AIAA	American Institute of Aeronautics and Astronautics
ALT	Approach and Landing Tests
AN	Normal Acceleration
AOA	Abort-Once-Around
APU	Auxiliary Power Unit
ASI	Aero-Stick Inputs
AX	Longitudinal Acceleration
BET	Best Estimated Trajectory
BFCS	Backup-Flight Control System
CAS	Command Augmentation System
CFD	Computational Fluid Dynamics
CG	Center of Gravity
CM	Coefficient of Pitching Moment
CP	Center of Pressure
CSE	Catalytic Surface Experiment
DFI	Developmental Flight Instrumentation
DoD	Department of Defense
DTO	Developmental Test Objectives
ELV	Expendable Launch Vehicle
FAD	Flight Assessment Delta
FPS	Feet Per Second
GEDA	Goodyear Electronic Differential Analyzer
GPC	General Purpose Computer
GNC	Guidance Navigation and Control
HAC	Heading Alignment Circle or Cone
HALO	Hypersonic Air Launch Option
HARV	High Angle of Attack Research Vehicle
HiMAT	Highly Maneuverable Aircraft Technology
HiRAP	High Resolution Accelerometer Package
HRSI	High Temperature Reusable Surface Insulation
HYFLEX	Hypersonic Flight Experiment
IFR	Instrument Flight Rules
IMU	Inertial Measurement Unit

LDEF	Long Duration Exposure Facility
LLRV	Lunar Landing Research Vehicle
LRSI	Low-Temperature Surface Insulation
MADS	Modular Auxiliary Data System
MMLE	Modified Maximum Likelihood Estimator
MMU	Manned Maneuvering Unit
NACA	National Advisory Committee for Aeronautics
NASP	National Aero-Space Plane (X-30)
NHFRF	National Hypersonic Flight Research Facility
NLS	National Launch System
NOAA	National Oceanic and Atmospheric Administration
OARE	Orbiter Acceleration Research Experiment
OSC	Orbital Science Corporation
OEX	Orbiter Experiment
OFT	Orbital Flight Test (see also STS)
OI	Operational Instrumentation
OMS	Orbital Maneuvering System
PCM	Pulse Code Modulation
PHYSX	Pegasus Hypersonic Experiment
PIO	Pilot Induced Oscillation
POPU	Push-Over Pull-Up
PSF	Pounds per Square Foot
PTI	Programmed Test Input
RCC	Reinforced Carbon-Carbon
RCS	Reaction Control System
RLV	Reusable Launch Vehicle
RPRV	Remote Piloted Research Vehicle
RTLS	Return-To-Launch-Site
SAS	Stability Augmentation System
SCA	Shuttle Carrier Aircraft
SEADS	Shuttle Experimental Air Data System
SILTS	Shuttle Infrared Leeside Temperature Sensing Experiment
SRB	Shuttle Rocket Booster
SRV	Spin Research Vehicle
SSME	Space Shuttle Main Engine
STA	Shuttle Training Aircraft
STS	Shuttle Transportation System
SUMS	Shuttle Upper Atmosphere Mass Spectrometer
TDRS	Telecommunication Data Relay System
TEB	Tri Ethyl Borate
TGH	Tile Gap Heating
TM	Telemeter
TPS	Thermal Protection System
UAV	Unmanned Aerial Vehicle

UCLA	University of California Los Angeles
USC	University of Southern California
VFR	Visual Flight Rules

References

1. Thompson, Milton O. and Curtis Peebles, Flying Without Wings: NASA Lifting Bodies and the Birth of the Space Shuttle (Smithsonian Institution Press, Washington, D. C. and London UK, 1999).

2. Dryden, Hugh L., "General Background of the X-15 Research-Airplane Project." In Research-Airplane-Committee Report on Conference on the Progress of the X-15 Project, pp. xvii-xix, 1956.

3. Jenkins, Dennis R. and Tony R. Landis, Hypersonic: The Story Of The North American X-15, Specialty Press, North Branch MN. 2003.

4. Review of the X-15 Program, NASA Technical Note D-1278, compiled by Joseph Weil, June 1962.

5. Progress of the X-15 Research Airplane Program, NASA SP-90, Flight Research Center, Edwards Air Force Base, October 7, 1965.

6. Stillwell, Wendell H., X-15 Research Results: With a Selected Bibliography NASA SP-60, 1965.

7. Thompson, Milton O., At the Edge of Space: The X-15 Flight Program, Smithsonian Institution Press, 1992.

8. Powers, Sheryll Goecke, Women in Flight Research at NASA Dryden Flight Research Center from 1946 to 1995. NASA Monographs in Aerospace History, Number 6, 1997.

9. Study of the Ejector Ramjet Engine for X-15 Propulsion, January 3, 1967, compiled by the Marquardt Corporation, Van Nuys, California.

10. Taylor, Lawrence W. Jr., G. H. Robinson, and Kenneth W. Iliff, "A Review of Lateral Directional Handling Qualities Criteria as Applied to the X-15." NASA SP-90. Progress of the X-15 Research Airplane Program, October 7, 1965, pp 45-60.

11. Banner, Richard D., Albert E. Kuhl, and Robert D. Quinn, Preliminary Results of Aerodynamic Heating Studies on the X-15 Airplane, NASA TM-X-638, 1962.

12. Quinn, Robert D. and Albert E. Kuhl, Comparison of Flight-Measured and Calculated Turbulent Heat Transfer on the X-15 Airplane at Mach Numbers from 2.5 to 6.0 at Low Angles of Attack. NASA TM-X-939, 1964.

13. Banas, Ronald P., Comparison of Measured and Calculated Turbulent Heat Transfer in a Uniform and Nonuniform Flow Field on the X-15 Upper Vertical Fin at Mach Numbers of 4.2 and 5.3. NASA TM-X-1136, 1965.

14. Quinn, Robert D. and Murray Palitz, Comparison of Measured and Calculated Turbulent Heat Transfer on the X-15 Airplane at Angles of Attack up to 19 degrees, NASA TM-X-1291, 1964

15. Iliff, Kenneth W. and Mary F. Shafer, "A Comparison of Hypersonic Flight and Prediction Results." AIAA PAPER 93-0311. AIAA, 31st Aerospace Sciences Meeting and Exhibit, Reno, NV, Jan. 11-14, 1993

16. Petersen, Commander Forrest S., USN, Herman A. Rediess, and Joseph Weil, Lateral-Directional Control Characteristics of the X-15 Airplane, NASA TM-X-726, March 1962.

17. Yancey, Roxanah B., Flight Measurements of Stability and Control Derivatives of the X-15 Research Airplane to a Mach Number of 6.02 and an Angle of Attack of 25 degrees. NASA TN D-2532, November 1964.

18. Horton, Victor W., Richard C. Eldredge, and Richard E. Klein, Flight-Determined Low-Speed Lift and Drag Characteristics of the Lightweight M2-F1 Lifting Body. NASA TN D-3021, September 1965.

19. Smith, Harriet J., Evaluation of the Lateral-Directional Stability and Control Characteristics of the Lightweight M2-F1 Lifting Body at Low Speeds. NASA TN D-3022, September 1965.

20. Keener, Earl R. and Jack J. Brownson, Wind-Tunnel Investigation of the Aerodynamic Characteristics of the M2-F2 Lifting-Body Entry Configuration at Transonic and Supersonic Mach Numbers. NASA TM X-2511, April 1972.

21. Waltman, Gene L., Black Magic and Gremlins: Analog Flight Simulations at NASA's Flight Research Center, Monographs in Aerospace History #20, NASA SP-2000-4520.

22. Ashkenas, Irving L., and Duane T. McRuer, Approximate Airframe Transfer Functions and Application to Single Sensor Control Systems. WADC-TR-58-82, (ASTIA No. AD 151025), Wright Air Dev. Center, U. S. Air Force, June 1958.

23. Taylor, Lawrence W., Jr., and Kenneth W. Iliff, Fixed-Base Simulator Pilot Rating Surveys for Predicting Lateral-Directional Handling Qualities and Pilot Rating Variability. NASA TN D-5358, August 1969.

24. Meeker, J. I., Evaluation of Lateral-Directional Handling Qualities of Piloted Re-Entry Vehicles Utilizing Fixed-Base and In-Flight Evaluations. NASA CR-778, May 1967.

25. Evaluation of Lateral-Directional Handling Qualities of Piloted Re-Entry Vehicles Utilizing Fixed-Base and In-flight Evaluations, Cornell Aeronautical Laboratory, Inc., CAL No. TC-1921-F-5, Final Report for Contract No. AF 33(615)-1253, January 1966.

26. Painter, Weneth D. and Berwin M. Kock, Operational Experiences and Characteristics of the M2-F2 Lifting Body Flight Control System. NASA TM X-1809, June 1969.

27. Taylor, Lawrence W., Jr., and George B. Merrick, X-15 Airplane Stability Augmentation System. NASA TN D-1157, March 1962

28. Iliff, Kenneth W. and Lawrence W. Taylor Jr., Determination of Stability Derivatives from Flight Data Using a Newton-Raphson Minimization Technique. NASA TN D-6579, March 1972.

29. Reed, R. Dale, Wingless Flight, The Lifting Body Story. NASA SP-4220, The NASA History Series, 1997.

30. Taylor, Lawrence W., Jr. and Kenneth W. Iliff, "A Modified Newton-Raphson Method for Determining Stability Derivatives from Flight Data." Second International Conference on Computing Methods in Optimization Problems, San Remo, Italy, September 9-13, 1968.

31. Taylor, Lawrence W., Jr., Kenneth W. Iliff, and Bruce G. Powers, "A Comparison of Newton-Raphson and Other Methods for Determining Stability Derivatives from Flight Data," 3rd Technical Workshop on Dynamic Stability Problems, NASA Ames Research Center, Nov. 4-7, 1968.

32. Becker, John V., "The X-15 Program in Retrospect," Third Eugen Sänger Memorial Lecture, Deutsch Gesselschaft für Luft und Raumfahrt, December 1968. Also in Raumfahrtforschung, Vol. 13, 1969, p. 45.

33. Watts, Joe D., Flight Experience with Shock Impingement and Interference Heating on the X-15-2 Research Airplane. NASA TM X-1669, October 1968.

34. Brown, David, "Internet began 30 years ago at UCLA" and Barry M. Leiner, Vinton G. Cerf, David D. Clark, Jon Postel, Lawrence G, Roberts, Stephen Wolff contributors to "A History of the Internet" Ingeniare, Spring 1999, pp. 19-20.

UCLA School of Engineering and Applied Science magazine, published by Public Information Office at UCLA.

35. Sim, Alex G., Results of a Feasibility Study Using the Newton-Raphson Digital Computer Program to Identify Lifting Body Derivatives from Flight Data. NASA TM X-56017, October 1973.

36. Sim, Alex G., Flight-Determined Stability and Control Characteristics of the M2-F3 Lifting Body Vehicle. NASA TN D-7511, December 1973.

37. Sim, Alex G., A Correlation Between Flight-Determined Derivatives and Wind-Tunnel Data for The X-24B Research Aircraft. NASA SX-3371, March 1976.

38. Kirsten, Paul W., Wind Tunnel and Flight Test Stability and Control Derivatives for the X-24A Lifting Body. FTC-TD-71-7, Technology Document No. 71-7, April 1972.

39. Nagy, Christopher J. and Paul W. Kirsten, Handling Qualities and Stability Derivatives of the X-24B Research Aircraft. AFFTC-TR-76-8, March 1976.

40. Thompson, Milton O., Flight Research: Problems Encountered And What They Should Teach Us, Monographs of Aerospace History #22, NASA SP-2000-4522. 2000.

41. Holleman, Euclid C., Stability and Control Characteristics of the M2-F2 Lifting Body Measured During 16 Glide Flights. NASA TM X-1593, June 1968.

42. Richardson, David F., Analysis of the Approach, Flare and Landing Characteristics of the X-24A Lifting Body. FTC-TD-71-9, Technology Document No. 71-9, July 1972.

43. Stuart, John L., Analysis of the Approach, Flare, and Landing Characteristics of the X-24B Research Aircraft. AFFTC-TR-76-9, November 1977.

44. Taylor, Lawrence W., Jr., Harriet J. Smith, and Kenneth W. Iliff, "A Comparison of Minimum Time Profiles for the F-104; Using Balakrishnan's Epsilon Technique and the Energy Method". Symposium on Optimization, Nice, France, June 29- July 5, 1969.

45. Weil, Joseph, and Bruce G. Powers, Correlation of Predicted and Flight Derived Stability and Control Derivatives - With Particular Application to Tailless Delta Wing Configurations. NASA TM-81361, July 1981.

46. Gamble, Joe D., Douglas R. Cooke, Jimmy M. Underwood, Howard W. Stone, Jr., and Donald C. Schlosser, "The Development and Application of Aerodynamic Uncertainties; and Flight Test Verification for the Space Shuttle Orbiter." NASA Conference Publication 2342 Part 1, Space Shuttle Technical Conference, 1985.

47. Fulton, Fitzhugh L., Jr., "Shuttle Carrier Aircraft Flight Test", The Society Of Experimental Test Pilots, Twenty First Symposium Proceedings, October 12-15, 1977.

48. Andrews, W.H., "Space Shuttle Orbiter Approach and Landing Program Status," AIAA 77-1204, AIAA Aircraft Systems & Technology Meeting, Seattle, Washington, August 1977.

49. Haise, Fred W., Jr., "Approach and Landing Test Program Status". The Society of Experimental Test Pilots, Twenty First Symposium Proceedings, October 12-15, 1977.

50. Tomayko, James E., Computers Take Flight: A History of NASA's Pioneering Digital Fly-By-Wire Project, The NASA History Series, NASA SP-2000-4224, 2000.

51. Iliff, Kenneth W., Richard E. Maine, and Mary F. Shafer, "Subsonic Stability and Control Derivatives for an Unpowered, Remotely Piloted 3/8-Scale F-15 Airplane Model Obtained from Flight Test," NASATN-D-8136, Jan. 1976.

52. Iliff, Kenneth W. and Mary F. Shafer, Extraction of Stability and Control Derivatives from Orbiter Flight Data. NASA Technical Memorandum 4500, June 1993.

53. Engle, Joseph H. and Richard H. Truly "Approach and Landing Test Supplemental Report". The Society of Experimental Test Pilots, Twenty First Symposium Proceedings, October 1977.

54. Brand, Vance D., "Return to Earth in the Space Shuttle". The Society of Experimental Test Pilots Twenty-first Symposium Proceedings, October 1977.

55. Smith, John W. and John W. Edwards. Design of A Nonlinear Adaptive Filter for Suppression of Shuttle Pilot-Induced Oscillation Tendencies. NASA Technical Memorandum 81349, April 1980.

56. Powers, Bruce G., "An Adaptive Stick-Gain to Reduce Pilot-Induced Oscillation Tendencies." AIAA paper 82-4078, AIAA Journal of Guidance Vol.5, No. 2,

March-April 1982.

57. Hooks, Ivy, David Homan, and Paul Romere, "Aerodynamic Challenges of ALT". Space Shuttle Technical Conference, NASA Conference Publication 2342 Part 1, 1985.

57A. Meyer, Robert R., Jr., Calvin R. Jarvis, and Jack Barneburg, In Flight Aerodynamic Load Testing of the Shuttle Thermal Protection System. AIAA Paper No. 81-2468, November 1981.

58. Iliff, K.W., R. E. Maine, and D. R. Cooke, Selected Stability and Control Derivatives from the First Space Shuttle Entry. AIAA Paper 81-2451, AIAA, SETP, SFTE, SAE, ITEA, and IEEE, First Flight Testing Conference, Las Vegas, NV, November 1981.

59. Maine, R. E. and K W. Iliff, Selected Stability and Control Derivatives from the First Three Space Shuttle Entries. AIAA Paper 82-1318, AIAA, 9th, Atmospheric Flight Mechanics Conference, San Diego, August 1982.

60. Iliff, Kenneth W. and Mary F. Shafer, Space Shuttle Hypersonic Aerodynamic and Aerothermodynamic Flight Research and the Comparison to Ground Test Results. NASA TM-4499. June 1993.

61. Bouslog, S.A., M.Y. An, L.N. Hartmann, and S.M. Derry, Review of Boundary Layer Transition Flight Data on the Space Shuttle Orbiter. AIAA Paper 91-0741, January 1991.

62. Trujillo, Bianca M., Robert R. Meyer, Jr., and Paul M. Sawko, In-Flight Load Testing of Advanced Shuttle Thermal Protection Systems. AIAA Paper 83-2704 at The AIAA 2nd Flight Test Conference, Las Vegas, Nevada, November 1983. Also published as NASA Technical Memorandum 86024, December 1983.

63. Moes, Timothy R., and Robert R. Meyer, Jr., In-Flight Investigation of Shuttle Tile Pressure Orifice Installations. NASA Technical Memorandum 4219, 1990.

64. Meyer, Robert R. Jr., and Jack Barneburg, In-Flight Rain Damage Tests of the Shuttle Thermal Protection System. NASA Technical Memorandum 100438, May 1988. Also presented as AIAA 88-2137 at the AIAA 4th Flight Test Conference May 1988.

65. Shuttle Performance: Lessons Learned. NASA Conference Publication 2283 Part 1 and 2, 1983, Compiled by James P. Arrington and Jim J. Jones.

66. Holloway, Paul F. and David A. Throckmorton, "Shuttle Orbiter Experiments - Use of an Operational Vehicle for Advancement and Validation of Space Systems Design Technologies". Space Systems Design and Development Testing. AGARD-CP-561, AGARD Conference Proceedings 561, March 1995.

67. Blanchard, Robert C., K.T. Larman, and M. Barrett, The High Resolution Accelerometer Package (HiRAP) Flight Experiment Summary for the First 10 Flights. NASA Reference Publication 1267, April 1992.

68. Blanchard, Robert C., Kevin T. Larman, and Christina D. Moats, Rarefied-Flow
Shuttle Aerodynamics Flight Model. NASA Technical Memorandum 107698, February 1993.

69. Larson, Terry J., and Paul M. Siemers, III, Subsonic Tests of an All-Flush-Pressure-Orifice Air Data System. NASA Technical Paper 1871, June 1981.

70. Henry, M.W., H. Wolf, and Paul M. Siemers, III An Evaluation of Shuttle Entry Air Data System (SEADS) Flight Pressures: Comparisons With Wind Tunnel and Theoretical Predictions. AIAA Paper 88-2052, May 1988.

71. Wolf. H., M.W Henry, and Paul M. Siemers, III, Shuttle Entry Air Data System (SEADS): Optimization of Preflight Algorithms Based on Flight Results. AIAA Paper 88-2053, May 1988.

72. Moss, James N., and Graeme A. Bird, "Monte Carlo Simulations in Support of the Shuttle Upper Atmospheric Mass Spectrometer Experiment". J. Thermophysics. Vol. 2, No. 2, April 1988.

73. Blanchard, R. C., R. J. Duckett, and E. W. Hinson, "The Shuttle Upper Atmosphere Mass Spectrometer Experiment." Journal of Spacecraft and Rockets. Vol. 21, No. 2, March-April 1984, pp. 202-208.

74. Blanchard, R C., T. A. Ozoroski, and J. Y. Nicholson, SUMS Experiment Flight Results on STS-35. NASA TM-107738, February 1993.

75. Blanchard, R. C., J. Y. Nicholson, and J. R. Ritter, "STS-40 Orbital Acceleration Research Experiment Flight Results During Typical Sleep Period". Microgravity Science and Technology. Vol. 5 No. 2, pp. 86-93, 1992.

76. Blanchard, Robert C., John Y. Nicholson, and James R. Ritter, Preliminary OARE Absolute Acceleration Measurements on STS-50. NASA Technical Memorandum 107724, February 1993.

77. Stewart, David A., John V. Rakich, and Martin J. Lanfranco, "Catalytic Surface Effects on Space Shuttle Thermal Protection System During Earth Entry on Flights STS-2 through STS-5," in Shuttle Performance: Lessons Learned, NASA CP-2283, pt. 2, 1983, pp.827-846.

78. Rakich, J. V., D. A. Stewart, and M. J. Lanfranco, "Results of a Flight Experiment on the Catalytic Efficiency of the Space Shuttle Heat Shield". Third Joint Thermophysics, Fluids, Plasma and Heat Transfer Conference, St. Louis, MO, AIAA Paper 82-0944, June 1982.

79. Pitts, W. C. and M. S. Murbach, Flight Measurements of Tile Gap Heating on the Space Shuttle. AIAA-82-0840, June 1982.

80. Throckmorton, David A., E. Vincent Zoby, James C. Dunavant, and David L. Myrick, Shuttle Infrared Leeside Temperature Sensing (SILTS) Experiment - STS-28 Preliminary Results. AIAA Paper 90-1741, June 1990.

81. Throckmorton, David A., E. Vincent Zoby, James C. Dunavant, and David L. Myrick, Shuttle Infrared Leeside Temperature Sensing (SILTS) Experiment - STS-35 and STS-40 Preliminary Results. AIAA Paper 92-0126, January 1992.

82. Muylaert. J., L. Walpot, P. Rostand, M. Rapuc, G. Brauckmann, J. Paulson, D. Trockmorton, and K. Weilmuenster, "Extrapolation From Wind Tunnel to Flight: Shuttle Orbiter Aerodynamics." Hypersonic Experimental and Computational Capability, Improvement and Validation. AGARD-AR-319 Vol. II, December 1998.

83. Carter, John F. and Christopher J. Nagy, The NASA Landing Gear Test Airplane. NASA Technical Memorandum 4703, June 1995.

84. Kolodziej, Paul, Jeffery V. Bowles, and Cathy Roberts, Optimizing Hypersonic Sharp Body Concepts from a Thermal Protection System Perspective. AIAA Paper 98-1610, 1998.

85. Iliff, Kenneth W. and Kon-Sheng Charles Wang, X-29 Lateral-Directional Stability and Control Derivatives Extracted From High-Angle-of-Attack Flight Data. NASA TP 3664, December 1996.

86. Iliff, Kenneth W. and Kon-Sheng Charles Wang, Flight-Determined, Subsonic, Lateral-Directional Stability and Control Derivatives of the Thrust-Vectoring F-18 High-Angle-of-Attach Research Vehicle (HARV), and Comparisons to the Basic F-18 and Predicted Derivatives. NASA TP 1999-206573. January 1999.

87. Wright, Robert L., and Ernest W. Zoby, Flight Boundary Layer Transition Measurements on a Slender Cone at Mach 20. AIAA 77-719, June 1977.

88. Proceedings of the X-15 First Flight 30th Anniversary Celebration. NASA Conference Publication 3105, June 1989.

89. Bertelrud, Arild, Geva de la Tova, Philip J. Hamory, Ronald Young, Gregory K. Noffz, Michael Dodson, Sharon S. Graves, John K. Diamond, James E. Bartlett, Robert Noack, and David Knoblock, Pegasus(r) Wing-Glove Experiment to Document Hypersonic Crossflow Transition-Measurement System and Selected Flight Results. NASA TM 2000-209016, January 2000.

90. Moes, Timothy R. and Kenneth Iliff, Stability and Control Estimation Flight Test Results for the SR-71 Aircraft With Externally Mounted Experiments. NASA TP-2002-210718, June 2002.

91. Jackson, E. Bruce, Christopher I. Cruz, and W. A. Ragsdale, Real-Time Simulation Model of the HL-20 Lifting Body. NASA Technical Memorandum 107580, July 1992.

92. Jackson, E. Bruce and Christopher I. Cruz, Preliminary Subsonic Aerodynamic Model for Simulation Studies of the HL-20 Lifting Body. NASA Technical Memorandum 4302, 1992.

93. Saltzman, Edwin J., K. Charles Wang, and Kenneth W. Iliff, Aerodynamic Assessment of Flight-Determined Subsonic Lift and Drag Characteristics of Seven Lifting-Body and Wing-Body Reentry Vehicle Configurations. NASA TP-2002-209032, November 2002.

Index

Adams, Michael "Mike"   26, 113, 114

Aerodynamic Data Book   149-151, 198, 228, 235-237, 239, 240, 244, 254, 255, 271, 278, 279, 289, 290

Aircraft

AeroCommander   72, 73

B-29   xiv, 339, 259

B-50   339, 359

B-52   xv, xvi, 7, 14, 15, 18, 27, 28, 66, 79, 82-84, 91, 117, 128, 141, 164, 165, 176, 304, 310-313, 334, 339, 344, 359

C-47/DC-3/R4D   35, 43, 45-50, 65, 81

Convair 990   278

D-21   28, 112, 141, 315, 316

D-558-II   xv, xvi, 24, 124, 339, 345

F-8 Digital Fly-By-Wire   1, 87

F-15   116, 117, 119, 140, 190

F-15 RPRV   9, 140, 162, 177, 196, 229, 303

F-104   16, 18, 75, 84, 88, 92, 97, 129, 130, 190, 270

F-111   136, 137

Gossamer Albatross   342, 343

Gossamer Condor   343

HALO   313-321, 323-326, 327, 328, 336, 337, 346, 359

Helios   343, 344

HiMAT   229

HL-10   59, 62-65, 71-77, 79, 80, 96-98, 105, 106, 121-125, 127, 129, 134, 135, 333, 334, 363

HL-20   333, 334, 336

Hyper III   147

L-1011   311, 344

Lenticular shape   57, 58, 62

M1-L   57, 62

M2-F1   ix, 25, 31-55, 57-60, 63, 65, 67, 70, 79-84, 92-94, 96, 97, 123, 124, 126, 130, 171, 172, 176, 250, 362

M2-F2   ix, 25, 57-60, 62-69, 70-77, 79-97, 99, 102, 105, 21, 123, 124, 126-129, 131, 171-173, 176, 186, 189, 250, 362

M2-F3   126, 127, 129, 131, 141, 363

M-21	28, 141, 315
National Aerospace Plane	x, 305-310, 323-325, 327, 332, 333, 336, 337, 344, 354, 355
NF-104	53
Piaggio	65-67.
747 Shuttle Carrier Aircraft	153-156, 162-165, 168, 171-176, 182, 183, 185, 190, 250, 295, 296, 304, 314, 320, 321, 339, 356, 359
Shuttle Training Aircraft	161, 173, 227, 228, 242
SR-71	x, 28, 113, 140, 141, 307, 314-321, 323-325, 331, 332, 359
SV-5	59, 63, 64, 74, 79, 96, 124, 125
SV-5J	79, 124
T-33	71, 72, 81
T-38	123, 173
Tu-22	100
X-1	xiv, xv, xvi, 24, 53, 339, 345
X-2	xv, xvi, 339, 345
X-15	ix, xv, xvi, 1-29, 51, 70, 88, 105-114, 124, 133, 137, 138, 141, 147, 164, 169, 172, 186, 192, 193, 250, 303-305, 339, 340-342, 345, 346, 354, 365
X-20 Dyna-Soar	xvi, 7, 9, 28, 51, 64, 112, 153, 341, 342
X-23A	64, 125, 139, 334-336
X-24A	79, 117, 121-131, 139, 140, 333-336, 364
X-24B	127-131, 147, 182, 265, 364
X-24C/NHFRF	336, 337
X-33	156, 157, 328-333, 336, 337, 344, 346, 354, 355, 356, 359
X-37	299, 300
X-38	300, 334-336
X-40	299
X-43	300
XB-70	ix, 84, 85, 95, 96, 105, 146, 191
YF-12	140, 141
Aldrin, Buzz	72, 119
Allen, Joseph "Joe"	252
Allen, Mary	279
Apollo program	xvii, 32, 170, 308
Apollo 8	72, 119
Apollo 11	xi, 72, 117-120
Apollo 13	160
Apollo 17	290

Apollo 18   290
Apollo 19   290
Apollo Soyuz Test Project   290
Approach and Landing Test   81, 149, 151, 153-156, 160-165, 167-
                                     187, 227 ,228, 229, 236, 241, 242, 250,
                                     269, 284, 339

Apt, Milburn G. "Mel"   xv, 23
Armstrong, Johnny   11
Armstrong, Neil A.   19-21, 25, 64, 72
Arnaiz, Henry   310-313
Bacon, Donald "Don"   76, 77
Balakrishnan, A.V.   115
Barton, Rick   158, 195, 239
Baunbach, Joseph "Joe"   156, 239
Bayle, Guy   156
Bikle, Paul   19, 32, 51, 53, 79, 81, 144, 163, 341,
                                     346

Blanchard, Bob   272, 273
Bobko, Karol   260
Bode plots   33
Boundary layer   9, 193, 219, 220, 226, 235, 250, 256,
                                     260, 298, 310, 316, 320

Bowles, Jeff   317
Brand, Vance   252, 264, 284, 289, 290, 296
Brandenstein, Daniel   263
Bresina, John   197, 329
Buning, Pieter   271
Bush, President George H.W.   287, 326
Carman, Robert   133
Carter, Al   155, 204, 209, 213, 270
Cernan, Eugene "Gene"   290
Charles, Prince of Wales   184, 185
Collins, Michael "Mike"   72
Computer programming   32, 33, 57, 61, 62, 64, 221-223
Compton, Hal   239
Computational fluid dynamics   61, 193-195, 225, 271, 272, 297, 298,
                                     300, 301, 310

Conrad, Charles "Pete"   329
Cooke, Doug   158, 159, 195, 239, 326
Cooper-Harper scale   5, 69, 70, 71
Covey, Richard "Dick"   278, 284, 289
Crippen, Robert "Bob"   201-203, 208, 209, 261, 264, 268, 269,
                                     284, 293

Cronkite, Walter   169

Cross, Carl	84
Crossfield, R. Scott	13, 20, 23, 124
Culp, Mick	279
D'Agostino, Joe	168
Dana, William "Bill"	ix, x, 15, 24, 26, 46, 47, 81, 122, 126, 141, 326, 354
Day, Richard	xv, 204
Drake, Hubert "Jake"	133
Dryden, Hugh L.	346, 351
Duke, E. Lee	197
Durrett, John	79
Eggers, Alfred "Al"	31, 32
Eldredge, Richard "Dick"	31, 35, 42, 55
Enderes, Larry	xii, xiii
Engle, Joe	8, 16, 22, 25, 26, 72, 81, 160, 161, 179, 181, 187, 229, 230, 232, 233, 235-237, 254, 270, 283, 284, 290
Evans, Ronald	290
Faget, Maxime "Max"	146
Flyovers	18, 314, 315
Fullerton, Gordon	160, 161, 173, 181, 182, 185, 240, 250, 269, 270, 283
Fullerton, Marie	250
Fulton, Fitzhugh	176
Gamble, Joe	239
Gagarin, Yuri	xvii, 339
Garn, Senator Jake	296
Gemini program	xvii, 32, 170
Gentry, Jerauld "Jerry"	71, 81, 82, 93, 105, 121, 123
Gibson Edward "Hoot"	264
Gillam, Isaac "Ike"	169
Glenn, John	296
Goldin, Daniel	325-327
Gordon, Richard "Dick"	290
Greenfield, Lowell	xii, 61, 62, 64, 74, 76, 119, 120
Gregory, Frederick	292, 294
Grimes, Jim	xiii
Grissom, Virgil "Gus"	xvii
Haise, Fred	xi, 65, 66, 70-72, 81, 92, 93, 97, 118, 119, 160, 161, 172-174, 176, 181, 182, 184-186, 241, 242, 248, 269, 290
Hart, Terry	268
Harper, Bob	71
Hartsfield, Henry "Hank"	245, 251

Hauck, Frederick "Rick"	261, 278, 284
Herman, P.J.	61
Hoag, Peter	122, 127, 329
Holleman, Euclid "Ed"	1, 3
Home made rockets	ix, xii, xiii
Horton, Timothy "Tim"	204
Horton, Victor "Vic"	31, 49, 55
Intourist travel agency	99-101
Iowa State University	ix, xiii, 32, 33, 61
Ishmael, Steven	314, 315, 320, 327, 328, 329
Johnson, President Lyndon B.	53
Keener, Earl	59
Kempel, Robert "Bob"	80, 94, 97
Kirsten, Paul	239
Kittrell, Jack	206
Klein, Richard "Dick"	31, 55
Knight, William "Pete"	8, 15, 16, 26, 110, 192
Knipe, Dave	195, 239
Kock, Berwin	94
Kolf, Jack	11, 15
Komarov, Vladimir	102
Larson, Terry	272
Layton, Garrison "Garry"	97, 341
Lenoir, William "Bill"	252
Lousma, Jack	240, 242, 247, 269
Libbe, Steve	195
Lifting body program	xvii, 29, 31-55, 57-77, 79-98, 102, 105, 106, 121-131, 328-336
Low, George	348
Lytton, Lee	98-102
Maine, Rich	158, 159, 162, 175, 204-205, 209-213, 221, 239, 249, 278
Mallick, Donald	52, 128
Manke, John	127
Marquardt	5, 28, 112, 141, 315
Marshall, Laurie	278
Martin	63, 64, 122, 124-126
Matranga, Gene	1
Mattingly, T.K.	245, 247, 248, 251, 284
Maximum likelihood estimator	ix, 54, 94, 95, 105, 159, 221-223
McKay, John "Jack"	7, 8, 21-25, 46
McKay, James "Jim"	23, 24
McTigue, John	65
Nagel, Steven	290-292

Nelson, William "Bill"	296
Nelson, George "Pinky"	266, 268
Mercury program	xvii, 32, 169, 296
Meyerson, Rob	195
Mir space station	285, 353
Modified maximum likelihood estimator	ix, 221
Moore, Archie	204
Musick, Richard "Dick"	62
Northrop	65-67, 73, 75, 76, 122
Overmyer, Robert	252
Painter, Weneth "Wen"	73, 80, 94, 97
Parameter identification	54, 121, 155, 156, 162, 196, 221-223, 242, 270, 278
Pegasus booster	310-313, 323, 324, 325, 336, 344, 346
Pelly, Ron	279
Perry, John	12
Peterson, Bruce	51, 52, 55, 65, 71, 81, 97, 102, 105
Pilot-induced oscillation	36, 38, 39, 86-89, 105, 121, 123, 124, 186, 187, 241, 242
Pontiac tow vehicle	33, 34, 40, 45
Powers, Bruce	150, 204
Reagan, President Ronald	245-251, 260
Reed, R. Dale	29, 31-33, 35, 42-44, 57, 58, 63, 65, 74, 97
Rich, Ben	324
Richardson, Dave	239
Ride, Sally	261
Romere, Paul	195
Root locus technique	33, 35-39, 58, 67-69
Rushworth, Robert "Bob"	8, 18, 21-23, 25
Ryan, Bertha	31, 33, 35, 42, 43, 58, 74
Saltzman, Edwin	317
Scherer, Lee,	144, 162, 163
Schilling, Lawrence "Larry"	197
Schlosser, Don	156
Schmidt, Harrison "Jack"	290
Scobee, Francis "Dick"	129, 264, 265, 267-270, 294
Scott, David	162, 163, 169, 170, 185
Scramjet/ramjet	4, 5, 7-9, 28, 29, 92, 106-114, 141, 300, 301, 305-307, 309-314, 318-321, 324, 325, 336, 337
Shafer, Mary	120, 264, 265, 267, 268, 293
Shaw, Brewster	263
Shepard, Alan	xvii

Siemers, Paul   272
Sim, Alexander "Alex"   121, 158
Simulators   xi, 12, 24, 33-35, 37-39, 43, 60-62, 64,
                                                 66-70, 74-77, 79, 80, 89, 90, 142, 143,
                                                 161, 196-198, 230, 329
Skylab space station   133, 353
Slayton, Donald "Deke"   162, 175
Slide rule   3, 58
Smith, Harriet   31, 33, 35, 42, 54, 55, 58
Smith, John   61
Sorlie, Donald   81, 93
Soyuz 1   102
Space Shuttle data   157-160, 175-182, 184-187, 189-198,
                                                 205-217, 219-229, 232-237, 242-244,
                                                 246-256, 259-261, 264, 269-281, 283,
                                                 297-301, 372-379
Space Shuttle orbital missions   81, 124, 168, 201-208, 229-236, 240-
                                                 248, 252-254, 260-275, 277-281, 283-
                                                 296
Space Shuttle program   81, 129-131, 133-151, 153-165, 189-
                                                 198, 201-217, 219-237, 239-256, 259-
                                                 281, 283-301, 308-310, 345, 365
Sputnik I   xii, xvi, 308
Strip charts   13, 14, 47, 50, 62, 175, 203, 205, 206,
                                                 209, 210, 231, 232
Stewart, Dave   273, 274
Swigert, Jack   290
Syvertson, Clarence   32, 42
Szalai, Kenneth   305
Taylor, Lawrence "Larry"   9, 10, 25, 38, 39, 41, 43, 53, 57, 68, 71,
                                                 73, 80, 94, 95, 105, 120, 121
Tereshkova, Valentina   252
Thompson, Milton   8, 22, 25, 26, 34, 35, 37-53, 59, 60, 65-
                                                 71, 81-94, 97, 123, 141-143, 145, 148,
                                                 149, 162-164, 186, 196-198, 204, 206,
                                                 208, 247, 262, 304, 313
Throckmorton, Dave   274-276
Titov, Gherman   xvii
Travels of Ken Iliff   98-102
Truly, Richard "Dick"   160, 161, 179, 181, 229, 232, 235, 261,
                                                 262, 309, 310
Underwood, Jimmy   195
Unger, Charles   239
University of California at Los Angeles   114-116, 120, 121, 133, 142

University of Southern California	114
Van Hoften, James "Ox"	266, 268
Von Braun, Wernher	340
Voskhod 1	252
Vostok program	xvi, xvii
Walker, Joseph	21-24, 46, 84
Weil, Joe	150, 204
Weitz, Paul	260
White, Alvin "Al"	84
White, Robert "Bob"	20, 25
Williams, Walter	346
Wilson, Warren	11
Wolowicz, Chester	54
Yancey, Roxanah	54
Yeager, Charles "Chuck"	xiv, 51-53, 66, 67, 79, 234
Young, John	187, 201-203, 207-209,227-229, 263
Zubrin, Robert	354
Zuniga, Fanny	279

About The Authors

Kenneth W. Iliff

Dr. Iliff is known widely for his seminal contributions to aircraft parameter estimation. This is the field of determining the coefficients of differential equations by analyzing the response of the system. In aeronautics, this method is used to extract estimates of the aerodynamic, structural and performance parameters from flight data.

He was awarded the Kelly Johnson Award by the Society of Flight Test Engineers (permanently housed in the Smithsonian Air and Space Museum). He was elected a Fellow of the American Institute of Aeronautics & Astronautics (AIAA). He received the AIAA Dryden Lectureship in Research Medal in recognition of significant individual contribution to aeronautics and astronautics research. Other awards include the NASA Exceptional Scientific Achievement Medal, which is NASA's highest scientific award, the NASA Outstanding Leadership Medal, and the Milt Thompson Lifetime Acheivement Award. He has authored or coauthored over 81 national and international peer-reviewed technical papers, journal articles, and reports.

He attended Iowa State University, earning a B.S. in Mathematics and a B.S. in Aerospace Engineering; the University of Southern California, earning an M.S. in Mechanical Engineering; and the University of California, Los Angeles, earning a Ph.D. in Electrical Engineering and a M.E. in Engineering Management.

Dr. Iliff began work at the Dryden Flight Research Center in 1962 and conducted research on a wide variety of aeronautical topics. He was the Chief of the Fluid & Flight Mechanics Branch before becoming Dryden's Chief Scientist in 1994. He was an Adjunct Associate Professor in the UCLA School of Engineering from 1986 to 2003.

He has analyzed flight data from 25 first flights to support initial flight envelope expansion. Throughout his career he has analyzed flight data for 90 distinctly different aircraft configurations. These include most of the advanced research aircraft (such as the lifting bodies, X-15, XB-70, X-29, and Space Shuttle) that have been flown by NASA in the past 40 years. He retired from NASA in 2003.

Curtis Peebles

Curtis Peebles has been working at the NASA Dryden Flight Research Center History Office since November 2000 and is employed by Analytical Services & Materials, Inc. A freelance writer since 1977, he has written 12 books, including *Dark Eagles, The Corona Project, Flying Without Wings* (with the late

NASA research pilot Milt Thompson), and *Asteroids: A History,* as well as more than forty articles on various aspects of Cold War aerospace history. Peebles is a Fellow of the British Interplanetary Society and a member of the Flight Test Historical Foundation.

.

ISBN 978-1-78039-371-1

90000

For sale by the Superintendent of Documents, U.S. Government Printing Office
Internet: bookstore.gpo.gov   Phone: toll free (866) 512-1800;   DC area (202) 512-1800
Fax: (202) 512-2250 Mail: Stop SSOP, Washington, DC 20402-0001

ISBN 978-1-78039-371-1